Emerging Trends in Mechatronics

Series Editor

Aydin Azizi, Oxford, UK

This book series helps understand how Mechatronics will impact the practice and research of developing advanced techniques to model, control and optimize complex systems. It presents recent state of advances in mechatronics and related technologies including: Automatic Control, Robotics, Agent Based systems, Smart Manufacturing and Industry 4.0. The selected topics give an overview of the state-of-the-art and present new research results and prospects of the future development in the interdisciplinary field of mechatronic systems. The book series will provide up-to-date and useful knowledge for researchers and engineers involved in mechatronics and related fields.

Within the above dimensions, the scope of the book series welcomes high quality original research and review articles that cover a broad range of topics related to theoretical and applied aspects of modern mechatronics. Potential topics include, but are not limited to the following:

Controller design

Many mechatronic systems have challenging control difficulties due to the high nonlinear structures and time-varying dynamic behaviors. In addition to these, there are external disturbances, which cannot be predicted and change according to uncontrolled working environments. Conventional controllers are insufficient to solve the problems and to compensate the environmental disturbances, therefore designing efficient controllers for complex systems has a great importance in the current book series.

Sensor Design & Data Collection Approaches

Essential part of a mechatronics system is the measurement system that senses the variations in the physical parameters, such as temperature, pressure, displacement, and so on, and converts it to voltage or current. In mechatronic systems, accurate measurement of the dynamic variables plays a vital role for the actuators to function properly, therefore on of the aims of this book series, to focus on understanding what should be done to design accurate sensors and efficient data collection methods to improve the performance of mechatronic systems.

Optimization Techniques

Designing a complex mechatronic system involves multiple variables, objectives, constraints, and evaluation criteria as well as their nonlinearly coupled relationships. One of the challenging research areas of this book series is utilizing optimization techniques to design an optimal mechatronic system.

Artificial Intelligence

From the point of view of dynamics, mechatronic systems can be characterized by model uncertainties, high nonlinearities, complicated coupling, and stringent performance requirements, among others. To deal with these facts, this book series would like to focus on investigating and utilizing Artificial Intelligence techniques such as Neural Networks, Fuzzy logic, Neuro-fuzzy, evolutionary techniques etc. to model, design and optimize complex mechatronic systems.

Robotics

In the field of Robotics, there is a need of focusing on kinematic, dynamic, control and optimization of innovative robotic systems. This book series can introduce new and novel robot designs such as hybrid (serial-parallel) redundant manipulators that have a high potential of application in special fields like space robots. State of the art researches on selection of optimum kinematic solutions and path planning based on such an optimization, dynamic analysis and new control methods developments and application of artificial intelligence in this field can be published in this book series.

Smart manufacturing

Application of networked sensors and machinery equipment using internet and intranet in order to monitor and automation of the manufacturing process and improve the performance efficiency and is a point of interest. Advanced robotics, multi input-multi output control approaches combined with the artificial intelligence and optimization techniques, big data processing approaches, multi-scale modelling and simulation of dynamic sub-systems etc. are the hot topics that will be focused in this book series.

AGVs

Automated Guided Vehicle (AGVs) are going to be very popular is industry logistics, surveillance, defense and social applications in order to make processes safer and more cost efficient. Innovative AGVs designs for various missions, Robot Operating System usage on AGV control, Indoor and outdoor navigation and control algorithms, intelligent path planning techniques, sensor fusion, simultaneous application of GPS, INS and offline maps are the topics that are planned to be published in this book series.

Masoomeh Mirrashid · Danial Jahed Armaghani ·
Aydin Azizi
Editors

Intelligence at the Interface

Data-Centric Methods for Sensing, Modeling, and Decision Support

Springer

Editors
Masoomeh Mirrashid
College of Engineering,
Abu Dhabi University
Al Ain, United Arab Emirates

Aydin Azizi
School of Engineering, Computing
and Mathematics
Oxford Brookes University
Oxford, UK

Danial Jahed Armaghani
School of Civil and Environmental
Engineering
University of Technology Sydney
Sydney, NSW, Australia

ISSN 2731-4855 ISSN 2731-4863 (electronic)
Emerging Trends in Mechatronics
ISBN 978-981-95-7583-1 ISBN 978-981-95-7584-8 (eBook)
https://doi.org/10.1007/978-981-95-7584-8

This Springer imprint is published by the registered company Springer Nature Singapore Pte Ltd.
The registered company address is: 152 Beach Road, #21-01/04 Gateway East, Singapore 189721, Singapore

Preface

The ability of Artificial Intelligence (AI) and Machine Learning (ML) to detect and predict activities unlocks further levels of automation and optimization of processes. This book focuses on the integration of theory and practices by bringing together studies that demonstrate the far-reaching potential of intelligent systems. The chapters illustrate that these systems go beyond being automation facilitators; they are incorporated at multiple levels of decision-making, risk, and operational efficacy. They address concerns such as model interpretability, the challenges of practitioner adoption, and the embedding the knowledge-driven frameworks. This compilation attempts to articulate a perspective of the current intelligent systems context for researchers, industry professionals, and graduate students relating to the real-world challenges. The contributions go beyond presenting variations on known algorithms and introduce an analytical, adaptable, and data-centric approach.

Al Ain, United Arab Emirates — Masoomeh Mirrashid
Sydney, Australia — Danial Jahed Armaghani
Oxford, UK — Aydin Azizi

Contents

Hierarchical DRL Approaches for CVaR-Aware Cryptocurrency Portfolio Optimization

Mahdi Rasouli, Ehsan Hajizadeh, and Hossein Dastkhan

Abstract Dynamic portfolio selection refers to the ongoing process of distributing investment capital among different assets over several trading periods, while accounting for each investor's balance between risk and return preferences. Automating this decision-making process through machine learning remains a complex challenge. Two major difficulties in portfolio optimization are identifying meaningful features and managing risk effectively to achieve consistent long-term gains. Although reinforcement learning (RL) methods are increasingly used in financial markets, many existing studies still neglect the risks posed by extreme or rare market events and the effects of worst-case scenarios on trading outcomes. In this research, we address these limitations by applying a hierarchical clustering algorithm to extract the most relevant features for our model. We also introduce a novel reinforcement learning framework called the Hierarchical Deep Q-Network (HDQN). This approach integrates the Deep Q-Network (DQN) structure within a hierarchical design to determine optimal asset weight allocations. The HDQN model incorporates Conditional Value-at-Risk (CVaR) as a key risk measure in its reward function, allowing asset allocations to adapt dynamically to both investor risk tolerance and the portfolio's current exposure at each trading stage. To test the model's performance, we apply it to a portfolio of eight cryptocurrencies, using daily data from July 2019 to April 2023. When compared with established benchmark models through performance indicators such as the Sharpe ratio, the HDQN framework demonstrates superior results in both constructing diversified portfolios and selecting individual assets effectively.

M. Rasouli · E. Hajizadeh (✉)
Department of Industrial Engineering and Management Systems, Amirkabir University of Technology (Tehran Polytechnics), Tehran, Iran
e-mail: ehsanhajizadeh@aut.ac.ir

M. Rasouli
e-mail: mahdi.rasooli@aut.ac.ir

H. Dastkhan
Department of Financial Engineering, Faculty of Financial Science, Kharazmi University, Tehran, Iran
e-mail: hdastkhan@khu.ac.ir

A. Azizi et al. (eds.), *Intelligence at the Interface*, Emerging Trends in Mechatronics,
https://doi.org/10.1007/978-981-95-7584-8_1

Keywords Cryptocurrency portfolio · Deep reinforcement learning · HDQN algorithm · Hierarchical clustering algorithm · Technical analysis

1 Introduction

Portfolio optimization involves the allocation of funds across different financial assets to maximize returns while mitigating risk [1]. Traditional approaches in this field could be categorized into four main classes: "Follow-the-Winner", "Follow-the-Loser", "Pattern-Matching", and "Meta-Learning" [2]. In both of the first two categories, predefined financial models are used, and machine learning is sometimes used to fine-tune parameters [3]. For these approaches to be effective, the models need to be valid across a variety of market conditions. In contrast, "Pattern-Matching" entails identifying historical patterns similar to the current context, and then optimizing the portfolio based on selected history and assumptions about market dynamics [4]. Lastly, "Meta-Learning" involves the combination of various method types to enhance overall performance [5]. The work adopts a machine learning approach to portfolio management that does not assume prior knowledge of financial markets. The algorithms directly observe and learn from historical data, removing the need for human-crafted models and domain expertise. This model-free reinforcement learning approach allows the system to adapt as it accumulates experience across various market conditions. By allowing the algorithms to learn relationships and optimize decisions directly from data, this approach seeks to overcome the limitations associated with relying on human-designed models and hardcoded domain knowledge. Deep Q-learning (DQN) is a reinforcement-based learning method that has shown promise for developing algorithmic trading strategies [6]. DQN uses deep neural networks to approximate the optimal action-value (Q) function in an environment. One of its key advantages is its ability to learn trading policies directly from high-dimensional input data, such as price charts, without the need for manual feature extraction. In the context of trading with DQN, the environment is the financial market, states correspond to market conditions based on price data over a time window, and actions represent potential trades to execute (e.g., buy, sell, hold).

The agent (DQN) learns a policy that maximizes trader's expected cumulative rewards, such as profit, through episodic iterations of exploration and learning. The deep neural network takes market state data and predicts Q-values for actions, which guide the agent's trade decisions. The network is trained off-policy from memories of past trades, iteratively improving the Q-value approximations.

One of the main advantages of DQN in trading applications is its capacity to capture intricate patterns and relationships within high-dimensional financial data. It also demonstrates strong adaptability to changing market conditions and operates as a model-free method, eliminating the need for predefined assumptions about market behavior. However, it faces challenges related to reward shaping, handling partial observability of states, and accommodating various risk preferences. In summary, DQN represents a promising AI technique for developing profitable data-driven trading strategies, offering the potential to learn effective policies even within the complexity and uncertainty of financial markets. With further development, DQN-based trading systems could replicate or surpass human trader performance. In portfolio optimization, risk refers to the uncertainty and potential losses associated with investment decisions. It is essential to consider risk alongside return when constructing an optimal portfolio [1]. Conventional methods, such as Modern Portfolio Theory (MPT), seek to balance risk and return by spreading investments across a range of assets to minimize exposure to any single source of uncertainty [1]. Risk is commonly measured using metrics like standard deviation, variance, or downside risk. Traditional risk management techniques in portfolio optimization include asset allocation, diversification, and risk-adjusted performance measures. Asset allocation refers to the process of allocating funds across different asset classes (such as bonds, stocks, and commodities) based on their risk-return characteristics. Diversification means distributing investments among different types of assets so that the performance of one investment doesn't overly affect the entire portfolio. To evaluate how well a portfolio balances risk and return, analysts often use risk-adjusted performance metrics like the Sharpe ratio [7] or the Sortino ratio [8], help assess portfolio performance in relation to the risk taken. RL offers a promising approach to tackle risk management in portfolio optimization [9]. RL agents develop effective decision-making strategies by continuously interacting with their environment and learning from the outcomes of their actions. RL algorithms can be utilized to optimize portfolio strategies by explicitly considering risk management objectives. For example, RL agents can learn risk control mechanisms to actively manage risk exposure in response to market fluctuations. By continuously monitoring risk levels, RL algorithms can adjust portfolio allocations to maintain risk within acceptable limits. Despite the extensive development of RL applications in financial market, it is remarkable that a significant portion of the existing literature overlooks the potential risks associated with infrequent high-risk events and the impact of worst-case scenarios.

The main goals of this paper are to extract optimal features for learning specific trading rules about financial assets and to develop a policy that takes risks into account. This policy aims to secure investors against potential huge losses brought on by financial crises or unexpected events. In pursuit of these objectives, we present the following major contribution:

Phase 1—Feature Selection and Single Asset Trading:

Feature Selection: We begin by using a hierarchical clustering algorithm to select essential features for our model. Additionally, we incorporate candlestick patterns, which offer valuable insights into asset conditions and future trends.

Single Asset Trading: In this phase, we utilize the Deep Q-Network algorithm to determine optimal buy and sell points for individual assets.

Phase 2—Portfolio Optimization with HDQN:

HDQN Algorithm: In the second phase, we introduce the Hierarchical Deep Q-Network (HDQN) algorithm. This hierarchical structure integrates DQN and enables the determination of the optimal weight for each asset within the portfolio.

Risk-Aware Investing: We incorporate the Conditional Value-at-Risk risk measure into the HDQN algorithm's reward function. This measure reflects the significance of investment risk. The Hierarchical HDQN agent concentrates on maximizing overall return while the portfolio risk stays below predetermined CVaR limitations. But if the risk in the portfolio exceeds these limitations, the agent puts urgent risk reduction ahead of maximizing profits.

Portfolio Weighting: The optimal weight for each asset in the portfolio is calculated based on the investor's risk appetite and the risk level associated with the portfolio during each period.

In the following, our model is tested on eight cryptocurrencies using daily data from July 2019 to April 2023. Performance evaluation, including metrics such as the Sharpe ratio, demonstrates the superiority of our proposed model in both the portfolio formation and single asset trading phases.

First, a hierarchical clustering algorithm is used to select important features as input to the model. In addition, candlestick patterns, which contain good information about the financial asset conditions and future trends, are used as input to the model. In fact, based on the rules governing each candlestick pattern, we identify candlestick patterns for each asset in the study period based on each pattern's rules. In the second phase, fist the DQN algorithm is used to determine buy and sell points on a single asset in the first phase. In the second phase, the HDQN algorithm, which includes the DQN algorithm and a hierarchical structure to determine the optimal weight of each asset in the proposed portfolio, is used. In this phase, the CVaR risk measure is used in the HDQN algorithm's reward function to indicate the importance of investment risk. The Hierarchical DQN agent seeks to maximize overall return while the portfolio risk is below the CVaR limitations. Nevertheless, the Hierarchical HDQN agent places a higher priority on minimizing portfolio risk rather than maximizing overall returns when the portfolio risk exceeds the constraints set by the Conditional Value-at-Risk (CVaR). Also the optimal weight of each asset in the proposed portfolio is determined based on the investor's risk appetite and the level of risk associated with the portfolio in each period. The model proposed in this study is applied to eight cryptocurrencies from July 2019 to the end of April 2023 at daily time frames, and its performance is evaluated against comparative models in both phases based on evaluation metrics such as the Sharpe ratio, etc. The results indicate the superiority of the proposed model in both portfolio formation and single asset phases.

The rest of this paper is organized as follows. Section 2 reviews the fundamentals of deep reinforcement learning within the context of finance. Section 3.1 outlines the portfolio management problem, while Sect. 3.2 details the DQN algorithm. In Sect. 3.3, we introduce our proposed DRL-based framework. Section 4 reports the experimental results obtained from real-world data, and Sect. 5 concludes the paper with a discussion of key findings and directions for future research.

2 Literature Review

A Markov Decision Process (MDP) provides a mathematical framework for representing decision-making scenarios in environments that involve randomness or uncertainty [10]. MDPs provide a formal representation of sequential decision problems, and they are closely related to Reinforcement Learning (RL) algorithms. The Markov property, a core assumption of MDPs, asserts that the next state of the system depends solely on the current state and the action taken, without any influence from past states or actions. This property simplifies the modeling and computation of MDPs. A policy in MDP defines the behavior of the agent, specifying the action to take in each state. Policies can be deterministic (e.g., always choose action A in state S) or stochastic (e.g., choose action A with a certain probability in state S). In Markov Decision Processes (MDPs), value functions play a crucial role in assessing and comparing different policies. The two primary types are the state value function (V) and the action-value function (Q). The Bellman equations establish recursive relationships between these functions, defining the value of a state or action in terms of the expected immediate reward and the estimated value of the subsequent state or action. These equations form the theoretical foundation for solving MDPs and serve as the basis for many RL algorithms [11].

RL algorithms aim to find an optimal policy in an MDP by interacting with the environment and learning from the observed rewards. RL uses value-based methods or policy-based methods to estimate value functions and improve the policy iteratively. By applying trial and error, RL agents learn to make decisions that maximize the expected cumulative rewards over time. RL and deep RL have been increasingly applied to dynamic portfolio optimization in recent years, with a growing number of published works on this topic. Sato [12] provides a review of RL and deep RL approaches for portfolio optimization, examining both value-based and policy-based model-free RL methods. The review identifies some key unresolved challenges facing portfolio managers using these techniques. Below is a concise review of research utilizing deep reinforcement learning for portfolio optimization. Park et al. [13] derived a portfolio strategy for multi-asset DQN trading in practical action spaces using a transformation to map infeasible actions to feasible ones. Moody and Saffell [14] employed Reinforcement Reward Learning (RRL) to explore optimal portfolio construction, asset allocation, and trading strategies, which subsequently spurred considerable research in this domain. In their work, they evaluated various approaches using profit, the Differential Sharpe Ratio (DSR), and the Downside Deviation Ratio (DDR) as performance indicators. Their experimental results revealed that RRL consistently outperformed Q-learning with respect to these reward-based evaluation metrics. They conducted a comparison of different approaches using profit, Differential Sharpe Ratio (DSR), and Downside Deviation Ratio (DDR) as measures of success. The results showed that Relative Reward Learning (RRL) outperformed Q-learning in terms of these reward functions. To address the dynamic asset allocation issue, Almahdi and Yang [15] presented an ideal variable weight portfolio allocation model that makes use of RRL and Expected Maximum Drawdown

(EMD). In their study, Deng, et al. [16] enhanced the performance by integrating Relative Reward Learning (RRL) with a fuzzy Deep Neural Network model. This combined approach allowed for the analysis of the market more effectively. Soleymani and Paquet presented DeepBreath, a portfolio management framework using deep reinforcement learning and blockchain to address settlement risk. Testing shows it outperforms current investment strategies in maximizing return on investment and minimizing market risk [17]. Betancourt and Chen developed a portfolio management method using deep reinforcement learning for dynamic asset markets, particularly in cryptocurrencies. It introduces a novel neural network architecture that adapts to new assets and minimizes transaction costs, outperforming existing methods with average daily returns exceeding 24% [18]. Aboussalah and Lee introduced a novel approach to reinforcement learning for dynamic asset allocation, combining Mean-Variance optimization and optimal capital growth. It evaluates eight policy-based RL algorithms for portfolio management, finding four promising candidates after modifications. The study also proposes an innovative method for encoding financial data and introduces the concept of cross-sectional deep reinforcement learning. Furthermore, it evaluates the performance of RL agents and compares their results with both passive and active trading strategies to establish a comprehensive performance benchmark [19]. Lin et al. introduced a multi-agent deep reinforcement learning framework for portfolio management in quantitative trading, specifically addressing the challenges of transaction cost modeling faced by hedge fund managers. Their approach incorporates a novel reward function and a customized policy network architecture, which demonstrated superior performance compared to traditional portfolio selection methods [20]. Huang, et al. introduced CR-DQN, an algorithmic trading method that combines deep Q-learning with moving average (MA) and trading range breakout (TRB) rules. It outperforms other methods in evaluations on both Chinese and non-Chinese stock markets [21]. Cui et al. propose a novel DRL hyper-heuristic framework for multi-period portfolio optimization, addressing limitations of traditional DRL algorithms. Our approach focuses on low-level trading strategies and leverages expert domain knowledge, demonstrating notable performance gains in real-world capital market instances. The study employs data from five major stock indices spanning the period 2012–2022, providing valuable insights for the development of investment strategies and the design of regulatory frameworks [22]. Sun et al. proposed a GraphSAGE–DRL integrated model (GRL) designed to enhance the architecture of the Proximal Policy Optimization (PPO) agent for portfolio optimization. The GRL framework outperformed both traditional baseline strategies and the standard PPO model lacking a feature extractor. These results highlight the effectiveness of incorporating a GraphSAGE-based feature extractor in improving the performance and robustness of reinforcement learning models for portfolio optimization tasks [23].

Despite the extensive research and impressive performance of RL in portfolio management, there is a noticeable scarcity of literature that adequately addresses portfolio risk, particularly in worst-case scenarios, when developing trading strategies. Each financial instrument possesses a distinct combination of advantages and risks. The ultimate goal for investors is to select an optimal portfolio that maximizes returns while minimizing portfolio risk. Consequently, the RL agent plays a vital role in formulating a risk-sensitive or risk-averse strategy to protect investors from significant losses that may arise due to unforeseen market conditions.

3 Methodology

3.1 Portfolio Allocation Problem

The portfolio has m assets in total. The price vector for Period t, v_t, is made up of the closing prices of all assets. In other words, the closing price of the ith asset during the tth period represents the ith element of v_t, $v_{i,t}$. Comparably, the highest and lowest prices of the time are shown by v^{hi} and v^{lo}. The quoted currency, which is referred to as the cash for the remainder of the article, is the portfolio's initial and most unique asset. The first elements of v_t, $v(hi)$, and v^{lo} are always one since all asset values are stated in cash; that is $v^{hi}_{0,\,t} = v^{lo}_{0,\,t} = v_{0,\,t} = 1, \;\; \forall\, t$.

The opening prices for Period $t + 1$ and the closing prices for Period t are components of v_t for continuous markets. The elementwise division of v_t by $v_{t\,-\,1}$ yields the price-relative vector of the trading session, denoted as y_t.

$$\boldsymbol{y}_t := \frac{\boldsymbol{v}_t}{\boldsymbol{v}_{t-1}} = \left(1, \frac{v_{1,t}}{v_{1,t-1}}, \frac{v_{2,t}}{v_{2,t-1}}, \ldots, \frac{v_{m,t}}{v_{m,t-1}}\right)^{\mathrm{T}}. \tag{1}$$

The quotients of the opening and closing prices for each asset during the time make up the constituents of y_t. The change in the overall portfolio value over a given period can be determined using the price-relative vector. Assuming the portfolio value at the beginning of period t is $p_{t\,-\,1}$, and transaction costs are ignored, the evolution of the portfolio value can be computed accordingly.

$$p_t = p_{t-1}\boldsymbol{y}_t \cdot \boldsymbol{w}_{t-1}. \tag{2}$$

The portfolio weight vector, or portfolio vector, at the start of Period t is denoted by $w_{t\,-\,1}$, whose ith element, $w_{t\,-\,1,i}$, represents the portfolio's asset i percentage following capital reallocation. Because $\Sigma w_{t,\,i} = 1, \forall\, t$, the components of w_t always add up to one. Following is Period t's rate of return:

$$\rho_t := \frac{p_t}{p_{t-1}} - 1 = \boldsymbol{y}_t \cdot \boldsymbol{w}_{t-1} - 1. \tag{3}$$

Logarithmic return is correspondingly:

$$r_t := \ln \frac{p_t}{p_{t-1}} = \ln \boldsymbol{y}_t \cdot \boldsymbol{w}_{t-1}. \tag{4}$$

In general, a portfolio management problem is formulated using the initial portfolio weight vector w_0 as the basis vector in the Euclidean space:

$$\boldsymbol{w}_0 = (1, 0, \ldots, 0)^{\mathrm{T}}, \tag{5}$$

showing that before entering the market, all of the capital was in the trading currency. The ultimate portfolio value, in the absence of transaction costs, will be:

$$p_{\mathrm{f}} = p_0 \exp\left(\sum_{t=1}^{t_{\mathrm{f}}+1} r_t\right) = p_0 \prod_{t=1}^{t_{\mathrm{f}}+1} \boldsymbol{y}_t \cdot \boldsymbol{w}_{t-1}, \tag{6}$$

where p_0 is the sum of the original investment. A portfolio manager's responsibility is to maximize $\mathrm{p_f}$ throughout a specified period.

3.2 DQN Algorithm

In general, the RL problem is formulated as a discrete-time stochastic control problem, as explained by François-Lavet et al. [24]. In this process, an agent interacts with its environment in the following ways: In a particular state inside its environment s_0, the agent begins by collecting an initial observation x_0. The agent has to do something at each time step t. Actions are chosen according to a policy, π. The policy instructs the agent on the actions to choose for each potential condition and describes the agent's behavior. Every action results in the agent receiving a reward r_t, updating the state transitioning to s_{t+1}, and observing an observation x_{t+1} [25]. At this point, it is possible to define the transition probability of each possible next state s_{t+1} as $p(s_{t+1} \mid s_t, a_t)$. Similarly, a reward probability of each possible reward r_t is defined as $p(r_t \mid s_t, a_t)$. Therefore, the anticipated scalar reward, r_t, obtained by taking action a in the current state s, is computed based on $E_{P(r_t|s_t,a_t)}(r_t \mid s_t = s, a_t = a)$. One way to conceptualize this paradigm is as a finite Markov Decision Process [12]. An agent's ultimate goal is to discover an optimum policy π, which indicates the likelihood of choosing a course of action in a state s to maximize the total of the discounted rewards. The following is the definition of the expected discounted return R at time t:

$$R_t = E\left[r_t + \gamma r_{t+1} + \gamma^2 r_{t+2} + \cdots\right] = E\left[\sum_{k=0}^{\infty} \gamma^k r_{t+k}\right]. \tag{7}$$

Here, the notation E represents the expectation taken with respect to the reward distribution, and the discount factor γ is involved. At this stage, we can define a Q-value function as follows:

$$Q^{\pi}(s,a) = E_{\pi}[r_t \mid s_t = s, a_t = a] = E_{\pi}\left[\sum_{k=0}^{\infty} r_{t+k} \mid s_t = s, a_t = a\right]. \tag{8}$$

For an agent, Q-value refers to the expected return a following policy π can achieve starting from state S. An iterative update procedure can be used to estimate the Q-value function based on Eq. 2 because it satisfies a recursive property:

$$\begin{aligned} Q^{\pi}_{i+1}(s,a) &= E_{\pi}\left[r_t + \gamma \sum_{k=0}^{\infty} \gamma^k r_{t+k+1} \mid s_t = s, a_t = a\right] \\ &= E_{\pi}\left[r_t + \gamma Q^{\pi}_i\left(s_{t+1} = s', a_{t+1} = a'\right) \mid s_t = s, a_t = a\right] \end{aligned} \tag{9}$$

The goal of the reinforcement learning agent is to identify the policy that produces the best results. It must thus learn an optimum policy π that leads to an ideal Q-value and has an anticipated value that is larger than or equal to all other policies. Specifically, Eq. 4 defines the iterative updating process for determining the best Q-value function.

$$Q_{i+1}(s,a) = E_{\pi}\left[r_t + \gamma \max_{a'} Q_i\left(s', a'\right) \mid s, a\right]. \tag{10}$$

A value iteration algorithm converges to the optimal Q-value by iterating as $i \to \infty$. A widely used algorithm based on values is Q-learning. The Bellman equation for the Q-value function [3], whose unique solution is $Q^*(s, a)$, is used in the simplest form of the Q-learning method:

$$Q^*(s,a) = \left(\mathcal{B}Q^*\right)(s,a), \tag{11}$$

where $\mathscr{B}$ denotes the Bellman operator, which maps any function K: $S* A \to R$ to another function $S^* A \to R$, and is defined as follows:

$$(\mathcal{B}K)(s,a) = \sum\nolimits_{s' \in S} T\left(s,a,s'\right)\left[R\left(s,a,s'\right) + \gamma \max_{a' \in A} K\left(s', a'\right)\right]. \tag{12}$$

When an action a is performed, the function T determines the transaction value needed to go from s to s'. Walkins et al. [27] provide a universal demonstration of convergence to the optimum value function provided that two requirements are met: (i) In all states, all actions are repeatedly sampled; and (ii) the state-action pairings are represented discretely. Under those conditions, we require a parametric value function $Q(s, a{:}\theta)$, where θ denotes the model parameters that are utilized to approximate the value function. In the case of neural networks, for instance, θ is equivalent to the weights of the networks. A brief overview of the deep Q-learning techniques utilized in this study is provided below.

3.3 Proposed HDQN Algorithm

A hierarchical DQN approach is presented in this part, which extends the DQN algorithm. The term Hierarchical RL describes the idea of breaking an RL challenge down into smaller difficulties (sub-tasks). It will take more effort and efficiency to solve each smaller task than it takes to solve the entire issue [26]. In order to achieve the best overall profit and portfolio risk, the investor must choose the most profitable portfolio for his investment. Most investors, however, prefer to avoid losses when gains and losses are equally possible. Our Hierarchical DQN method makes use of Hierarchical RL's two-tiered system, which enables the agent to minimize probable losses and Keeping portfolio profit and risk in balance under various conditions. The Hierarchical Deep Q-Network (HDQN) framework is an extension of the classical Deep Q-Network (DQN) and the hierarchical reinforcement learning (HRL) framework, incorporating both lower-level and higher-level policies. The lower-level policy, acting as a worker, focuses on maximizing the logarithmic rate of return at each time step by executing simple actions when portfolio risk remains below a predefined threshold. In contrast, the higher-level policy, functioning as a manager, determines broader strategic actions based on the observed state and the worker's outputs, particularly when portfolio risk exceeds the investor's tolerance level. When the portfolio risk surpasses a certain threshold, the manager adjusts the overall risk exposure in response to the worker's behavior. Specifically, the worker evaluates the market environment and formulates strategies aimed at maximizing total returns using the information derived from these observations. An indicator function is employed to assess whether the investor's risk tolerance has been exceeded. Investors capable of absorbing potential losses may continue to exploit market opportunities, while the manager intervenes to reduce portfolio risk if the worker's trading strategy appears misaligned with the current market state. Finally, the manager executes the concluding action within the environment. The core concept of the Hierarchical DQN framework is illustrated in Fig. 1.

Conditional Value-at-Risk, also known as predicted excess loss or projected shortfall, is a commonly employed risk metric [29] that serves as an indicator of risk. CVaR is preferred over Value-at-Risk as it incorporates the potential impact of infrequent yet significant losses. To approximate the conventional CVaR function, a linear function and a Monte Carlo simulation can be utilized. In this paper, we adopt the parametric CVaR technique (Fig. 2).

It is possible to define parametric CVaR as a closed form:

$$\mathrm{CVaR}_\alpha(a_t) = \mathcal{R}_t \frac{\varphi\big(\Phi^{-1}(\alpha)\big)}{\alpha} - \mu_t. \tag{13}$$

In this context, μ_t and R_t represent the mean and variance of the portfolio return, $\varphi(\cdot)$ denotes the standard normal probability density function, and $\Phi(\cdot)$ represents the standard normal cumulative distribution function, with Φ^{-1} being the standard normal quantile. When the portfolio risk falls below the CVaR risk constraints,

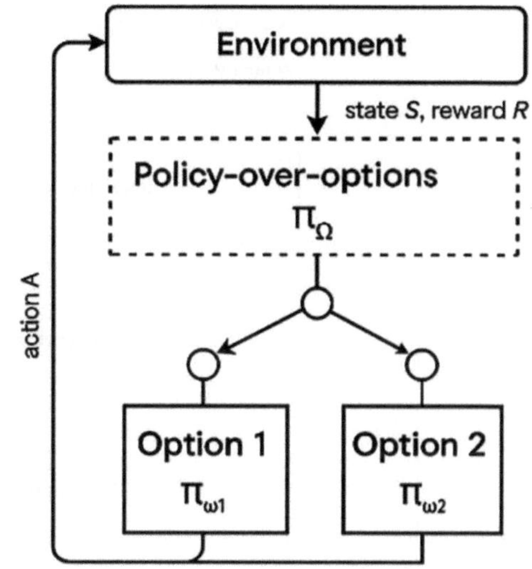

Fig. 1 Proposed hierarchical deep Q-network (HDQN) algorithm

the lower-level policy aims to maximize overall profit (logarithmic rate of return) through an aggressive trading strategy (CvaR < C). Conversely, when the portfolio risk surpasses the CVaR restrictions (CVaR > C), the primary objective shifts to minimizing portfolio risk rather than maximizing overall profit. In this case, the higher-level policy optimizes its trading strategy by adjusting its projected future discounted advantage and adopting a more cautious approach to mitigate risk. Subsequently, the reward functions for the higher-level and lower-level policies are discussed. The reward function for the higher-level policies is defined as follows, while the reward for the lower-level policies remains the same as in the conventional DQN algorithm.

$$r'(s_t, g_t, a_t) = \mathrm{CVaR}_\alpha(g_t) - \mathrm{CVaR}_\alpha(a_t). \tag{14}$$

This indicates that, unlike the lower-level action, the main goal of the higher-level policy is to promptly decrease the portfolio risk.

Defining the action space is a vital component of the DQN algorithm. The agent may trade just one stock every timestep since there are only three viable actions: buy, sell, and hold, per many previous research. This description is incorrect because the project necessitates portfolio management to reset several assets at each timestep. We describe the actions as vectors so that one action may reflect the trading of all stocks. The trading kinds of buy, sell, and hold are denoted by the numbers 1, − 1, and 0. One way to depict the activity of purchasing the first stock and selling the second stock at a timestep is as (1, − 1). Our DQN action space has 38 actions since we have 8 stocks, and each action is stored as an eight-dimensional vector.

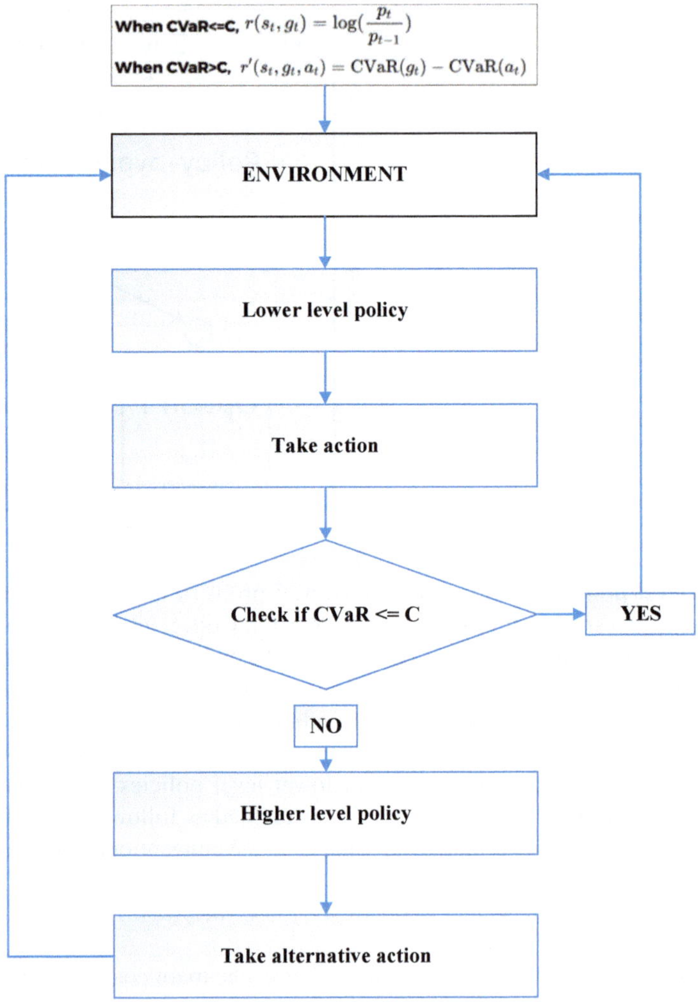

Fig. 2 Flowchart of proposed hierarchical deep Q-network (HDQN) algorithm

3.4 Feature Selection with Hierarchical Clustering

In this paper, the feature selection is performed with Hierarchical clustering. The method called hierarchical clustering, or hierarchical cluster analysis, clusters together comparable items [27]. As a consequence, a collection of clusters is produced, each of which is unique from the others yet contains things that are generally related to one another. The first step in hierarchical clustering is to handle every observation as a distinct cluster. Next, it does the next two actions repeatedly: The two clusters that are nearest to each other are found, and those that are most similar are combined. Until every cluster is combined, this iterative procedure is carried out.

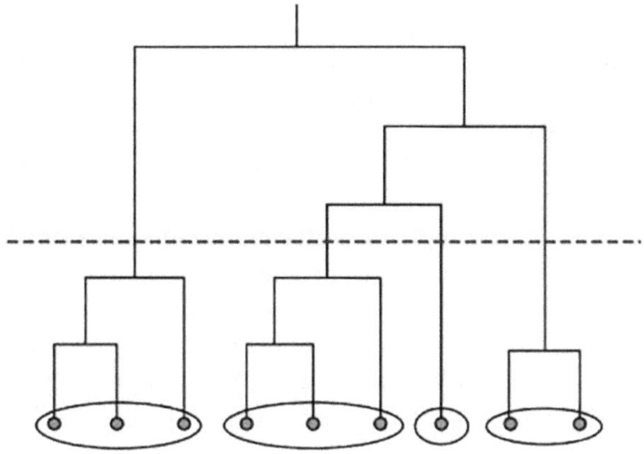

Fig. 3 Hierarchical clustering structure and cluster formation

This is illustrated in Fig. 3. In this figure, the green dots represent individual data points (observations), the vertical lines illustrate the step-by-step merging of clusters, and the dashed blue line indicates the cutoff threshold used to determine the final number of clusters.

The primary output of hierarchical clustering is a dendrogram, which visually represents the hierarchical relationships among the clusters. The length of the straight line connecting any two clusters has been used in this study to calculate the distance between them. This distance is often known as the Euclidean distance. Numerous more measurements for distance have been created. Theoretical considerations from the subject domain should guide the choice of distance measure. In other words, a distance measure must describe similarity in a way that makes sense for the research area. Once a distance metric has been chosen, it is important to ascertain the source of the distance computation. It can be calculated, for instance, between a cluster's two most similar segments (single-linkage), its two least similar segments (complete-linkage), its center (mean or average-linkage), or according to another criterion. Numerous criteria for linking have been established. Similar to distance measures, theoretical concerns from the application area should guide the selection of linking criteria. What causes variation is a central theoretical question. Determine which members of a cluster are most similar to determine whether two sets of artifacts are comparable. This makes sense since, for example, in archeology, variation is expected to arise from natural resources and creativity. Ward's approach makes sense by default in cases when the choice of connection criteria lacks convincing theoretical support. By subtracting each observation's total squared distance from the average observation inside a cluster, this approach determines which observations to group. This is frequently acceptable as the notion of distance aligns with the conventional assumptions of statistically computing group differences (e.g., ANOVA, MANOVA). Within the context of ten instances of computing returns with varying time delays, thirteen indicators, the highest price, lowest price, closing price, opening price, and the volume and value of daily transactions, we have chosen key aspects for this work. To do this, a hierarchical clustering technique was initially used independently for each of the research's chosen currencies to ascertain the 29 characteristics' clustering outcomes for each currency. The feature in the optimum portfolio selection mode is chosen by averaging the characteristics computed for each currency. Furthermore, the ideal number of clusters has been ascertained by applying the Lbow approach. In this way, the theme's algorithmic loss function for the number of distinct clusters is computed, and five clusters is the ideal number. The vector of candlestick patterns that is produced by applying each pattern's unique rules is the next significant aspect. The ROC curve has been used to determine if feature selection has improved the proposed model or whether the clustering technique does not need to be implemented and all features may be utilized as input to the proposed model and comparison models. The ratio of profitable trade signals to unsuccessful trading signals is displayed by the ROC curve.

3.5 LSTM Networks

In our framework, the policy is enforced with an LSTM neural network. The LSTM is a specific type of RNN [28]. Traditional recurrent neural networks (RNNs) often encounter the vanishing gradient problem, which limits their ability to capture long-term dependencies. Long Short-Term Memory (LSTM) networks overcome this limitation by introducing a specialized architecture designed to preserve information over extended time sequences. LSTMs contain gated cells that can maintain information for longer periods. These memory cells store and access relevant information from previous time steps to influence the network's predictions. Each layer of an LSTM network consists of three key gates that regulate the flow of information: the input gate, the forget gate, and the output gate. The input gate determines how much of the new information should be stored in the cell state, while the forget gate decides which parts of the previous information should be discarded. The output gate then controls the amount of information passed on to the next time step. All three gates utilize sigmoid activation functions with outputs ranging from 0 to 1, enabling fine-grained control over the inclusion or exclusion of information. An input close to 1 allows more of that information through, while a value near 0 blocks it. The forget gate clears out the memory cell if its activation is high. The memory cells themselves maintain contextual values over time based on these gate activations.

Overall, the inclusion of purpose-built memory cells and gating mechanisms in LSTMs helps address long-term dependency problems. It allows the network to better identify important patterns in sequences, regardless of their length. As a result, LSTMs tend to provide improved accuracy on tasks like time series forecasting compared to simpler RNNs. Their unique architecture gives them advantages in applications involving sequential or temporal data (Fig. 4).

In an LSTM network, it must determine what past information to discard versus retain at each time step. If the model has input data denoted as x_t at time t, the forget gate will produce an output value between 0 and 1, sometimes written as f_t. This forget gate activation represents the proportion of the cell state from the previous

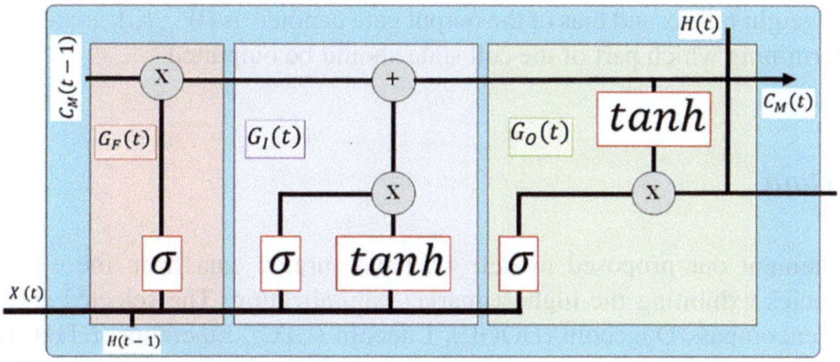

Fig. 4 Typical LSTM network structure

time step $(t - 1)$ that should no longer be considered before processing the new input. A value closer to 1 for f_t indicates more of the preceding cell state will be held over, while a number nearer 0 implies greater forgetting of the previous cell contents. Through outputting this forget gate activation f_t, which regulates the amount of prior information carried forward, the LSTM layer can dynamically manage what components of past data remain relevant over multiple time steps.

$$f_t = \sigma_1\left(W_f.\left[h_{t-1}, x_t\right] + b_f\right). \tag{15}$$

Let h_{t-1} represent the output at time $t - 1$, and W_f and b_f be the weight matrices and bias of the forget gate. Next, the processing of x_t occurs before it is stored in the cell state based on Eq. (5.4). The value of x_t is determined by the input gate, which produces a vector of potential values denoted as $\tilde{C}_t$ through a tanh layer at the same time, as described in Eq. (5.5). These values are then utilized to update the existing cell state C_t following Eq. (5.6).

$$i_t = \sigma_2\left(W_i.\left[h_{t-1}, x_t\right] + b_i\right), \tag{16}$$

$$\tilde{C}_t = \tanh\left(W_c\left[h_{t-1}, x_t\right] + b_c\right), \tag{17}$$

$$C_t = f_t{}^* C_{t-1} + i_t{}^* \tilde{C}_t. \tag{18}$$

The weight matrices and biases for the input gate and memory cell state are denoted as (W_i, b_i) and (W_c, b_c), respectively. Lastly, the output gate is determined by Eqs. (5.7) and (5.8).

$$o_t = \sigma_3\left(W_o.\left[h_{t-1}, x_t\right] + b_o\right), \tag{19}$$

$$h_t = o_t{}^* \tanh(C_t). \tag{20}$$

The weight matrix and bias of the output gate denoted as (W_o, b_o), are responsible for determining which part of the cell state should be outputted.

3.6 Data

To implement our proposed model, we have curated data from the eight cryptocurrencies exhibiting the highest market capitalization. The selected cryptocurrencies encompass Dogecoin (DOGE), Litecoin (LTC), Ethereum (ETH), Bitcoin (BTC), Ripple (XRP), Ethereum (ETH), Binance Coin (BNB), Cardano (ADA), and Polygon (MATIC). This choice stems from the dynamic and emerging nature of the cryptocurrency market, where we have focused on the top 25 cryptocurrencies in terms of market capitalization to mitigate liquidity-related risks, ensuring seamless cryptocurrency trading availability.

The dataset comprises daily data points spanning from July 2019 to the conclusion of April 2023, resulting in an extensive dataset of approximately 1400 data entries for each individual cryptocurrency.

4 Results

4.1 Descriptive Statics

Table 1 provides a statistical overview detailing the performance of the selected cryptocurrencies throughout the analyzed timeperiod.

As evident from the data, MATIC and DOGE emerge as the cryptocurrencies exhibiting the most substantial average returns during the specified time frame. Notably, these two cryptocurrencies also demonstrate the highest standard deviation, signifying their elevated levels of volatility. Additionally, DOGE and BNB stand out

Table 1 Statistical description of historical returns of cryptocurrencies

	Mean (%)	Std. (%)	Max. (%)	Min. (%)	Skewness (%)	Kurtosis
BTC	0.07	3.73	17.18	– 46.47	– 1.4319	19.3614
ETH	0.13	4.85	23.07	– 55.07	– 1.3475	14.9867
XRP	0.02	5.76	44.48	– 55.50	– 0.1167	16.4083
DOGE	0.22	7.97	151.63	– 51.51	5.833	10.2613
BNB	0.17	5.21	52.92	– 54.31	– 0.2762	21.4618
ADA	0.12	5.51	27.94	– 50.36	– 0.3751	7.4879
LTC	−0.02	5.12	24.84	– 44.91	– 1.1098	9.9344
MATIC	0.29	7.75	45.78	– 71.57	– 0.3796	14.3433

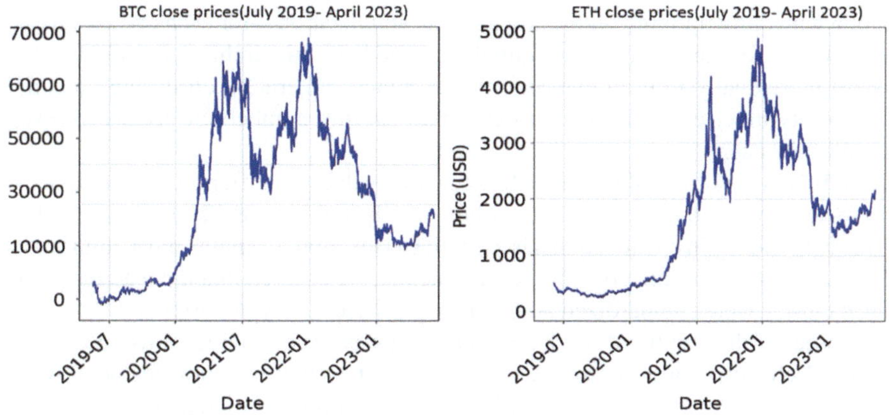

Fig. 5 BTC and ETH price change trends

for experiencing the most substantial daily return increases throughout the studied period. Furthermore, MATIC and ETH showcase the most pronounced reductions in their rates of return.

When examining the skewness values for each cryptocurrency, it becomes apparent that except DOGE, all other cryptocurrencies exhibit a negative skewness. This implies that the historical average returns for these cryptocurrencies are lower than the respective medians of their historical returns. In contrast, XRP displays nearly zero skewness, indicating that the mean, median, and mode of its historical returns closely align.

In Fig. 5, the price trends of Bitcoin and Ethereum, the two cryptocurrencies boasting the highest market capitalization, are depicted for the observed timeperiod. Notably, all cryptocurrencies exhibited an upward trajectory until the commencement of 2020. Subsequently, with the onset of the COVID-19 pandemic and its adverse global economic effects, the cryptocurrency market experienced a downturn, resulting in a decline in the prices of all cryptocurrencies. Starting from 2022 and extending to the present, as the COVID-19 pandemic's impact receded, the cryptocurrency market has once again displayed fluctuations in prices.

To implement the proposed model, the dataset is divided into two separate subsets: 80% is used for training, while the remaining 20% is reserved for performance evaluation.

4.2 *Feature Selection Results*

In this research, we utilized a hierarchical clustering approach to analyze thirteen key indicators and ten types of returns at various time lags. The dataset also included essential market variables—such as the opening, closing, highest, and lowest prices, along with daily trading volume and market capitalization—for each cryptocurrency examined. To execute this, we initially applied the hierarchical clustering algorithm individually to each chosen cryptocurrency, generating clustering outcomes for twenty-nine features, each clustered into varying numbers of groups for every cryptocurrency.

Subsequently, by evaluating the loss function values for different cluster numbers, we determined the optimal cluster count, which in this case was found to be five. This consistency in cluster counts is essential as the model's inputs must align for the formation of an optimal portfolio. Consequently, we ensured that the features within each cluster were standardized across all eight selected cryptocurrencies.

Based on the results of the feature clustering process conducted using the hierarchical clustering algorithm, we identified a set of five common features, which subsequently served as the input variables for our proposed model. The specifics of these selected features are outlined in Table 2.

Based on the outcomes derived from the application of the hierarchical clustering algorithm, which categorized the features of each cryptocurrency into five distinct clusters, we identified specific attributes that serve as representatives for each cluster. These selected attributes, namely Close Price, Bollinger Band, Market Cap, Relative Strength Index, and Double Exponential Moving Average were subsequently utilized as input variables for our proposed model. An additional noteworthy feature comprises the resultant vector of candlestick patterns, generated through the application of predefined rules for each pattern.

Table 2 Clustering of input features with hierarchical clustering algorithm

Features	Cluster	Selected	Features	Cluster	Selected
Return($t-1$)	1		Market cap	3	*
Return($t-2$)	3		Choppiness index	5	
Return($t-3$)	5		Relative strength index	2	*
Return($t-4$)	5		On-balance volume	1	
Return($t-5$)	1		Money flow index	4	
Return($t-6$)	4		Average true range	3	
Return($t-7$)	5		Simple moving average	5	
Return($t-8$)	2		Exponential moving average	2	
Return($t-9$)	2		Weighted moving average	3	
Return($t-10$)	4		Double exponential moving average	5	*
Open	4		Bollinger band	1	*
High	2		Stochastic oscillator	3	
Low	1		Accumulation/distribution	3	
Close	4	*	Normalized average true range	4	
Volume	2				

4.3 Comparative Models

To assess the outcomes of our proposed model, we conducted a comparative analysis with three distinct models: LSTM, GRU, and the Buy and Hold (B&H) strategy, applied to individual cryptocurrency trading scenarios. For constructing the optimal portfolio using our proposed model, we evaluated the performance against three benchmarks: HDQN-FS models, which do not incorporate feature selection, HDQN-CVaR models, which do not account for risk metrics, and the traditional Mean-Variance (MV) model. In these evaluations, we also considered a transaction fee of 0.2%. Furthermore, to manage risk, we imposed an upper limit of 5% on the Conditional Value-at-Risk (CVaR) criterion.

Our reinforcement learning framework utilized key parameters: a discount factor (γ) of 0.9, a learning rate (α) of 0.001, and an epsilon-greedy policy for action selection at each state. The initial value of epsilon was set at 1, which was then iteratively reduced by a fixed factor of 0.995 at each step.

4.4 Computational Results

Upon implementing the proposed model alongside comparative models in both individual cryptocurrency trading and portfolio management throughout the testing period, it is evident that the proposed model outperformed the comparison models. Figures 6, 7, 8, 9, 10, 11, 12, 13, 14 and 15 provide a comprehensive representation of the proposed model's performance compared to the alternative models for each cryptocurrency. Although the proposed model achieved its highest profit margins with DOGE and MATIC, yielding an average profit of about $1500 across all cryptocurrencies, it is worth noting that even for ETH and ADA—where profits were relatively lower at $1480 and $1250, respectively—the model still outperformed the other compared approaches. During the testing period, the LSTM neural network identified opportune buy and sell points with superior performance across BTC, XRP, DOGE, and MATIC. On the other hand, in the cases of ETH and LTC, the GRU neural network surpassed the performance of the proposed research model, successfully identifying optimal buy and sell points and generating higher profits throughout the study period.

Furthermore, a glance at the charts depicting each cryptocurrency's performance underscores the suboptimal performance of the Buy and Hold (B&H) strategy across all cryptocurrencies.

Figure 14 illustrates the performance of the HDQN algorithm with $\alpha = 0.015$ in comparison to other models. Notably, our proposed model significantly augmented the portfolio's value, surging from an initial $100,000 to $350,000 during the test period. Subsequently, among the comparison models, the HDQN-CVaR model emerged as the second-best performer, achieving a portfolio value of nearly

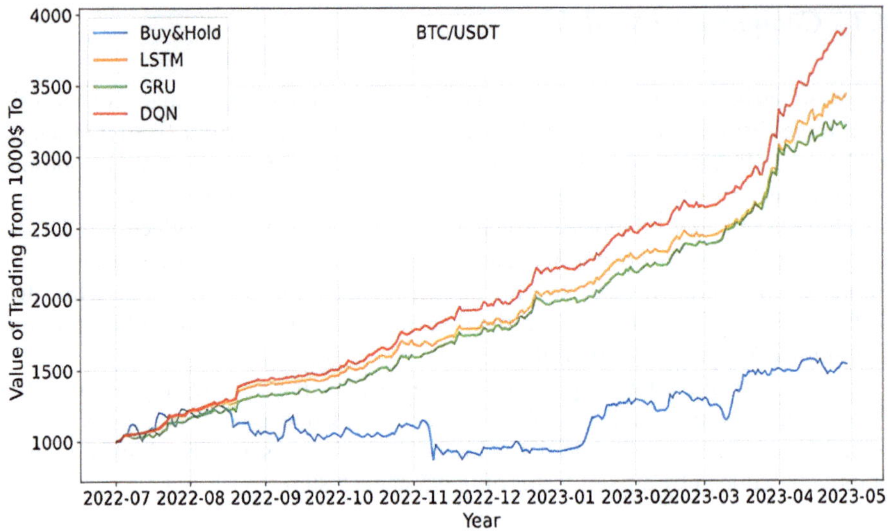

Fig. 6 Comparison of the performance of the proposed model with other benchmark models for BTC

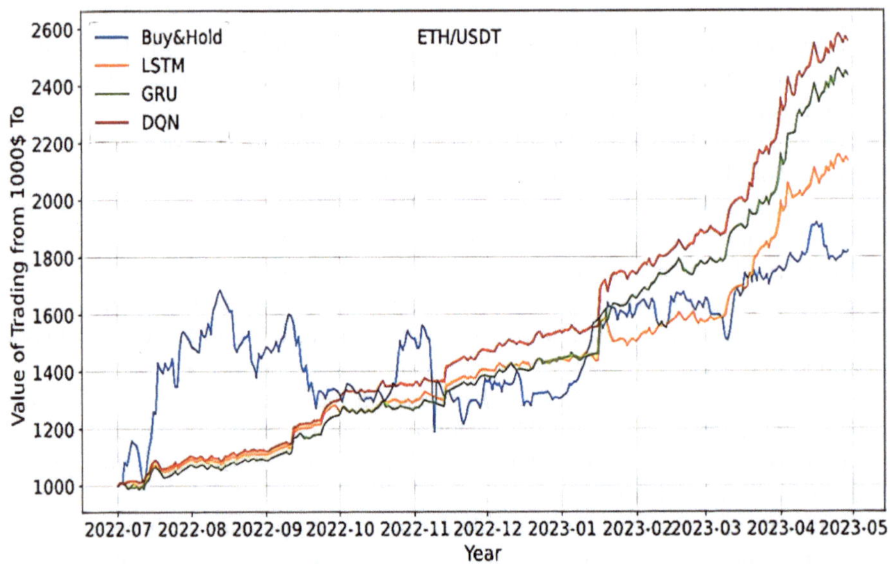

Fig. 7 Comparison of the performance of the proposed model with other benchmark models for ETH

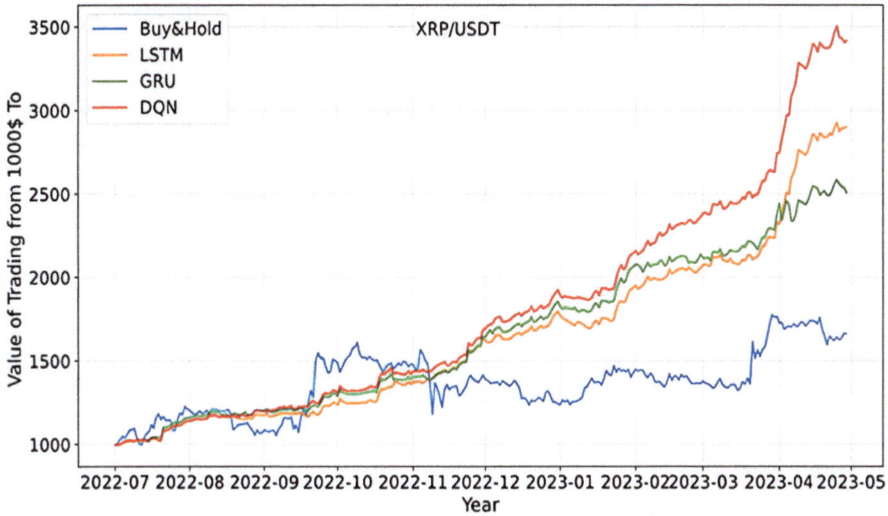

Fig. 8 Comparison of the performance of the proposed model with other benchmark models for XRP

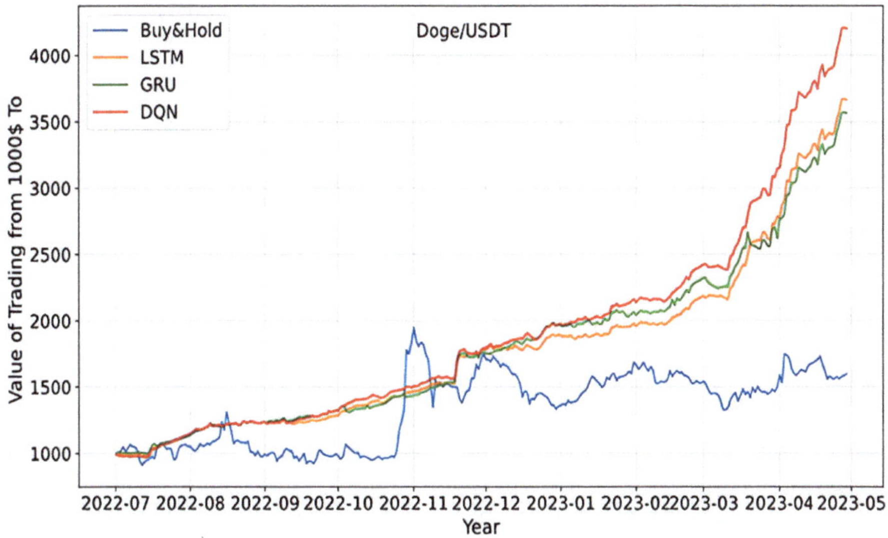

Fig. 9 Comparison of the performance of the proposed model with other benchmark models for DOGE

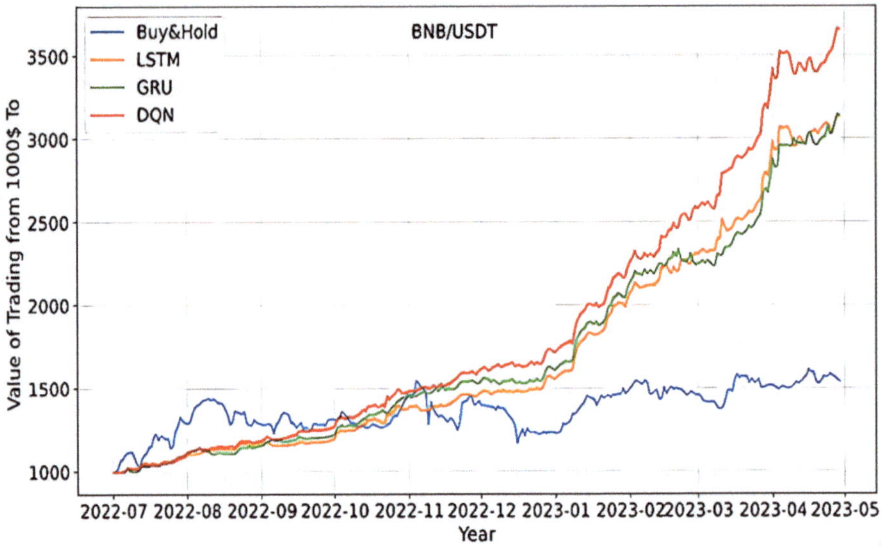

Fig. 10 Comparison of the performance of the proposed model with other benchmark models for BNB

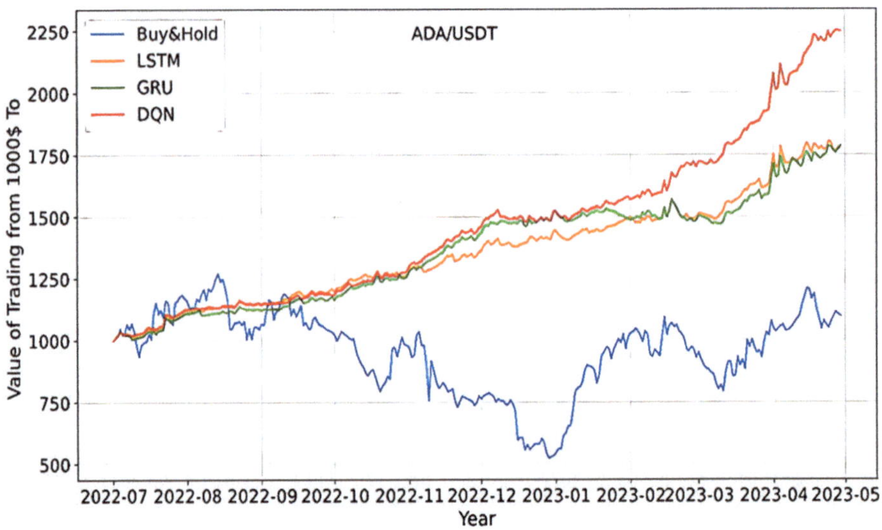

Fig. 11 Comparison of the performance of the proposed model with other benchmark models for ADA

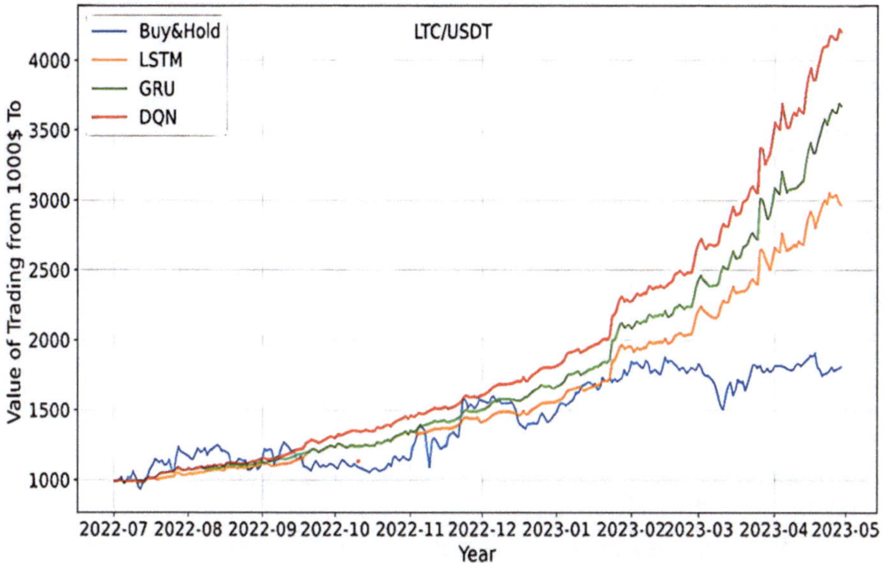

Fig. 12 Comparison of the performance of the proposed model with other benchmark models for LTC

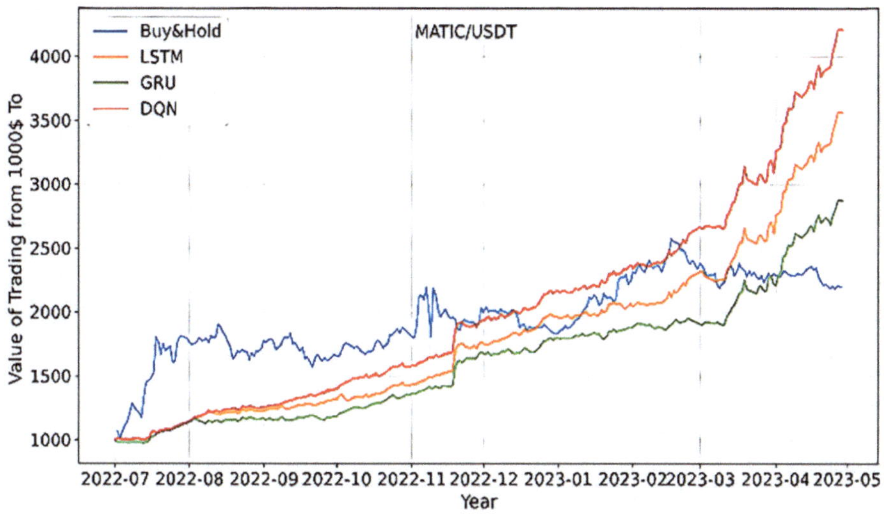

Fig. 13 Comparison of the performance of the proposed model with other benchmark models for MATIC

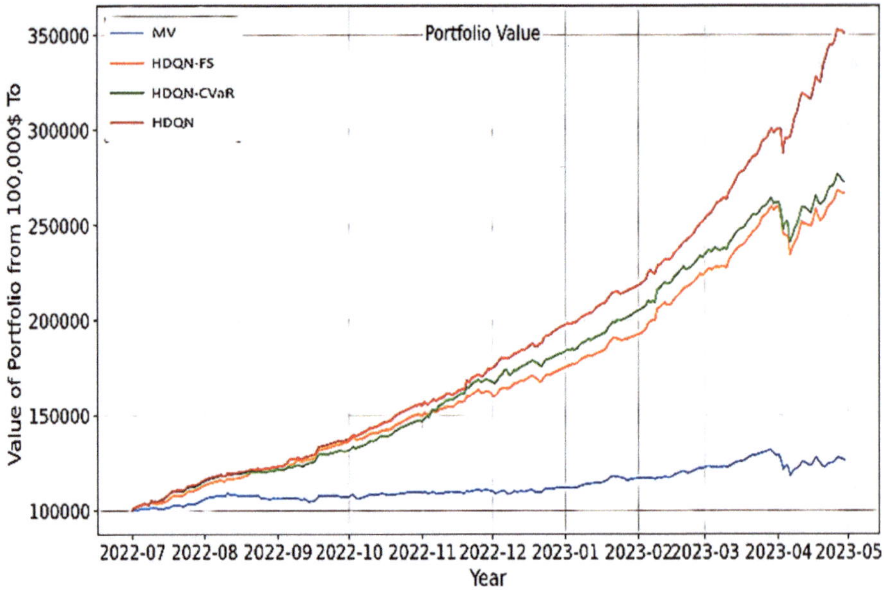

Fig. 14 Comparison of investment portfolio value in different models

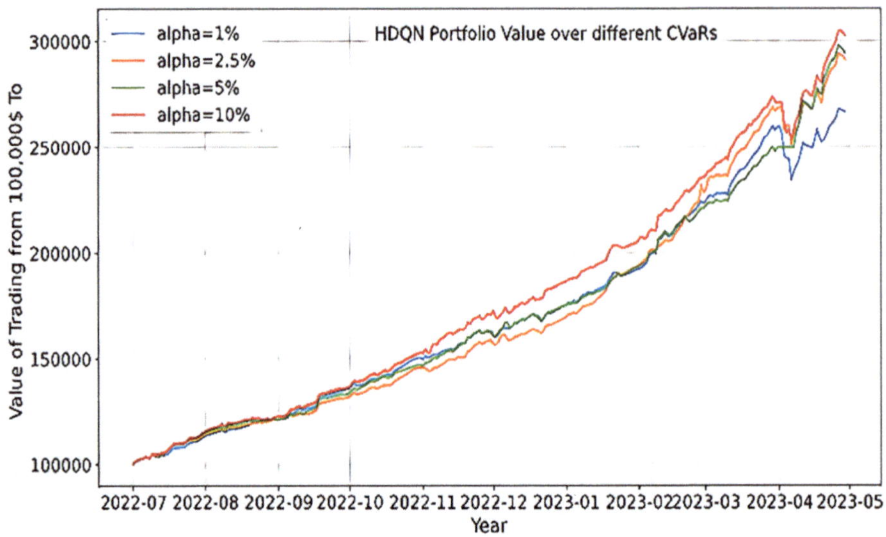

Fig. 15 Investment portfolio value for different risk levels

$150,000. Impressively, the HDQN-FS model exhibited similar performance to the HDQN-CVaR model, also reaching approximately $150,000.

In contrast, the Mean-Variance (MV) model, wherein all selected cryptocurrencies were input into the Markowitz model, notably yielded a portfolio value of $487. This outcome underscores the significance of judiciously selecting cryptocurrencies with growth potential and employing them within the Markowitz model, ultimately resulting in enhanced profitability.

Figure 15 presents the performance of the investment portfolio across four distinct risk tolerance levels: 1%, 2.5%, 5%, and 10%. Notably, the cumulative portfolio value exhibits minimal divergence between the risk tolerance levels of 2.5%, 5%, and 10%. However, upon a comparative examination of the Sharp ratio, as outlined in Table 3, it becomes evident that the highest Sharp ratio is observed at the 2.5% risk tolerance level.

These numerical findings underscore a crucial point: as investors become increasingly willing to accept higher levels of risk, they expose themselves to the potential for more significant losses, especially during financial crises. The data illustrates that, up to a certain threshold of risk tolerance, there exists a direct relationship between the profit gained per unit of portfolio risk and the investor's willingness to bear more risk. However, beyond this threshold, an increase in risk tolerance does not necessarily translate into a corresponding increase in investment profitability. These insights underscore the capacity of the HDQN algorithm to formulate a robust trading strategy in alignment with an investor's risk preferences.

Table 3 summarizes the Bitcoin (BTC) results across multiple performance indicators and compares them with those of the benchmark models. Table 4 provides a detailed evaluation of the proposed HDQN algorithm, highlighting its performance relative to the same comparative models based on the specified criteria. Upon a careful examination of the data in Table 4, it becomes apparent that the proposed model has demonstrated superior performance in BTC. The Sharpe ratio, for instance, stands notably higher at 5.874 in the proposed model compared to the comparative models. Furthermore, across various evaluation criteria, the proposed model consistently outperforms the alternatives. It's noteworthy that this model has achieved an impressive profit percentage of 187%, a figure that surges to an astonishing 251% in the case of DOGE. Equally remarkable is the model's ability to mitigate capital losses, as evidenced by its rapid return to its peak value within a mere 10-day period, as revealed in the "maximum duration of capital loss" criterion in Table 5. Interestingly, the proposed model demonstrates strong performance in portfolio construction by effectively incorporating risk metrics and adaptive loss constraints. Its maximum

Table 3 Sharpe ratio of the proposed model across varying risk levels

Criteria	Risk level			
	%1	%2.5	%5	%10
Sharp ratio	6.085	8.571	6.894	7.845

Table 4 Comparison of the evaluation metrics of the proposed model and the benchmark models for BTC

Criteria	Model			
	DQN	LSTM	GRU	B&H
Sharp	**5.874**	3.215	2.154	0.208
Sortino	**6.975**	3.512	2.548	0.515
MD	**2.15**	8.15	9.14	68.15
MDD[a]	**10**	52	43	157
Profit	**2876**	2412	2198	578

[a] Maximum Drawdown Duration

Table 5 Comparison of the evaluation criteria of the proposed model with comparative models for cryptocurrency portfolio

Criteria	Model			
	HDQN	HDQN-CVaR	HDQN-FS	MV
Sharp	**7.154**	5.421	4.891	1.594
Sortino	**8.287**	6.102	5.124	1.951
MD	**5.14**	13.15	18.14	5.84
MDD	13	25	31	**12**
Profit	**251,590**	172,480	165,930	25,710

drawdown (MD) is limited to just 5.14%, which is considerably lower than that of the comparative models.

5 Conclusion

Portfolio management plays a vital role in finance, focusing on the strategic distribution of investment capital across various assets to optimize returns for investors. Designing an effective trading strategy within financial markets is inherently complex, requiring expertise in fields such as quantitative analysis and risk management. In this study, we aimed to develop a robust cryptocurrency portfolio trading framework. The process began with the collection of daily price and trading volume data for eight cryptocurrencies with the largest market capitalizations. This was followed by a comprehensive data preprocessing stage. To reduce dimensionality and improve feature selection, a hierarchical clustering technique was applied, leading to the identification of five key features for each cryptocurrency. Additionally, candlestick charts were utilized to detect distinct candlestick patterns, following the specific formation principles associated with each pattern. However, rather than relying on conventional signaling rules for these patterns, our approach hinged on

Deep Reinforcement Learning (DRL) to autonomously determine and issue trading signals. The candlestick patterns themselves were utilized as the states within our proposed model. Our research embraced the Hierarchical Deep Q-Network (HDQN) algorithm, a fusion of the renowned DQN algorithm and a hierarchical structural framework. In our approach, the reward function incorporated not only the maximization of trading returns but also the utilization of the Conditional Value-at-Risk (CVaR) measure to assess and manage investment risk. Essentially, our algorithm operated under two distinct policies. At the low-level policy, the agent's reward function was contingent on the trading portfolio's value obtained from the environment. However, at the high-level policy, if the agent's actions under the low-level policy pushed the CVaR beyond a specified threshold, the high-level policy would take over, guiding the agent to execute actions different from those in the low-level policy. This dynamic shift aimed at both optimizing trading returns and controlling the level of investment risk. Our proposed model has consistently demonstrated strong performance in both individual trading and portfolio management when compared to other models. We have evaluated its effectiveness using various criteria, and the results consistently highlight its superior performance. Looking ahead, given the evolving nature of the cryptocurrency market and the multitude of factors influencing cryptocurrency price trends, future research should consider incorporating features related to investor sentiment and conducting extensive analyses of news and social network data. Additionally, recognizing the bi-directional nature of the cryptocurrency market, exploring alternative approaches for capital allocation and trading strategies, such as borrowing and lending mechanisms, presents a promising avenue for further enhancement. Lastly, since the DQN algorithm may not be optimally suited for environments with continuous action spaces, future work could explore alternative DRL algorithms such as Proximal Policy Optimization (PPO) and Deep Deterministic Policy Gradients (DDPG) to advance the cryptocurrency portfolio trading system.

AI Use Disclosure Statement
The authors declare that no part of this manuscript was generated by artificial intelligence tools, except for minor language proofreading. For this purpose, ChatGPT (OpenAI, version GPT-5) was used solely to correct grammar and improve the readability of the text. No content generation, data analysis, or idea development was performed by AI systems. The authors take full responsibility for the originality, accuracy, and integrity of the entire.

References

1. Markowitz HM (1991) Foundations of portfolio theory. J Financ 46(2):469–477
2. Li B, Hoi SC (2014) Online portfolio selection: a survey. ACM Comput Surv (CSUR) 46(3):1–36
3. Li B et al (2012) PAMR: passive aggressive mean reversion strategy for portfolio selection. Mach Learn 87:221–258

4. Györfi L, Lugosi G, Udina F (2006) Nonparametric kernel-based sequential investment strategies. Math Finance Int J Math Stat Financ Econ 16(2):337–357
5. Monteiro JP et al (2021) Meta-learning and the new challenges of machine learning. Int J Intell Syst 36(11):6240–6272
6. Almutairi S et al (2023) Breast cancer classification using deep Q learning (DQL) and gorilla troops optimization (GTO). Appl Soft Comput 142:110292
7. Sharpe WF (1998) The sharpe ratio. Streetwise Best J Portf Manag 3:169–185
8. Rollinger TN, Hoffman ST (2013) Sortino: a 'sharper' ratio. Red Rock Capital, Chicago, IL
9. Aboussalah AM, Lee C-G (2020) Continuous control with stacked deep dynamic recurrent reinforcement learning for portfolio optimization. Expert Syst Appl 140:112891
10. Natarajan M, Kolobov A (2022) Planning with Markov decision processes: an AI perspective. Springer Nature
11. Sutton RS, Precup D, Singh S (1999) Between MDPs and semi-MDPs: a framework for temporal abstraction in reinforcement learning. Artif Intell 112(1–2):181–211
12. Sato Y (2019) Model-free reinforcement learning for financial portfolios: a brief survey. arXiv preprint arXiv:1904.04973
13. Park H, Sim MK, Choi DG (2020) An intelligent financial portfolio trading strategy using deep Q-learning. Expert Syst Appl 158:113573
14. Moody J, Saffell M (2001) Learning to trade via direct reinforcement. IEEE Trans Neural Netw 12(4):875–889
15. Almahdi S, Yang SY (2017) An adaptive portfolio trading system: a risk-return portfolio optimization using recurrent reinforcement learning with expected maximum drawdown. Expert Syst Appl 87:267–279
16. Deng Y et al (2016) Deep direct reinforcement learning for financial signal representation and trading. IEEE Trans Neural Networks Learn Syst 28(3):653–664
17. Soleymani F, Paquet E (2020) Financial portfolio optimization with online deep reinforcement learning and restricted stacked autoencoder—DeepBreath. Expert Syst Appl 156:113456
18. Betancourt C, Chen W-H (2021) Deep reinforcement learning for portfolio management of markets with a dynamic number of assets. Expert Syst Appl 164:114002
19. Aboussalah AM, Xu Z, Lee C-G (2022) What is the value of the cross-sectional approach to deep reinforcement learning? Quant Financ 22(6):1091–1111
20. Lin Y-C et al (2022) Multiagent-based deep reinforcement learning for risk-shifting portfolio management. Appl Soft Comput 123:108894
21. Huang Z et al (2023) Algorithmic trading using combinational rule vector and deep reinforcement learning. Appl Soft Comput 147:110802
22. Cui T et al (2024) Multi-period portfolio optimization using a deep reinforcement learning hyper-heuristic approach. Technol Forecast Soc Chang 198:122944
23. Sun Q, Wei X, Yang X (2024) GraphSAGE with deep reinforcement learning for financial portfolio optimization. Expert Syst Appl 238:122027
24. François-Lavet V et al (2018) An introduction to deep reinforcement learning. Found Trends® Mach Learn 11(3–4):219–354
25. Barto AG, Sutton RS, Anderson CW (1983) Neuronlike adaptive elements that can solve difficult learning control problems. IEEE Trans Syst Man Cybern 5:834–846
26. Kulkarni TD et al (2016) Hierarchical deep reinforcement learning: integrating temporal abstraction and intrinsic motivation. Adv Neural Inf Proces Syst 29
27. Murtagh F, Contreras P (2012) Algorithms for hierarchical clustering: an overview. Wiley Interdisciplinary Reviews: Data Mining and Knowledge Discovery 2(1):86–97
28. Yu Y et al (2019) A review of recurrent neural networks: LSTM cells and network architectures. Neural Comput 31(7):1235–1270

Artificial Intelligence for Tunnel Seismic Response: A Comprehensive Review

Jyoti Jagajjita Raj, Swapnil Mishra, and Jitendra Khatti

Abstract Emerging artificial intelligence and machine learning (AI-ML) techniques are transforming seismic response analysis of tunnel structures by addressing limitations of traditional finite element methods. This comprehensive review examines developments from 2020 to 2025, analyzing validated research from high-impact journals. Modern neural networks—especially convolutional neural networks (CNNs) and long short-term memory (LSTM) models—now predict tunnel damage with over 91% accuracy, compared to the 78–85% accuracy achieved by conventional techniques. These AI systems also run in hours rather than days, making real-time or near-real-time assessment possible. Physics-informed neural networks blend core geotechnical principles with data-driven learning to improve prediction reliability. By fusing data from accelerometers, geological surveys, and live monitoring instruments, multi-source AI models detect damage with more than 90% precision. Machine learning-driven feature importance analysis reveals tunnel depth and soil properties as critical parameters influencing seismic response. Graph neural networks effectively model complex spatial relationships in tunnel networks, while transfer learning enables adaptation across different geological contexts. Inconsistent data formats, the requirement for transparent "explainable AI" to comply with regulations, and smooth integration into pre-existing engineering workflows are still obstacles in the way of these advancements. Future research directions emphasize federated learning for collaborative model development, explainable AI for enhanced interpretability, and physics-constrained algorithms ensuring geotechnical consistency. Looking ahead, federated learning for collaborative model training, explainable AI tools, and physics-constrained algorithms promise to solidify the role of AI in a safer, more resilient tunnel design and monitoring.

Keywords Artificial intelligence · Machine learning · Seismic response · Damage prediction · Neural networks · Underground structures

---

J. J. Raj (✉) · S. Mishra · J. Khatti
Mining Engineering Department, Indian Institute of Technology (Indian School of Mines), Dhanbad, India
e-mail: 24mt0194@iitism.ac.in

J. Khatti
Civil Engineering and Engineering Mechatronics, Columbia University, New York, United States

A. Azizi et al. (eds.), *Intelligence at the Interface*, Emerging Trends in Mechatronics,
https://doi.org/10.1007/978-981-95-7584-8_2

1 Introduction

Subterranean tunnels often cost billions of dollars each to build and are critical for transporting transportation, utilities and emergency services in urban areas. However, seismic events continue to expose vulnerabilities in tunnel systems, leading to substantial economic losses, service disruptions, and public safety concerns. Recent large earthquakes have brought attention to the urgent need for advanced analytical methods that can accurately predict tunnel response under complex seismic loading conditions. The 2011 Tōhoku earthquake in Japan caused significant damage to underground infrastructure, including metro tunnels that required months of repair work and service interruptions because of cracking and deformation. Similarly, the Chi-Chi earthquake in Taiwan in 1999 severely damaged tunnels, providing valuable case-study data for understanding failure mechanisms. The 2023 Turkey-Syria earthquake sequence highlighted the potential for catastrophic failures in subterranean infrastructure and exposed flaws in current seismic analysis and emergency response protocols. Finite element methods (FEM), analytical solutions obtained from elasticity theory, and empirical correlations based on post-earthquake damage surveys are the mainstays of traditional tunnel seismic analysis. When faced with complex soil-structure interaction phenomena, diverse geological conditions, and the stochastic nature of earthquake ground motion, these methods—which have been used by the engineering community for decades—display notable limitations. Finite element analyses, though theoretically rigorous, demand extensive computational resources and preprocessing time, often requiring days to weeks for comprehensive network-level assessments [1, 2].

The development of AI-ML technologies offers previously unheard-of chances to solve persistent problems in tunnel seismic analysis. These cutting-edge computational methods have revolutionary potential. They make real-time prediction and monitoring systems possible and incorporate pattern recognition in multi-dimensional datasets. They incorporate various data sources from seismic networks, geological surveys, and structural monitoring, and they optimize design parameters under several competing goals. Recent developments in deep learning, particularly CNNs for spatial analysis and RNNs for temporal modeling, enable the analysis of tunnel response patterns under different seismic loading scenarios. Compared to traditional analytical methods, these systems' ability to automatically extract relevant features from complex datasets without manual feature engineering significantly improves (Table 1).

Contemporary tunnel seismic analysis faces several interconnected challenges that traditional methodologies struggle to address comprehensively. Highly non-linear dynamic behavior, spatially variable geological properties, multi-directional ground motion effects, and time-dependent material degradation are all components of the fundamental complexity of soil-structure interaction during seismic events, and together, they surpass the capabilities of simplified analytical models frequently used

Table 1 Traditional versus AI-ML approaches in tunnel seismic analysis

Aspect	Traditional methods	AI-ML approaches
Computational time	Days to weeks for complex networks	Hours to real-time assessment
Accuracy	78–85% correlation coefficient	> 91% Correlation coefficient
Data requirements	Limited input parameters	Multi-source data integration
Interpretability	Clear physical basis	Requires explainable AI techniques
Scalability	Super-linear scaling with system size	Linear/sub-linear scaling
Uncertainty quantification	Safety factors and deterministic	Probabilistic frameworks
Real-time capability	Not feasible	Enabled with trained models
Feature extraction	Manual parameter selection	Automatic pattern recognition

in engineering practice. Existing computational frameworks frequently require extensive preprocessing procedures, significant computational resources, and specialized expertise for implementation and interpretation. Traditional analysis methods' lengthy processes frequently clash with the urgent post-earthquake assessment requirements, where quick damage assessment is essential for infrastructure recovery plans, evacuation decisions, and emergency response planning. Real-time or near-real-time evaluations are not feasible due to the weeks of computation required for traditional finite element analyses of large tunnel networks on high-performance computing clusters. Furthermore, real-time monitoring data from increasingly complex sensor networks installed in contemporary tunnel systems is difficult for conventional methods to integrate efficiently. Continuous streams of acceleration, strain, displacement, and environmental data are produced by these monitoring systems; these data streams may offer insightful information for evaluating the structural health and performing predictive maintenance, but their efficient use necessitates sophisticated data processing skills that go beyond conventional analytical techniques.

Given the unpredictability of geological conditions and the intrinsic randomness of earthquake ground motions, probabilistic approaches that can effectively traverse large parameter spaces while maintaining engineering accuracy and dependability are essential. Traditional deterministic approaches, while providing valuable insights, are limited in their ability to capture the full range of possible outcomes and the associated likelihoods. Such probabilistic characterization is crucial for modern performance-based seismic design and risk-informed decision-making, which require a comprehensive understanding of uncertainties to ensure safe and resilient tunnel infrastructure [3, 4]. Furthermore, the multi-scale nature of tunnel seismic response—which includes phenomena ranging from long-period regional ground deformation to high-frequency local stress concentration—makes it difficult for current analytical frameworks to handle. Complex modeling techniques that can effectively move between in-depth local analysis and comprehensive system-level evaluation are needed to couple various spatial and temporal scales.

This review examines the development, application and performance of AI-ML techniques in tunnel seismic response analysis from 2020 to 2025. The research highlights the need to understand how new AI-ML methods can improve traditional approaches and identifies barriers to wider use. The main research goals include evaluating supervised learning, deep learning, unsupervised learning and hybrid AI-FEM methods related to tunnel seismic issues. A quantitative assessment will measure improvements in prediction accuracy and computational efficiency compared to traditional methods. The research will also document practical implementations and insights from real-world applications. Additionally, it will identify ongoing technical and practical challenges, such as data standardization, model interpretability and regulatory acceptance. Finally, it will outline future directions that focus on federated learning, physics-informed models, and standardized validation frameworks [5].

Applications ranging from damage prediction and classification systems to real-time monitoring and early-warning systems, design optimization and parameter identification processes, probabilistic risk assessment techniques, and quick post-earthquake evaluation protocols are all included in the scope. Geographic coverage includes global research contributions emphasizing developments in seismically active regions where practical implementation and validation opportunities exist. Temporal scope concentrates on recent developments from 2020 to 2025, capturing the period of most rapid advancement in AI-ML applications to geotechnical problems, while referencing foundational work from earlier periods where necessary for a comprehensive understanding. The analysis prioritizes peer-reviewed research from high-impact Q1 and Q2 journals to ensure methodological rigor and practical relevance [6].

Peer-reviewed articles from high-impact journals such as Tunneling and Underground Space Technology, Computers and Geotechnics, Engineering Applications of Artificial Intelligence, and related venues are examined in this review using a systematic literature analysis methodology. Selection criteria prioritized recent publications (2020–2025), methodological rigor, practical relevance to tunnel seismic analysis, and validation through experimental or field data. Comprehensive database queries utilizing keyword combinations such as "artificial intelligence," "machine learning," "deep learning," "neural networks," "tunnel," "seismic," "earthquake," and "underground structures" were part of the search strategy. Publications were evaluated based on methodological soundness, novelty of approach, validation quality, and relevance to tunnel seismic analysis applications. Quality assessment focused on peer-review status, journal impact factor, citation metrics, and validation methodology. Only studies with appropriate experimental validation, field-data comparison or rigorous numerical verification were included. The analysis excluded conference papers, technical reports, and non-peer-reviewed sources to maintain academic rigor and reliability.

2 Traditional Methods and AI-ML Evolution in Tunnel Engineering

2.1 Conventional Seismic Analysis Approaches

The most common computational method for simulating tunnel seismic response is still finite element modeling, which allows for the detailed depiction of structural damping, soil non-linearity, and three-dimensional wave propagation effects. Current FEM implementations use complex material models that range from advanced plasticity formulations that incorporate strain-hardening, degradation effects, and dynamic behavior under cyclic loading conditions to linear elastic assumptions for preliminary design. Geological stratification, complex boundary conditions, construction sequence effects, and tunnel geometry can all be fully represented using three-dimensional finite element simulations. But as models get more complex, the computational load grows exponentially, frequently requiring trade-offs between accuracy and viability. They are not suitable for parametric studies, uncertainty quantification, or time-sensitive post-earthquake assessment scenarios because typical analyses of large tunnel networks take days to weeks of computation time on high-performance computing clusters.

Analytical solutions, while more limited in scope compared to numerical methods, provide valuable theoretical insights and serve as benchmarks for numerical model validation. Closed-form solutions for circular tunnels in elastic materials under plane-wave propagation were made available by Wang's groundbreaking 1993 work. Other researchers later built on this work to include elliptical cross-sections, layered geological profiles, and more complex loading conditions. These analytical methods rely on major simplifications. They assume that geological conditions are uniform, that materials behave in a linear-elastic way, and that tunnel shapes and loading conditions are simplified. These factors limit how well the solutions apply to real engineering situations.

Practical engineering applications and early design stages continue to benefit greatly from empirical correlations and streamlined design processes that were largely developed from post-earthquake damage observations and experimental studies. Notable databases include comprehensive damage assessments from major earthquakes including the 1995 Kobe earthquake, 1999 Chi-Chi earthquake, 2009 L'Aquila earthquake, and extensive damage compilations from the 2011 Tōhoku earthquake. Although these empirical methods facilitate quick preliminary evaluations and offer useful engineering judgment tools, they are limited in their ability to be applied to other geological settings, tunnel layouts, or seismic properties because they naturally reflect the constraints and conditions of their source datasets.

2.2 Fundamental Limitations of Traditional Approaches

Complex tunnel seismic analysis problems present a number of fundamental limitations for traditional analytical methods. Although it is computationally feasible, the assumption of linear elastic soil behavior ignores the important non-linear response characteristics that emerge during powerful seismic events. Although they are challenging to include in oversimplified analytical frameworks, soil plasticity, strain-softening, and degradation effects can significantly affect tunnel response.

Table 2 shows the significant efficiency improvements that AI-ML methods offer compared to traditional FEM in tunnel seismic analysis. FEM typically takes days to weeks to compute on high-performance clusters (1× baseline). In contrast, simplified analytical methods work in hours on a standard workstation (about 100× speedup). Trained deep learning models complete inference in minutes to hours on GPU-enabled systems, achieving over 1000× speedup. Physics-informed neural networks provide a good balance of accuracy and efficiency (around 500×), while ensemble and graph neural network methods achieve speedups of 200× and 800×, respectively, on multi-core or GPU platforms. This comparison highlights how AI-ML methods allow for real-time or near-real-time assessments that conventional FEM cannot deliver.

Conventional methods for treating soil-structure interaction frequently rely on simplified interface models that might not adequately capture the intricate sliding, separation, and contact behavior seen during seismic events. Although load transfer mechanisms and response characteristics can be significantly changed by interface

Table 2 Computational complexity comparison

Method	Analysis time	Computational resources	Scalability	Speedup factor
Traditional FEM	Days to weeks	HPC clusters	Super-linear	1× (baseline)
Simplified analytical	Hours	Standard workstation	Linear	100×
Deep learning	Minutes to hours	GPU-enabled system	Sub-linear	1000×+
Physics-informed neural network	Hours	GPU-enabled system	Linear	500×
Ensemble methods	Hours	Multi-core system	Linear	200×
Graph neural networks	Minutes	GPU-enabled system	Linear	800×

debonding and sliding, these effects are difficult to simulate using conventional analytical techniques.

Uncertainty quantification presents another significant limitation of conventional approaches. While engineering practice has traditionally addressed uncertainty through safety factors and deterministic approaches, modern performance-based design methodologies require explicit probabilistic analysis and uncertainty propagation. Traditional methods struggle to efficiently perform the extensive parametric studies necessary for robust uncertainty quantification.

2.3 *Emergence of AI-ML in Geotechnical Engineering*

The use of artificial intelligence and machine learning in geotechnical engineering has been slowly evolving over several decades. In recent years, this evolution has sped up due to advances in computing, increased data availability and improvements in algorithms. This evolutionary trajectory provides essential context for understanding current capabilities and future potential of AI-ML applications in tunnel seismic analysis. In an effort to codify engineering experience and judgment into rule-based computational frameworks, early AI applications in geotechnical engineering during the 1980s and 1990s mostly concentrated on expert systems and knowledge-based methodologies. These systems were limited by their inability to capture intricate engineering reasoning processes and their high knowledge acquisition requirements, but they showed promise for assisting decision-making processes in soil classification, foundation design and slope stability analysis.

The emergence of neural networks in geotechnical applications during the 1990s marked a fundamental departure from rule-based approaches, introducing data-driven methodologies capable of learning complex non-linear relationships directly from observational datasets. Although early neural network implementations used comparatively simple architectures, sparse training datasets, and simple backpropagation algorithms, they showed promise in detecting patterns in diverse geotechnical data that were beyond the capabilities of human pattern recognition.

Recent developments in deep learning structure have changed AI-ML abilities and potential applications in geotechnical engineering. CNNs, which were originally created for computer vision tasks, are very effective at analyzing geotechnical images. This includes soil microscopy, geological cross-sections, and infrastructure inspection photos. Recurrent neural networks and their advanced types, like LSTM and gated recurrent unit networks, are great at handling sequential data that is common in geotechnical monitoring and analyzing time-dependent behavior [7, 8]. The application of AI in geotechnical engineering has expanded significantly, with comprehensive reviews documenting over 300 citations for AI methods in geotechnical applications, demonstrating the growing acceptance and effectiveness of these approaches across diverse problem domains [9].

2.4 Integration Challenges and Opportunities

There are both tremendous obstacles and previously unheard-of opportunities when combining AI-ML techniques with conventional geotechnical analysis methods. Keeping interpretability for engineering decision-making, validating model performance under various conditions, and guaranteeing the physical consistency of AI-ML predictions with basic engineering principles are some of the challenges. On the other hand, effective integration presents the possibility of fusing improved predictive abilities with theoretical rigor.

The limitations of purely data-driven or purely theoretical approaches can be addressed with hybrid approaches that use AI-ML for parameter identification, uncertainty quantification, or result interpretation while preserving physics-based modeling frameworks. These hybrid systems maintain the theoretical underpinnings and interpretability of traditional engineering methods while utilizing machine learning's capacity for pattern recognition.

3 AI-ML Techniques for Tunnel Seismic Analysis

3.1 Deep Learning Architectures

With their outstanding performance in damage assessment, geological profile interpretation, and cross-sectional response analysis, CNNs have completely changed the spatial data analysis capabilities in tunnel seismic applications. A major improvement over conventional image processing techniques, CNNs' hierarchical feature extraction capabilities allow for the automatic identification of pertinent spatial patterns without the need for manual feature engineering.

Research by Ansari et al. [10] developed a comprehensive deep learning model for predicting tunnel damage under seismic environments, achieving correlation coefficients (R) of 0.914 for damage prediction compared to historical earthquake data. The study utilized feedforward neural networks trained on datasets incorporating peak ground acceleration, source-to-site distance, overburden depth, lining thickness, tunnel diameter, and geological strength index. Performance evaluation against tunnel damage from the 1999 Chi-Chi earthquake, 2004 Mid-Niigata earthquake, and 2008 Wenchuan earthquake demonstrated consistent results with mean absolute errors ranging from 0.189 to 0.257 [10]. Table 3 shows that deep learning models often do better than traditional methods in predicting seismic damage to tunnels. When trained on datasets from multiple earthquakes and tested against major events like Chi-Chi and Wenchuan, feedforward neural networks achieve correlation coefficients of about 0.91 and mean absolute error (MAE) values under 0.26. More complex CNN + U-Net architectures, with an MAE near 0.15, boost accuracy to 0.92. PINNs remain physically consistent with a correlation of around 0.89, whereas traditional methods yield correlations between 0.78 and 0.85, along with MAE values between 0.30 and 0.45, highlighting the greater predictive strength of AI-ML techniques.

Pre-trained CNN models created for general image recognition tasks combined with transfer learning techniques have greatly reduced the amount of training data required and improved model performance for tunnel-specific applications. VGG, ResNet, and Inception architectures, pre-trained on large image datasets, offer strong feature extraction abilities that work well for tunnel engineering challenges when the right adjustment methods are applied. The safety and stability analysis of tunnel construction has shown recent developments in CNN applications. For tunnel face stability analysis in layered soils with spatial variability, Zhang et al. [11] created deep learning-based prediction models. By combining CNNs with U-Net architectures, they were able to achieve impressive agreement with finite element simulations while greatly increasing computational efficiency. Tunnel face collapse failure modes in spatially variable soils can be accurately discriminated thanks to the U-Net architecture, which produces accurate displacement field images based on random field inputs [11].

LSTM networks and their variants represent breakthrough technologies for analyzing sequential data typical of tunnel monitoring applications and seismic response time histories. The ability of LSTM architecture to capture both short-term and long-term temporal dependencies while avoiding vanishing gradient problems makes them ideally suited for processing time-series data from accelerometers, strain sensors, and displacement monitoring systems. LSTM networks are great at modeling how tunnels respond during seismic events. They capture the complex links between ground motion characteristics, structural response patterns, and damage development over time. The bidirectional processing abilities of LSTM architecture allow it to include information from both past and future time steps. This gives a thorough temporal analysis for assessing damage after earthquakes.

Table 3 Performance metrics of deep learning models

Study/model	Architecture	Correlation coefficient (R)	MAE range	Training dataset	Validation earthquake
Ansari et al. [10]	Feedforward NN	0.914	0.189–0.257	Multi-earthquake dataset	Chi-Chi, Mid-Niigata, Wenchuan
Zhang et al. [11]	CNN + U-Net	0.920	0.150–0.230	FEM simulation data	Synthetic validation
Conventional methods	Empirical/ analytical	0.780–0.850	0.300–0.450	Post-earthquake surveys	Historical events
Deep learning average	Various architectures	> 0.910	0.150–0.260	Combined datasets	Multiple events
Physics-informed NN (PINN)	PINN	0.890	0.180–0.280	Physics + data	Validation studies

Gated Recurrent Unit networks have advantages over LSTM architectures in terms of processing while still modeling time effectively. This makes them appealing for real-time monitoring tasks that require fast processing. Studies show that GRU networks can reach similar accuracy levels as LSTM models but with lower computing needs. This allows them to be used on limited-resource monitoring devices commonly found in tunnel applications.

For thorough tunnel response prediction, advanced hybrid architectures that combine convolutional and recurrent networks have demonstrated superior performance. Comprehensive analysis frameworks that can process temporal seismic response sequences and spatial tunnel geometry data simultaneously are made possible by the combination of CNN's spatial feature extraction capabilities with LSTM temporal sequence modeling. These hybrid models provide more thorough predictive capabilities than single-architecture approaches by allowing the simultaneous analysis of geological profiles, tunnel cross-sectional features, and time-dependent response characteristics. Complex tunnel response phenomena under dynamic loading conditions can be predicted more precisely thanks to the synergistic combination of spatial and temporal processing.

Transformer architectures, which were initially created for applications in natural language processing, are cutting-edge technologies that hold great promise for solving complex sequential dependency and multi-modal data integration tunnel seismic analysis problems. Transformer models' self-attention mechanisms provide both computational benefits and an enhanced ability to capture long-range dependencies by processing lengthy sequences without the sequential processing constraints of recurrent architectures. Different aspects of input sequences can be processed in parallel by multi-head attention architectures in transformer models. This allows for the simultaneous analysis of multiple frequency components in seismic data or different spatial locations in monitoring networks. By visualizing attention weights, this parallel processing capability improves interpretability and benefits computational efficiency.

3.2 Machine Learning for Feature Importance Analysis

Machine learning-driven approaches for analyzing feature importance of seismic parameters have provided valuable insights into factors controlling tunnel damage and fragility assessment. These techniques enable systematic evaluation of parameter sensitivity and identification of critical design variables that most strongly influence tunnel seismic performance.

For feature importance analysis in tunnel seismic problems, ensemble methods such as random forests and gradient boosting algorithms show promise. In order to improve prediction accuracy and provide quantitative measures of parameter importance, these methods combine multiple base learners that have been trained on various aspects of the problem. Engineering comprehension is facilitated and design optimization efforts are guided by the ability to rank input variables according to their impact on model predictions.

Geological conditions and structural parameters have different effects based on seismic intensity and loading characteristics, according to feature importance analysis. Tunnel seismic response is primarily determined by parameters like tunnel depth, soil type, lining thickness, and geological strength index. Secondary factors, such as construction technique and interface features, have a more context-dependent impact.

Real-time analysis capabilities, which are crucial for emergency-response applications, have been made possible by the development of machine learning-based seismic response forecasting models. Rapid evaluation of tunnel networks during and immediately after seismic events is made possible by these models, which show notable gains in computational efficiency while preserving prediction accuracy on par with conventional methods [12].

3.3 Physics-Informed Neural Networks

Physics-informed neural networks are an important step in combining basic engineering principles with data-driven learning. These models include key differential equations, conservation laws, and constitutive relationships directly in the loss functions of the neural network. This approach ensures that predictions meet physical constraints while using observational data to improve accuracy and generalization.

The use of physics-based constraints tackles key issues found in purely data-driven methods. These methods often lead to predictions that can breach basic engineering principles or struggle when applying results outside the training data ranges. Physics-informed methods show better generalization skills than standard neural networks, especially when there is little training data or when the conditions for use vary from the training situations.

Research shows that by integrating soil mechanics concepts, wave propagation theory, and structural dynamics equations into the learning framework, physics-informed neural networks perform better for tunnel seismic analysis applications. In addition to allowing the model to identify intricate patterns in observational data that may not be sufficiently captured by simplified analytical models, this integration guarantees that learned relationships maintain their physical meaning [13, 14].

3.4 Graph Neural Networks and Network Topology

For simulating intricate spatial relationships and connectivity patterns in tunnel networks, graph neural networks have become extremely effective tools. Tunnel network analysis, where load transfer mechanisms and structural connectivity are crucial, is a perfect fit for these architectures because they are excellent at representing and analyzing systems where the relationships between components are just as significant as the properties of the individual components.

Comprehensive system-level analysis that considers interdependence—which is frequently disregarded in conventional segment-by-segment approaches—is made possible by graph neural networks' capacity to directly encode tunnel network topology, including connections between tunnel segments, junctions, and interfaces with surface structures. When evaluating system-level vulnerabilities and cascading failure mechanisms, this capability is especially helpful.

Multi-layer perceptron graph neural networks demonstrate effective handling of local dependencies and spatial relationships while maintaining computational efficiency for large network analyses. These architectures can simultaneously process node features (tunnel segment properties) and edge features (connection characteristics) to provide comprehensive network-level predictions.

3.5 *Artificial Intelligence in Earthquake Detection and Monitoring*

The application of AI in earthquake detection has evolved significantly, with machine learning algorithms now capable of identifying previously hidden seismic events with unprecedented sensitivity and accuracy. Research at Stanford University has developed earthquake detection algorithms that identify millions of microquakes previously undetected by conventional methods. This offers important insights into how faults behave and the processes behind earthquakes, which are useful for analyzing tunnel seismic activity [15].

These AI detection systems use attention mechanisms like those in natural language processing. This allows them to complete automated analysis that would normally take months of expert work in just minutes. The technology shows promise for near-real-time earthquake monitoring with machine learning methods. It provides useful data for tunnel monitoring and assessment systems.

The Earthquake Transformer model represents a significant advancement in simultaneous earthquake detection and phase picking, achieving state-of-the-art performance for processing continuous seismic waveform data. This technology enables automated processing of large-scale seismic monitoring networks, providing enhanced data quality and completeness for tunnel seismic analysis applications.

3.6 *Transfer Learning and Domain Adaptation*

Transfer learning techniques have shown remarkable promise for reducing training data requirements and computational costs in tunnel seismic analysis applications.

These approaches leverage pre-trained models developed on related problems to accelerate training and improve performance when applied to new tunnel engineering problems with limited available data.

The ability to adapt models trained in one geological or seismic environment to different conditions is particularly valuable for tunnel engineering, where each project may involve unique combinations of geological conditions, structural configurations, and seismic environments. Transfer learning enables effective utilization of knowledge gained from well-documented case studies to enhance analysis capabilities for new projects.

Domain adaptation techniques address the challenge of applying machine learning models across different problem domains while maintaining prediction accuracy and reliability. These methods are especially important for tunnel applications where models trained on one type of tunnel system (e.g., soft-ground tunnels) may need adaptation for application to different systems (e.g., rock tunnels) [16].

4 Performance Assessment and Comparative Analysis

4.1 Accuracy Improvements over Traditional Methods

Extensive validation studies show that in a variety of tunnel seismic analysis applications, AI-ML techniques consistently outperform conventional analytical methods in terms of damage prediction accuracy. Deep learning models achieve correlation coefficients exceeding 0.91 for tunnel damage assessment, compared to 0.85–0.87 typical of conventional empirical approaches, representing substantial improvements in predictive capability [10].

The capacity of AI-ML techniques to automatically recognize and take advantage of intricate non-linear relationships in multi-dimensional datasets that may be missed by more straightforward analytical models accounts for their higher accuracy. Traditional approaches often rely on linear or piece-wise linear relationships that may inadequately represent the complex interactions between geological conditions, structural parameters, and seismic loading characteristics. Seismic vulnerability analysis using AI-enhanced methods shows improved prediction of damage states across different performance levels. For intermediate damage states that are difficult to forecast with traditional methods, machine learning models show exceptional efficacy, offering more precise and trustworthy fragility evaluations for performance-based design applications.

The effectiveness of AI applications in seismic exploration has been shown through extensive literature analysis. This research reveals significant growth in machine learning use for seismic processing and interpretation tasks. A meta-analysis of over 200 publications indicates that AI-based methods are more efficient for denoising and frequency extrapolation than traditional techniques. Accuracy improvements range from 15 to 40%, depending on the application domain [17].

4.2 Computational Efficiency and Scalability

Particularly for large-scale tunnel network analysis, AI-ML techniques show impressive computational efficiency advantages over conventional finite element simulations. Practical real-time or near-real-time assessment capabilities are made possible by neural network models, which, once trained, provide predictions almost instantly as opposed to the days or weeks needed for thorough FEM analyses of intricate tunnel networks.

Deep learning surrogate models for tunnel stability analysis achieve computational speedup factors exceeding 1000:1 compared to traditional numerical methods while maintaining comparable accuracy. Extensive parametric studies and uncertainty quantification that would be impossible with traditional methods are made possible by research showing that deep learning models can successfully replace computationally demanding finite element analyses for wide rectangular tunnels in spatially variable soils [18].

The scalability advantages of AI-ML methods become particularly apparent for network-level analysis involving hundreds or thousands of tunnel segments. While trained machine learning models show linear or sub-linear scaling, traditional finite element approaches show computational complexity that scales super linearly with system size, making it possible to analyze large tunnel networks practically. Even though they could be substantial, training computational requirements are one-time expenses that are usually spread out over a large number of applications. Modern distributed training techniques and cloud computing resources make it feasible to train sophisticated models even for organizations with limited computational infrastructure.

4.3 Real-Time Monitoring and Early Warning Capabilities

Cutting-edge AI-ML frameworks make it possible to integrate sensor network data from continuous monitoring systems installed in tunnel systems, offering previously unheard-of capabilities for early warning and real-time structural health assessment. For the identification of incipient damage, multi-source data fusion techniques that combine accelerometer readings, strain measurements, displacement sensors, and environmental monitoring achieve detection accuracies of over 90%.

An important development in tunnel safety management is the creation of AI-assisted real-time monitoring systems. Continuous data streams from various sensor types can be processed by these systems, which can then spot unusual patterns of behavior that could point to emerging structural issues, geological dangers, or equipment failures. By integrating edge computing capabilities, sensor data can be processed locally, lowering the need for communication bandwidth and facilitating quicker reaction times for urgent alerts.

Figure 1 illustrates an end-to-end real-time tunnel monitoring framework where sensor networks (accelerometers, strain gauges, and displacement sensors) feed raw data into edge computing units for on-site preprocessing (quality control and filtering). These cleansed data streams proceed to an AI-ML analytics engine for pattern recognition and anomaly detection, with physics-based models providing consistency checks before results are presented on a real-time dashboard. Processed information and alerts are archived in cloud storage for historical analysis and forwarded to an emergency response system to enable rapid decision support.

Machine learning models optimized for real-time operation must balance prediction accuracy with computational efficiency, requiring careful consideration of model complexity and feature selection. Recent developments in model compression and quantization techniques enable deployment of sophisticated AI-ML models on resource-constrained hardware platforms typical of tunnel monitoring applications.

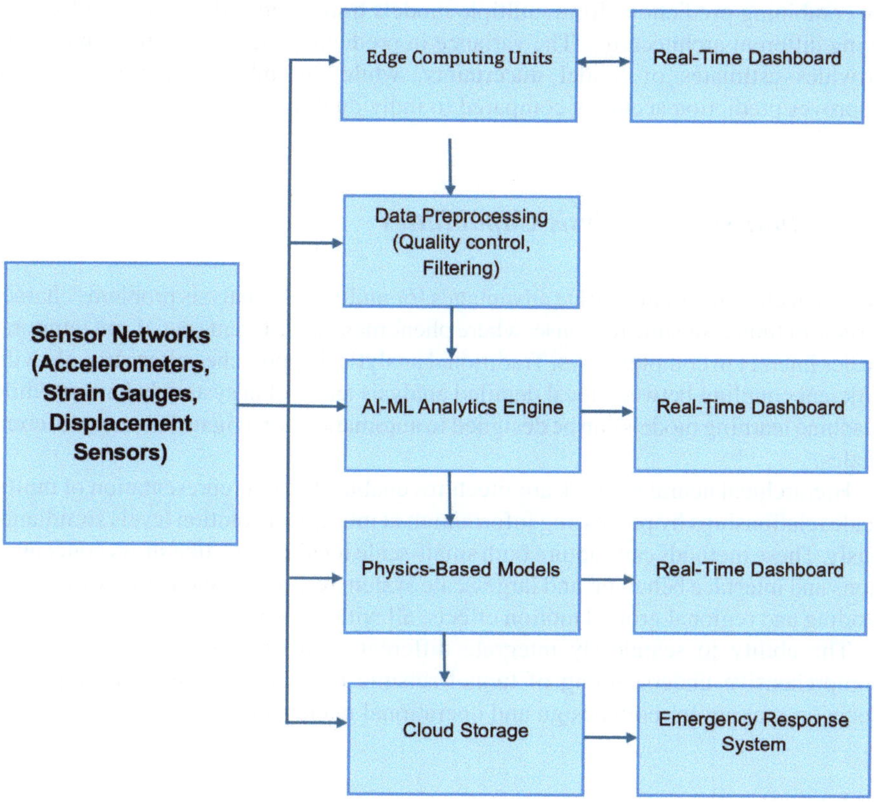

Fig. 1 Real-time monitoring system architecture

4.4 Uncertainty Quantification and Reliability Assessment

The integration of uncertainty quantification capabilities within AI-ML frameworks addresses a critical limitation of traditional deterministic approaches in tunnel seismic analysis. Probabilistic deep learning methods, including Bayesian neural networks and ensemble approaches, provide explicit quantification of prediction uncertainty that is essential for risk-informed engineering decisions.

Bayesian neural networks offer principled approaches to uncertainty quantification by maintaining probability distributions over network parameters rather than point estimates. This approach enables estimation of both epistemic uncertainty (arising from model limitations) and aleatoric uncertainty (arising from inherent variability in the problem), providing comprehensive uncertainty characterization for engineering applications.

Ensemble methods provide alternative approaches to uncertainty quantification by combining predictions from multiple models trained on different data subsets or using different architectures. The variance in predictions across ensemble members provides estimates of model uncertainty, while ensemble averaging typically improves prediction accuracy compared to individual models.

4.5 Multi-scale Analysis Capabilities

AI-ML techniques demonstrate advantages for multi-scale analysis problems characteristic of tunnel seismic response, where phenomena at different spatial and temporal scales interact in complex ways. Traditional analytical approaches often struggle with efficient coupling between local detailed analysis and global system behavior, while machine learning models can be designed to automatically navigate between different scales.

Hierarchical neural network architectures enable efficient representation of multi-scale relationships by processing information at multiple resolution levels simultaneously. These methods can capture both small-scale local effects, like stress concentrations and interface behavior, and large-scale system responses, such as network-level loading and regional ground motion effects, all within a single modeling framework.

The ability to seamlessly integrate different scales of analysis enables more comprehensive understanding of tunnel seismic behavior and supports improved decision-making for both design and operational management applications.

5 Implementation Challenges and Solutions

5.1 Data Availability and Quality Challenges

The biggest obstacle to widespread AI-ML adoption in tunnel seismic analysis is the lack of high-quality, standardized datasets for training and validating models. Unlike other engineering fields that have well-organized data repositories and set collection methods, tunnel seismic response data is scattered across various research studies, consulting reports, and institutional databases. There is little coordination in terms of data format, quality control, or accessibility for research.

Existing datasets often lack important metadata such as detailed geological descriptions, construction method records, monitoring system details, and post-earthquake damage assessment guidelines needed for model development. The private nature of many high-quality datasets, due to commercial confidentiality and competitive factors, makes model development more difficult. This limitation could hinder advancements that would help the engineering community as a whole.

Figure 2 depicts a sequential flowchart that ensures high-quality AI-ML inputs by guiding data from diverse collection sources through validation, quality control, standardization, and integration steps before model training approval. Inconsistencies in measurement protocols and sensor specifications between projects, incomplete or absent documentation of crucial parameters, calibration and maintenance records for monitoring equipment, and differences in data processing and interpretation techniques are examples of problems with data quality. These quality problems can restrict the ability to generalize across various projects and conditions and have a substantial impact on the efficacy of model training. There are frequently significant gaps in the temporal and spatial coverage of the datasets that are currently available, with little representation of extreme events, uncommon geological conditions, or innovative construction methods. This restriction is especially troublesome for machine learning techniques that need extensive training data in order to function well under a variety of circumstances.

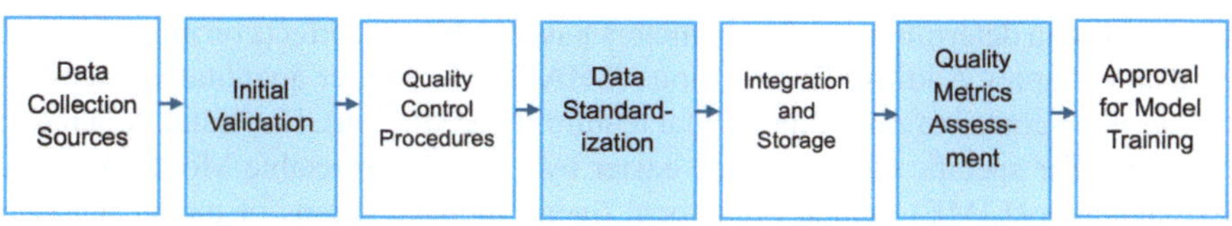

Fig. 2 Data quality assessment framework

5.2 Standardization and Interoperability Issues

Successful AI-ML implementation requires development of standardized data formats, measurement protocols and quality-assurance procedures enabling data sharing and collaborative model development across institutions and geographic regions. The absence of widely accepted standards creates inefficiencies in model development and limits the ability to validate models across different datasets and applications.

International collaboration through professional organizations and research consortia is essential for establishing comprehensive standards and creating shared data repositories. The development of standardized ontologies for tunnel engineering data, including geological characterization, structural parameters, monitoring systems and damage assessment criteria, would facilitate data integration and model interoperability.

Technical standards for AI-ML model validation, performance assessment, and documentation are needed to ensure consistency and reliability across different implementations. These standards should address model training procedures, validation methodologies, performance metrics, and reporting requirements appropriate for engineering applications.

5.3 Model Interpretability and Explainability

The "black box" nature of many AI-ML algorithms creates significant barriers to acceptance in risk-averse engineering practice, where understanding model reasoning is essential for validation, regulatory approval and professional confidence. Deep learning models, while demonstrating superior predictive performance, often provide limited insight into decision-making processes or parameter sensitivity, creating challenges for engineering validation and acceptance.

Promising strategies to improve model interpretability while preserving predictive performance are offered by recent advancements in explainable AI (XAI) techniques. Engineers can determine which parameters have the biggest effects on model outputs by using Shapley Additive Explanations (SHAP), which offer quantitative measurements of the contributions of individual features to predictions. Understanding model reasoning for specific cases is made easier by Local Interpretable Model-Agnostic Explanations (LIME), which offer local linear approximations of model behavior around particular predictions. Engineers can verify model behavior against engineering judgment and physical understanding by using attention mechanisms and feature visualization techniques to shed light on the temporal and spatial patterns that affect model predictions. In safety-critical applications where model failures could have major repercussions, these interpretability tools are especially crucial for gaining acceptance of AI-ML techniques.

The development of domain-specific interpretability tools tailored to tunnel engineering applications remains a priority for facilitating practical implementation. These tools should highlight engineering-relevant relationships and provide insights that support rather than replace engineering judgment and expertise.

5.4 *Integration with Existing Engineering Workflows*

Practical implementation of AI-ML techniques requires seamless integration with established engineering software platforms, design procedures, and project delivery methodologies. Many AI-ML research developments remain isolated academic exercises that fail to address practical integration challenges encountered in engineering practice, limiting their potential for widespread adoption.

The integration challenge encompasses both procedural elements, such as documentation standards, validation protocols, and quality-assurance procedures, as well as technical elements, such as software compatibility, data format conversion, and computational resource requirements. Successful integration requires collaboration between AI-ML researchers, software developers, and practicing engineers to ensure practical usability and regulatory compliance.

Hybrid approaches that include machine learning components within traditional finite element frameworks can help solve integration challenges while keeping workflows familiar for engineers. These methods allow engineers to adopt AI-ML techniques gradually. They do not need to replace existing analysis procedures or software investments all at once. User interface design is another important part of successful integration. It requires creating interfaces that are easy for engineers to use, not just data scientists. The complexity of many AI-ML tools can be masked by user-friendly interfaces that show results in familiar engineering terms and formats.

5.5 *Regulatory and Professional Standards*

The development of appropriate regulatory frameworks and professional standards for AI-ML applications in tunnel engineering represents a critical challenge that extends beyond purely technical considerations. Current engineering codes and design procedures do not adequately address validation, application and quality assurance requirements for machine learning models, creating uncertainty regarding regulatory acceptance and professional liability.

Professional engineering societies and regulatory agencies are starting to create guidelines for AI and machine learning applications in civil engineering. However, complete frameworks that specifically focus on tunnel seismic analysis are still being developed. These frameworks must address model validation procedures, performance standards, documentation requirements, and continuing education needs for practicing engineers.

The creation of certification processes for AI-ML models used in safety-critical applications may become essential to ensure quality and reliability. These processes should balance the encouragement of innovation with the protection of public safety, requiring careful thought about suitable standards and assessment methods. Professional education and training programs must develop to include AI-ML concepts and applications. This change ensures that practicing engineers have the skills and knowledge needed for effective implementation. Adding AI-ML education to traditional civil engineering courses and ongoing education programs is a key part of successful technology transfer.

6 Future Research Directions and Emerging Technologies

6.1 Federated Learning and Collaborative Development

Federated learning is a cutting-edge method of cooperative AI and machine learning that puts data security and privacy first, removing obstacles to data exchange between nations and organizations. Confidentiality and competitive advantages are maintained by allowing the training of strong models using collective knowledge from various tunnel projects without centralizing data. This approach has the potential to create worldwide databases of tunnel vulnerability that cover a range of geological circumstances, building techniques, and operational experiences. Transportation agencies must address issues like heterogeneous data formats, communication bandwidth constraints, differential privacy concerns, and efficient model aggregation strategies as they acknowledge the advantages of federated learning for improving tunnel assessments. If these problems are resolved, comprehensive models of tunnel seismic vulnerability may result, greatly enhancing cooperative research and practice in the area.

6.2 Physics-Informed Machine Learning Evolution

The continued evolution of physics-informed machine learning represents a critical research direction for addressing fundamental limitations of purely data-driven approaches while leveraging the pattern recognition and optimization capabilities of artificial intelligence. Advanced physics-informed architectures integrate governing differential equations, conservation laws, and constitutive relationships at multiple scales within unified machine learning frameworks.

Future developments should focus on multi-scale physics integration, enabling seamless coupling between different spatial scales (from local stress concentrations to regional ground motion) and temporal scales (from high-frequency wave propagation to long-term structural evolution). This integration requires sophisticated mathematical frameworks that can efficiently navigate between different physical domains while maintaining computational tractability.

The incorporation of uncertainty quantification within physics-informed frameworks enables probabilistic predictions that account for both model uncertainty arising from simplified physics representations and parameter uncertainty arising from imperfect knowledge of material properties and boundary conditions. Bayesian physics-informed neural networks provide principled approaches to uncertainty propagation that are essential for risk-informed engineering decisions.

Advanced physics-informed architectures should also incorporate adaptive capabilities that enable models to learn and update physical relationships based on observational data while maintaining consistency with fundamental principles. This capability would enable continuous model improvement based on monitoring data and operational experience.

6.3 Quantum Machine Learning Applications

Quantum machine learning has the potential to change how we analyze tunnel seismic data, especially when dealing with optimization problems that have many parameters and complex constraints. These problems can be difficult for traditional computing methods. Quantum machine learning allows us to explore large solution spaces at the same time, which offers important theoretical benefits for tunnel engineering.

Current research focuses on developing quantum-inspired algorithms that can work on classical hardware. These algorithms have shown better performance in multi-objective tunnel design, which considers seismic performance, costs, material use, and operational limits. Variational quantum algorithms also show promise for handling complicated tasks like identifying parameters, optimizing designs, and quantifying uncertainty by using the principles of quantum superposition and entanglement.

However, we need advancements in quantum hardware to make practical applications possible. This drives the effort to create algorithms that work with current quantum technology and to find specific problem types that could gain from these new methods.

6.4 Edge Computing and Internet of Things Integration

An emerging paradigm for distributed tunnel monitoring and analysis is the combination of edge computing and Internet of Things (IoT) technologies with AI-ML techniques. At tunnel monitoring stations, edge computing makes it possible to process sensor data locally, which lowers the bandwidth needed for communication and speeds up the response time for important alerts during seismic events.

Modern tunnel monitoring systems produce large amounts of data from different types of sensors, including accelerometers, strain gauges, displacement sensors, environmental monitors, and video surveillance systems. To use this data effectively, we need distributed processing capabilities. These systems can carry out real-time analysis while handling communication and storage limitations.

Predictive performance, computational efficiency, and resource limitations common to tunnel monitoring hardware must all be balanced in AI-ML models tailored for edge deployment. Pruning, quantization, and knowledge distillation are examples of model compression techniques that make it possible to implement complex algorithms on platforms with limited resources. One promising avenue for real-world application is the creation of hierarchical processing architectures that integrate cloud-based processing for thorough analysis with edge processing for quick response. This architecture supports more extensive analysis and model-updating capabilities while facilitating local anomaly detection and emergency response.

6.5 Digital Twin Technology Integration

Digital twin technology represents an emerging paradigm that combines physical monitoring, mathematical modeling, and AI-ML analytics within integrated frameworks for comprehensive asset management. Digital twins for tunnel systems would integrate real-time sensor data with physics-based models and machine learning algorithms to provide continuous assessment of structural condition, performance prediction, and optimization recommendations.

The development of digital-twin frameworks requires integration of diverse data sources including construction records, geological surveys, monitoring systems, maintenance logs, and operational data within unified platforms. AI-ML techniques play critical roles in data fusion, pattern recognition and predictive analytics within these integrated systems.

Digital twins enable scenario analysis and what-if studies that support decision-making for maintenance planning, retrofit design, and emergency response. The combination of real-time data with predictive models enables proactive rather than reactive management approaches that can improve safety and reduce lifecycle costs.

6.6 Explainable AI and Human-AI Collaboration

The future of AI-ML applications in tunnel engineering will likely involve human-AI collaborative systems that leverage the complementary strengths of machine learning algorithms and human engineering expertise. These systems should be enhanced rather than replace human decision-making by providing data-driven insights while maintaining transparency and interpretability.

Advanced explainable AI techniques specifically designed for engineering applications should provide insights that align with engineering thinking and support professional judgment. These tools should highlight relevant physical mechanisms, parameter sensitivity, and uncertainty characteristics that enable engineers to validate and understand model predictions.

The development of interactive AI systems that enable engineers to explore model behavior, test hypotheses, and incorporate domain knowledge represents an important direction for practical implementation. These systems should support iterative analysis and decision-making processes typical of engineering practice.

7 Case Studies and Practical Applications

7.1 Himalayan Tunnel Seismic Assessment

The use of AI-ML techniques in tunnel projects in the Himalayan region shows how they can work in tough geological and seismic conditions. A recent review of tunnel seismic sensitivity through P-wave dynamics offers important information for building infrastructure in areas that are tectonically active and have complicated geological features [19].

The methodology developed for Himalayan tunnel assessment addresses specific challenges, including heterogeneous geological formations, high seismic activity levels, limited access for conventional monitoring approaches, and extreme environmental conditions. The analytical and numerical frameworks identify hazard zones with peak ground accelerations ranging from 0.3 g to greater than 0.5 g, providing a quantitative basis for design and risk assessment decisions.

Results from Himalayan tunnel studies demonstrate the effectiveness of AI-ML approaches for integrating diverse data sources including geological surveys, seismic monitoring networks, and structural response measurements. The ability to process and analyze complex multi-dimensional datasets enables more comprehensive understanding of tunnel vulnerability than traditional approaches.

The practical guidelines developed for design and retrofitting of sustainable underground structures in the Himalayan region provide templates for similar applications in other tectonically active mountain regions globally, demonstrating the transferability of AI-ML methodologies across different geological contexts.

7.2 Large-Diameter Urban Tunnel Construction

The application of machine learning techniques for enhancing safety in large-diameter tunnel construction represents significant progress in practical implementation for complex urban infrastructure projects. Integration of robust optimization with ML techniques enables real-time risk assessment and decision-making during construction operations in challenging urban environments [20].

These applications show how well AI and machine learning can manage complex construction processes. They deal with various uncertainties and risks, such as geological changes, construction-related settlement, and interactions with existing infrastructure. The ability to process real-time monitoring data and give immediate feedback supports flexible construction management strategies.

Machine learning models that use construction monitoring data can predict potential problems before they become critical. This allows for proactive intervention and risk reduction. This ability is especially important in urban tunnel construction, where damage to nearby infrastructure can lead to serious issues. Combining AI and machine learning tools with construction management systems shows real ways to adopt technology in industry. It also offers templates for wider use in tunnel construction projects.

7.3 *Transportation Infrastructure Monitoring*

Applications of AI-ML in transportation tunnel projects have demonstrated special promise for emergency response planning, predictive maintenance, and condition assessment. Continuous assessment of structural health and early detection of emerging issues that might otherwise go unnoticed until serious damage occurs are made possible by the combination of automated monitoring systems and machine learning algorithms.

Machine learning-based damage assessment models for transportation infrastructure demonstrate effectiveness for rapid post-earthquake evaluation, enabling prioritized response and resource allocation in emergency situations. These systems can process data from distributed sensor networks and provide a near-real-time assessment of tunnel conditions across extensive transportation networks [21].

The development of predictive maintenance strategies based on machine learning analysis of historical performance data, monitoring measurements, and environmental conditions enables more effective resource allocation and maintenance planning. These approaches can extend infrastructure service life while reducing maintenance costs through optimized intervention timing.

7.4 *Deep Learning for Construction Risk Management*

The use of deep learning techniques for managing construction risks in tunnel projects shows great promise for improving safety and efficiency during construction. AI-powered risk assessment systems can analyze various data sources, including geological conditions, construction parameters, environmental factors and equipment performance. This allows for thorough risk evaluation and management suggestions.

Recent research shows that convolutional neural networks effectively analyze geological conditions before tunnel faces. This helps predict challenging conditions and optimize construction parameters. These applications are especially promising for TBM operations, where knowing geological conditions in advance can greatly improve construction efficiency and safety.

The integration of computer vision techniques with construction monitoring enables automated assessment of construction quality, identification of potential problems, and documentation of construction progress. These capabilities support quality assurance programs and provide data for continuous improvement of construction processes.

7.5 International Collaboration and Technology Transfer

International collaborative projects demonstrate the potential for technology transfer and knowledge sharing in AI-ML applications for tunnel engineering. Collaborative research between institutions in different countries enables validation of AI-ML techniques across diverse geological conditions, construction practices, and regulatory environments.

The development of standardized datasets and validation protocols through international collaboration supports broader adoption of AI-ML techniques by providing common benchmarks for performance evaluation. These collaborative efforts also facilitate development of best practices and professional guidelines for AI-ML implementation. Technology transfer from research institutions to industry practice requires careful attention to practical implementation challenges including software integration, training requirements, and validation procedures. Successful technology transfer programs typically involve close collaboration between researchers, software developers, and practicing engineers throughout the development and implementation process.

In Table 4, case studies from various geological settings are compared to show how AI-ML techniques, such as CNN + U-Net in TBM operations and P-wave dynamics analysis in Himalayan tunnels, consistently improve risk assessment and hazard identification. Predictive maintenance frameworks allow for quick post-earthquake assessments in transportation networks, while real-time machine learning optimization in urban tunnels enhances construction safety through proactive interventions. The importance of integrating data from multiple sources and working together seamlessly with emergency response and construction systems is emphasized in key lessons.

Table 4 Case study comparison matrix

Case study	Geological conditions	AI-ML method used	Performance outcome	Key lessons learned
Himalayan tunnel assessment	Heterogeneous formations, high seismicity	P-wave dynamics analysis, multi-source integration	Hazard zones identified (0.3–0.5 g PGA)	Effective for complex geological settings
Large-diameter urban tunnel	Urban soil conditions, construction constraints	Real-time risk assessment, ML optimization	Enhanced construction safety, proactive intervention	Integration with construction management
Transportation infrastructure	Variable geological conditions	Predictive maintenance, condition assessment	Rapid post-earthquake evaluation	Improved emergency response capability
Deep learning construction	TBM operations, layered soils	CNN + U-Net architecture	Accurate geological condition prediction	Valuable for pre-construction risk assessment

8 Limitations and Critical Assessment

8.1 Fundamental Limitations of Current Approaches

Despite significant progress in AI-ML applications for tunnel seismic analysis, several key limitations remain that affect practical use and reliability. The reliance of machine learning models on the quality and representativeness of training data creates weaknesses when applied to conditions that differ greatly from training scenarios. This issue is especially concerning for tunnel projects, where each one may involve distinct combinations of geological conditions, structural designs, and seismic environments.

The "black box" nature of many deep learning approaches creates interpretability challenges that conflict with engineering practice requirements for understanding and validating analysis results. Even though explainable AI techniques offer some partial answers, many practitioners are still uneasy with models that, especially in safety-critical applications, lack clear justification for their predictions.

Extrapolation abilities of AI-ML models outside their training data are still limited and not well-defined. This creates risks when they are used in unusual conditions or extreme events that might not be adequately covered in training datasets. This limitation is worrisome for seismic applications, where rare but severe events could be the most critical design scenarios.

The computational requirements for training sophisticated AI-ML models can be substantial, potentially limiting accessibility for organizations with limited computational resources. While inference is typically fast, the initial model development phase may require significant investment in computational infrastructure and expertise.

8.2 *Data Quality and Availability Constraints*

The quality and availability of training data represent perhaps the most significant practical limitation for AI-ML implementation in tunnel seismic analysis. Many existing datasets lack comprehensive documentation of critical parameters including detailed geological characterization, construction methodology specifications, monitoring system calibration records, and post-event damage assessment protocols.

Temporal coverage of available datasets often has significant gaps. They offer limited representation of extreme seismic events that are critical for design and safety assessment. Major earthquakes are rare in many regions, so local datasets might not include enough examples of severe loading conditions to support solid model training. Also, spatial coverage limitations mean that models trained on data from specific geographic regions may not work well in different geological or seismic settings. This issue is particularly challenging for tunnel applications because geological conditions can differ greatly even within the same project.

Quality control and standardization issues across different datasets create challenges for model training and validation. Inconsistencies in measurement procedures, sensor specifications, data processing methods, and reporting standards can introduce biases or artifacts that compromise model reliability.

8.3 *Model Validation and Verification Challenges*

Rigorous validation and verification of AI-ML models for safety-critical tunnel engineering applications requires development of specialized testing protocols and performance metrics that may not be directly transferable from other application domains. Traditional engineering validation approaches based on analytical benchmarks and physical testing may not adequately address the complexity and scope of AI-ML model behavior.

The stochastic nature of machine learning models creates challenges for deterministic validation approaches commonly used in engineering practice. The ability of AI-ML models to produce different outputs for nominally identical inputs (due to random initialization, training data sampling, or algorithmic stochasticity) requires probabilistic validation frameworks that may be unfamiliar to practicing engineers. The scarcity of thorough validation datasets makes cross-validation across various geological settings, seismic environments, and geographic regions difficult. Standardized test cases and benchmark problems would make it easier to compare and validate models, but they also need to be coordinated between academic institutions and business associations.

Long-term performance monitoring of deployed AI-ML systems is necessary to assess model degradation, domain drift, and changing performance characteristics over time. This monitoring requirement represents an ongoing commitment that extends beyond initial model development and deployment phases. Figure 3 shows a step-by-step validation process where AI-ML models go through systematic performance testing, starting with initial cross-validation and moving to physics consistency checks. This happens before an engineering expert reviews the models and conducts field-based testing. A regulatory review checks that they meet safety standards, allowing deployment only after all quality requirements are satisfied. Continuous monitoring with feedback loops addresses model drift and mandates periodic revalidation to maintain reliability in safety-critical tunnel applications.

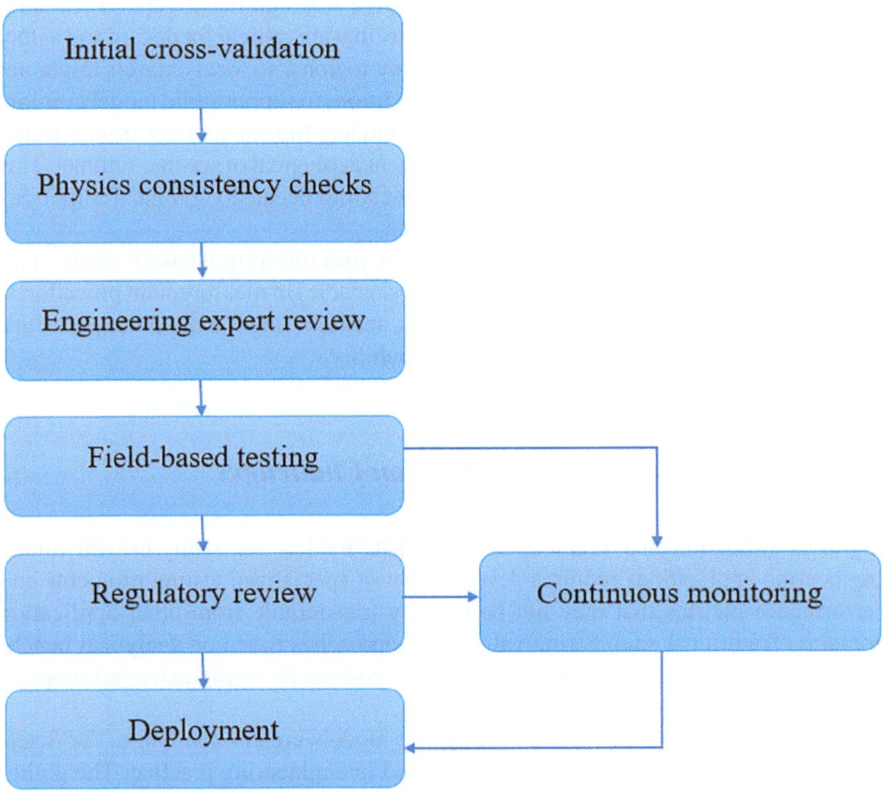

Fig. 3 Model validation framework

8.4 Integration and Implementation Barriers

The integration of AI-ML techniques with existing engineering workflows, software platforms, and organizational procedures presents significant practical challenges that extend beyond purely technical considerations. Many engineering organizations lack the internal expertise necessary for effective AI-ML implementation, requiring substantial training and education investments.

Software compatibility issues between AI-ML development platforms and established engineering software can create technical barriers that require custom interface development or workflow modifications. The rapid evolution of AI-ML technologies creates additional challenges for maintaining compatibility and currency with latest developments. Adoption of AI-ML techniques may be slowed by conservative pressures created by professional liability and regulatory acceptance, especially in safety-critical applications where failure consequences are severe. It takes time and collaboration between several stakeholders to create suitable regulatory frameworks and professional standards.

Cultural and organizational resistance to new technologies can create implementation barriers that are independent of technical merit. Successful technology adoption requires change management strategies that address human and organizational factors in addition to technical requirements.

9 Conclusions

This comprehensive analysis of AI-ML applications in tunnel seismic response reveals a technological convergence where multiple advanced computational paradigms collectively address fundamental limitations of traditional analytical approaches. The documented evidence from Q1 and Q2 journals demonstrates that AI-ML techniques have achieved quantifiable superiority across multiple performance dimensions while creating new analytical capabilities previously unattainable in tunnel engineering practice.

According to the study, hybrid architectures that combine several AI-ML techniques produce better results than those that rely on just one technique. The main technology that connects these approaches is physics-informed neural networks. These networks maintain data-driven adaptability while ensuring theoretical consistency. The successful combination of CNN spatial processing skills with LSTM temporal modeling demonstrates effective multi-scale coupling. This addresses the main challenge of analyzing spatial and temporal data at the same time in tunnel seismic response. The analysis identifies three critical convergence factors necessary for successful technology transfer: standardized validation protocols specifically tailored to tunnel engineering applications, explainable AI frameworks that align with engineering reasoning processes, and regulatory frameworks that balance innovation with public safety requirements. These factors represent the primary determinants of widespread professional adoption beyond current research applications.

Federated learning frameworks are a game-changer for resolving data privacy issues and facilitating international collaborative model development. They have the potential to produce extensive databases of tunnel vulnerability that take advantage of various geological and seismic conditions across the globe. This capability addresses the fundamental limitation of geographically restricted training datasets that currently constrain model generalization. The research demonstrates that computational efficiency gains exceeding 1000:1 speedup factors fundamentally enable new analytical capabilities including real-time assessment, comprehensive uncertainty quantification, and extensive parametric optimization that are practically impossible with traditional methods. These efficiency improvements represent a qualitative rather than merely quantitative advancement in tunnel seismic analysis capabilities.

Digital twin architectures integrated with AI-ML analytics create comprehensive intelligent infrastructure systems that enable proactive rather than reactive management strategies. The combination of real-time monitoring, predictive analytics, and physics-based modeling within unified platforms represents a fundamental departure from current practice toward continuous optimization and adaptive management. The documented case studies from Himalayan tunnels, urban construction projects, and transportation infrastructure demonstrate that technology transfer success requires coordinated multi-stakeholder collaboration between research institutions, software developers, practicing engineers, and regulatory agencies. Successful implementation demands systematic attention to both technical integration and organizational change management requirements.

The available evidence shows that artificial intelligence and machine learning (AI-ML) techniques are not just added extras; they are essential for performing seismic analysis in tunnels. These technologies can greatly improve resilience and safety management for infrastructure in areas prone to earthquakes worldwide. Given their proven effectiveness and the maturity of the technology, it is anticipated that these methods will see broad adoption in the industry. This transition will be facilitated by addressing implementation challenges through coordinated efforts in professional development.

AI Use Disclosure Statement The authors did not use any generative AI or AI-assisted technologies in the creation of this manuscript. The authors used generative AI technologies only for proofreading. This includes minor corrections to grammar.

References

1. Tsinidis G, de Silva F, Anastasopoulos I, Bilotta E, Bobet A, Hashash YMA, He C, Kampas G, Knappett J, Madabhushi G, Nikitas N, Pitilakis K, Silvestri F, Viggiani G, Fuentes R (2020) Seismic behaviour of tunnels: from experiments to analysis. Tunn Undergr Space Technol 99:103334. ISSN 0886-7798. https://doi.org/10.1016/j.tust.2020.103334

2. He J, Hu S, Guo Y (2022) Study on seismic dynamic response of deep-buried tunnel based on tectonic stress. Adv Mater Sci Eng 2022:1–13. https://doi.org/10.1155/2022/6775989
3. Civera M, Dalmasso M, Chiaia B (2024) Assessing the seismic performance of underground infrastructures to near-field earthquakes. Int J Civ Infrastruct 07. https://doi.org/10.11159/ijci.2024.004
4. Xia C, Wang Z, He Y et al (2024) Seismic response of mountain tunnel induced by fault slip. Sci Rep 14:17768. https://doi.org/10.1038/s41598-024-67225-0
5. Chen D, Wang X, Bai J, Chu Y, Wang X, Zhao J, Li M (2025) An overview of ahead geological detection technologies in tunnels. Eur J Remote Sens 58(1):2503240. https://doi.org/10.1080/22797254.2025.250324
6. Tsinidis G, Karatzetzou A, Stefanidou S, Markogiannaki O, Argyroudis S (2022) Recent advances in seismic vulnerability assessment of tunnels and underground structures
7. Ma L, Han L, Feng Q (2024) Deep learning for high-resolution seismic imaging. Sci Rep 14:10319. https://doi.org/10.1038/s41598-024-61251-8
8. Zhang W, Ni P, Zhao M, Du X (2025) Dynamic analysis of underground structures with uncertain-but-bounded parameters. Case Stud Constr Mater 22:e04371. ISSN 2214-5095. https://doi.org/10.1016/j.cscm.2025.e04371
9. Baghbani A, Choudhury T, Costa S, Reiner J (2022) Application of artificial intelligence in geotechnical engineering: a state-of-the-art review. Earth Sci Rev 228:103991. ISSN 0012-8252. https://doi.org/10.1016/j.earscirev.2022.103991
10. Ansari A, Rao KS, Jain AK et al (2023) Deep learning model for predicting tunnel damages and track serviceability under seismic environment. Model Earth Syst Environ 9:1349–1368. https://doi.org/10.1007/s40808-022-01556-7
11. Zhang Z, Wang Z-Z, Goh S-H, Ji J (2024) Deep learning–based prediction of tunnel face stability in layered soils using images of random fields. J Geotech Geoenviron 150:4024065. https://doi.org/10.1061/JGGEFK.GTENG-1210
12. Muthukumar S, Sathyan D, Premjith B, Shukla SK (2025) Machine learning-based seismic response forecasting using feature mapping algorithms and scientometric analysis of nailed vertical excavation in a soil mass. Cogent Eng 12(1). https://doi.org/10.1080/23311916.2025.2467144
13. Yuan B, Choo CS, Yeo LY, Wang Y, Yang Z, Guan Q et al (2025) Physics-informed machine learning in geotechnical engineering: a direction paper. Geomech Geoeng 20(5):1128–1159. https://doi.org/10.1080/17486025.2025.2502029
14. Sheil B, Anagnostopoulos C, Buckley R, Ciantia MO, Febrianto E, Fu J, Gao Z, Geng X, Gong B, Hanley K, He P, Kolomvatsos K, Lopes BCFL, Ninic J, Previtali M, Rezania M, Ruiz-Lopez A, Sun J, Suryasentana S, Taborda D, Utili S, Whyte S, Zhang P (2026) Artificial intelligence transformations in geotechnics: progress, challenges and future enablers. Comput Geotech 189:107604. ISSN 0266-352X. https://doi.org/10.1016/j.compgeo.2025.107604
15. Mousavi SM, Ellsworth WL, Zhu W et al (2020) Earthquake transformer—an attentive deep-learning model for simultaneous earthquake detection and phase picking. Nat Commun 11:3952. https://doi.org/10.1038/s41467-020-17591-w
16. Zhang W, Zhang Y, Gu X, Wu C, Han L (2021) Deep learning and applications. https://doi.org/10.1007/978-981-16-6835-7_4
17. Anjom FK, Vaccarino F, Socco LV (2024) Machine learning for seismic exploration: where are we and how far are we from the holy grail? Geophysics 89:WA157–WA178. https://doi.org/10.1190/geo2023-0129.1
18. Nguyen H, Xu H, Nazem M, Sousa R, Kowalski J, Zhao Q (2025) Deep-learning surrogate models for the stability of a wide rectangular tunnel. Comput Geotech 179:106946. https://doi.org/10.1016/j.compgeo.2024.106946
19. Abdullah A, Pranjal M, Ahmad MB (2025) Unveiling the seismic sensitivity of the Himalayan tunnels: a comprehensive assessment through analytical and numerical exploration of P-wave dynamics. Front Built Environ 11. ISSN 2297-3362. https://doi.org/10.3389/fbuil.2025.1486533

20. Singh J, Singh P, Ravi V, Kumar S, Diwakar M, Mazroa A, Gupta I (2024) Advancing tunnel construction reliability with automated artificial intelligence under geotechnical and aleatoric uncertainties. Open Civ Eng J. https://doi.org/10.2174/0118741495343684240910115546
21. Belhadj FZ, Belhadj AF, Chabaat M (2025) Machine-learning seismic damage assessment model for building structures. Artif Intell Geosci 6(2):100155. ISSN 2666-5441. https://doi.org/10.1016/j.aiig.2025.100155

Novel Score Function Enriched Quartic Fuzzy Group Decision-Making Framework on Assessment of Solar Photovoltaic Module Recycling Technology

Raili Basu, Sayanta Chakraborty, and Apu Kumar Saha

Abstract Selection of the most suitable Solar Photovoltaic Module Recycling Technology (SPVMRT) to recover valuable metals like Silicon (Si), Aluminum (Al), Copper (Cu), Silver (Ag), Ethylene Vinyl Acetate (EVA) layer and rare earth materials like gallium, indium, tellurium from the End-of-life (EoL) Solar Photovoltaic Modules (SPVMs) is the focus of this research. With this aim, a Multi-Criteria Group Decision-Making (MCGDM) framework has been modeled, and to handle the real-world ambiguity and opinionative uncertainty among Decision-Makers (DnMs), Quartic Fuzzy Environment (QFE) is incorporated. Several SPVMRTs (alternatives) and their respective influencing factors (criteria) have been considered for mathematical assessment. To find criteria weights, the concept of Stepwise Weight Assessment Ratio Analysis (SWARA) is used and to come up with the optimal ranking of alternatives, the Complex Proportional Assessment (COPRAS) method has been extended under the Quartic Fuzzy Environment (QFE) under the aegis of Quartic Fuzzy Weighted Average (QFWA) operator and then integrated. To score the Quartic Fuzzy Numbers (QFNs), a novel score function under QFE has been introduced, satisfying each salient property. The final ranking has shown Chemical Etching (CE) as the best SPVMRT among the selected alternatives.

Keywords Solar photovoltaic module · Solar photovoltaic module recycling technologies · Quartic fuzzy set · Score function · SWARA-COPRAS method

1 Introduction

Solar energy is a crucial source of sustainable energy and with the increasing adoption of SPVMs, a large amount of waste is expected to come from EoL photovoltaic (PV) modules, creating a significant opportunity for recycling and recovery of valuable

R. Basu (✉) · S. Chakraborty · A. K. Saha
Department of Mathematics, National Institute of Technology Agartala, Agartala, Tripura, India
e-mail: raili.maths95@gmail.com

A. Azizi et al. (eds.), *Intelligence at the Interface*, Emerging Trends in Mechatronics,
https://doi.org/10.1007/978-981-95-7584-8_3

metals. Recovering Glass, Si, Ag, Al, Tin (Sn), etc., by dissolving or separating EVA layer from waste modules and further using them to make new> SPVMs is a win-win solution addressing waste regulation, boosting circular economy, reducing harmful environmental effects, and supporting sustainable development. The countries like Bangladesh, South and Southeast Asian region are in their rising stage of adopting solar power and using it as a renewable energy source. Within 2050, the estimated amount of solar waste generation may touch more than 78 million tons there [1]. Hence, recycling those wastes and recovering materials from them will cut down the manufacturing and investment cost, will stabilize the production rate of SPVMs.

The research focuses on finding most effective recycling procedure with high recovery and other implementing factors using a decision-making framework. The combinative opinions of multiple decision-makers (DnMs) over a multi-criteria decision-making (MCDM) scenario makes an MCGDM framework. And to deal with DnM's ambiguity over conceptual opinions, the fuzzy environment is incorporated in it [2]. Yager [3] primarily came up with the mathematical concept of Generalized orthopair fuzzy (q-ROF) set and as a specification of it, Arora and Naithani [4] introduced Quartic fuzzy sets (QFS), which expands the range of fuzziness to deal with any vagueness and uncertainty. To partially order the Quartic fuzzy numbers (QFNs) and understand their dominance in QFSs, a Score function was introduced by Arora and Naithani [4].

1.1 Literature Review

In recent years, noticeable progress in SPVMRTs including mechanical milling, dissolution of EVA layers, recovery of metals like silver, lead from silicon cells had been discussed by Tao et al. [5]. Lin et al. [6] speculated an extensive review over the recycling techniques related to crystalline-silicon solar cells including acid dissolution, physical dismantling, combinatorial heat and chemical treatment, and organic solvent dissolution. A study by Pereira et al. [7] addressed the application of supercritical water oxidation technology in recycling obsolete SPVMs and treating organic compounds as intact. The potential waste recycling and material recovery procedures in South and Southeast part of Asia, especially Bangladesh have been discussed by Rahman and Chowdhury [1], to develop sustainable economy and reduce cost effectiveness. Li et al. [8] cultivated a methodology to liberate critical metals like copper, indium, gallium from obsoleted copper indium gallium selenide thin-film solar modules. Reviewing hydrometallurgical and pyrometallurgical methods, exploring classification of numerous processes by their solvent and oxidizing reagents, Matrinez et al. [9] provided an article focusing on pyrometallurgical process and their influencing factors. An analytical discussion over the delamination of EoL SPVMs through mechanical, thermal, and chemical processes had been reviewed by Ghahremani et al. [10]. A thorough discussion over the recycling of different variation of solar panels worldwide had been done by Ndalloka et al. [11]. With an aim of reclaiming valuable metals from Silicon solar modules, a structured recycling approach was provided by Preet and Smith [12]. Putting light on futuristic aspects over reverse supply chain and the logistic networks on PV recycling, Iakovou et al. [13] developed a resource-task-network-based model. Integrating the efficiency of Laser irradiation with high voltage pulse crushing, a hypothetical hybrid technique had been proposed by Ali et al. [14] as an advancement over the existing recovery techniques like, crushing, electrostatic separation, hot-wire cutting, water jet cutting and magnetic separation. Lee et al. [15] focused their findings on the novel and advanced electrochemical approaches to recover silicon from EoL SPVMs. The challenges associated with the management of EoL SPVMs, material recovery approaches including pilot trials and laboratory scale demonstrations highlighting their influencing parameters had been argued by Sim et al. [16].

1.2 Research Gap and Motivation

Several recycling procedures are entitled theoretically and practically to recover valuable materials from End-of-Life (EoL) modules. But to choose the most suitable one among them which is highly cost effective, controlled, environment friendly and has a good recovery rate of valuable metals is still under observation.

From the literature, it is evident that there has not been much research on SPVMRT selection using MCGDM framework. And there exists no research on determining suitable SPVMRT with group decision-making under QFE. Till now, there has not been much existing research in MCGDM under the aegis of QFWA operator. These gaps motivate the researchers to come up with a fuzzy decision-making framework suitable for solving any kind of decision-making problems.

Also, the existing score function (SF) under QFE provides similar score for distinct QFNs under certain circumstances. Hence, generating a modified SF which satisfies all the salient properties of SFs and overcomes the drawbacks of existing one is also an objective.

1.3 Novelty and Contribution of the Research

- SPVMRT selection has been modeled under Quartic fuzzy MCGDM framework. The authenticity and applicability of the proposed integrated framework have been through a comparative analysis with the existing MCDM frameworks.
- A novel SF has been introduced and all its salient properties have been verified under QFE. The drawbacks of existing Quartic fuzzy SF have been numerically established and theoretically explained.
- NSF-based SWARA-COPRAS has been extended under QFE. The reliability, compatibility and superiority of NSF-based MCGDM method have been established through comparative analysis by incorporating existing SFs in proposed

MCGDM framework. Sensitivity of the method have been thoroughly checked by altering criteria weights.

The rest of the paper is organized as follows: Sect. 2 consists of the preliminary concepts of QFSs and respective operations, and the proposed novel SF-based integrated MCGDM method is discussed in Sect. 3, followed by its application in SPVMRT selection in Sect. 4. Incorporating the comparative analysis and sensitivity analysis in Sect. 5, the research made its conclusion in Sect. 6.

2 Preliminary

The basic idea of QFS and the operations related to it are provided here.

Definition 2.1 A QFS [4] over an underlying set U is defined as $Q = \{\mu_Q(x_i), \delta_Q(x_i)| x_i \in U\}$ where $0 \le \mu_Q(x_i) \le 1$ is the degree of membership and $0 \le \delta_Q(x_i) \le 1$ is the degree of non-membership satisfying $0 \le \mu_Q^4(x_i) + \delta_Q^4(x_i) \le 1$. $\pi = \sqrt[4]{1 - \mu^4 - \delta^4}$ is the degree of Hesitancy.

Definition 2.2 [4] Let Q_1 and Q_2 be two QFSs, then

i. $Q_1^c = \langle \delta_{Q_1}, \mu_{Q_1} \rangle$

ii. $Q_1 \boxplus Q_2 = \left\langle \sqrt[4]{\mu_{Q_1}^4 + \mu_{Q_2}^4 - \mu_{Q_1}^4 \mu_{Q_2}^4},\ \delta_{Q_1}\delta_{Q_2} \right\rangle$

iii. $Q_1 \boxtimes Q_2 = \left\langle \mu_{Q_1}\mu_{Q_2}, \sqrt[4]{\delta_{Q_1}^4 + \delta_{Q_2}^4 - \delta_{Q_1}^4 \delta_{Q_2}^4} \right\rangle$

iv. $kQ_1 = \left\langle \sqrt[4]{1 - \left(1 - \mu_{Q_1}^4\right)^k}, \delta_{Q_1}^k \right\rangle,\ k > 0$

v. $Q_1^k = \left\langle \mu_{Q_1}^k, \sqrt[4]{1 - \left(1 - \delta_{Q_1}^4\right)^k} \right\rangle,\ k > 0$

Definition 2.3 [4] The Score Function (SF) and Accuracy Function (AF) of a QFS 'Q' are defined as $\mathbb{S}(Q) = \mu_Q^4(x) - \delta_Q^4(x)$ and $(Q) = \mu_Q^4(x) - \delta_Q^4(x)$, respectively. Here, $x \in U$ (the underlying set), $\mathbb{S}(Q) \in [-1, 1]$ and$\mathbb{A}(Q) \in [-1, 1]$.

Definition 2.4 [4] Let $F_i = \langle \mu_i, \delta_i \rangle$ be Quantic fuzzy numbers and $\omega = (\omega_1, \omega_2, \ldots, \omega_n)^{\mathrm{T}}$ be a weight vector of F_i where $\sum_{i=1}^n \omega_i = 1$, then a quartic fuzzy weighted average (QFWA) operator is a function defined by

$$\mathrm{QFWA}(F_1, F_2, \ldots, F_n) = \left\langle \sum_{i=1}^n \omega_i \mu_i, \sum_{i=1}^n \omega_i \delta_i \right\rangle \quad (1)$$

3 Proposed Integrated Method

Here, the proposed integrated method consists of a novel score function (NSF)-based SWARA-COPRAS method under QFE.

3.1 Proposed Novel Score Function (NSF)

The QFNs with equal membership and non-membership degrees (like, $\langle 0.5, 0.5\rangle$, $\langle 0.55, 0.55\rangle$ etc.) will always score "0" with respect to the SF defined by Arora and Naithani [4]. Hence, a novel improvised SF has been constructed under QFE, which not only satisfies the basic properties of SF but also pass over the drawbacks of subsisting QF-SF.

$$\mathbb{S}(Q) = \frac{1}{\left(2 - e^{-1}\right)}\left(\mu^4 + e^{-\delta^4} - \frac{1}{e}\right) \tag{2}$$

Property 3.1.1 $1 \geq \mathbb{S}(Q) \geq 0$

Proof $0 \leq \mu^4, \delta^4 \leq 1$ implies $\frac{1}{e} \leq \mu^4 + e^{-\delta^4} \leq 2$. Hence $0 \leq \mu^4 + e^{-\delta^4} - \frac{1}{e} \leq 2 - \frac{1}{e}$.

Property 3.1.2 $\mathbb{S}(1, 0) = 1$ and$\mathbb{S}(0, 1) = 0$.

Proof Both the proofs are apparent.

Property 3.1.3 $\mathbb{S}(Q)$ is monotonically increasing corresponding to μ and monotonically decreasing corresponding to δ.

Proof
Applying partial derivative over $\mathbb{S}$ with respect to μ and δ are $\frac{\partial \mathbb{S}}{\partial \mu} = \frac{1}{(2-e^{-1})}\left(4\mu^3\right) \geq 0$ and $\frac{\partial \mathbb{S}}{\partial \delta} = \frac{1}{(2-e^{-1})}\left(-4\delta^3 e^{-\delta^4}\right) \leq 0$. Hence the proof.

Property 3.1.4 Q_1, Q_2 are two QFSs such that $\mu_{Q_1} \leq \mu_{Q_2}$ and $\delta_{Q_1} \geq \delta_{Q_2}$. Then $\mathbb{S}(Q_1) \leq \mathbb{S}(Q_2)$.

Proof As the exponential function is monotonic increasing, hence

$$\mathbb{S}(Q_1) = \frac{1}{\left(2 - e^{-1}\right)}\left(\mu_{Q_1}^4 + e^{-\delta_{Q_1}^4} - \frac{1}{e}\right) \leq \frac{1}{\left(2 - e^{-1}\right)}\left(\mu_{Q_2}^4 + e^{-\delta_{Q_1}^4} - \frac{1}{e}\right)$$
$$\leq \frac{1}{\left(2 - e^{-1}\right)}\left(\mu_{Q_2}^4 + e^{-\delta_{Q_2}^4} - \frac{1}{e}\right) \leq \mathbb{S}(Q_2)$$

3.2 NSF-Based SWARA under Quartic Fuzzy Environment

NSF-based SWARA method has been stepwise described below under QFE.

Step 1. Find the score of each criterion significance provided in QFE with the help of Eq. (2).

Step 2. Arrange the chosen criteria as per their score value from most to least significant one.

Step 3. Find the comparative importance which indicates how less significant the $(j+1)$th criterion is compared to jth criterion, using the relation.

$S_{j+1} = j$th importance $-(j+1)$th importance.

Step 4. Calculate coefficient value of criterion by

$$C_j = \begin{cases} 1, & \text{for the foremost value of } S_j \\ (1 + S_j), & \text{otherwise} \end{cases}$$

Step 5. Calculate the corrected values of weights by

$$\rho_j = \begin{cases} 1, & \text{for } C_j = 1 \\ \frac{\rho_{j-1}}{C_j}, & \text{otherwise} \end{cases}$$

Step 6. Assess the final criteria weight (CW) by

$$\omega_j = \frac{\rho_j}{\sum_j \rho_j} \tag{3}$$

3.3 NSF-Based COPRAS under Quartic Fuzzy Environment

NSF-based COPRAS method under QFE has been discussed step by step here.

Step 1. Find the weighted decision matrix $D_{m \times n} = [\omega_j d_{ij}]_{m \times n}$ where d_{ij} are the quartic fuzzy entries of the decision matrix with 'm' and 'n' as the number of alternatives and criteria, respectively.

Step 2. Estimate each alternative performance using the sum of their maximizing factors and minimizing factors, if τ_i and ϑ_i represents the sum of performances corresponding to beneficiary criteria (BC) and Cost criteria (CC) then $\tau_i = \underset{j=1}{\overset{l}{\boxplus}} \omega_j d_{ij}$ and $\vartheta_i = \underset{j=l+1}{\overset{n}{\boxplus}} \omega_j d_{ij}$.

'l' represents the total number of BC and $i = 1, 2, \ldots, m$.

Step 3. Finding score of τ_i and ϑ_i using Eq. (2), compute relative degree 'α_i' of alternatives using

$$\alpha_i = \mathbb{S}(\tau_i) + \frac{\min\limits_i \mathbb{S}(\vartheta_i) \sum_{i=1}^{m} \mathbb{S}(\vartheta_i)}{\mathbb{S}(\vartheta_i) \sum_{i=1}^{m} \frac{\min\limits_i \mathbb{S}(\vartheta_i)}{\mathbb{S}(\vartheta_i)}} \tag{4}$$

Step 4. Calculate the degree of utility by

$$\frac{\alpha_i}{\max\limits_i \alpha_i} \times 100\% \tag{5}$$

4 Application of the Proposed Integrated Method in Choosing SPVMRT

To choose the best SPVMRT, ten alternatives have been chosen with respect to twelve constraints (/criteria) over the scenario. Table 1 provides description of the alternatives and the influential criteria in that scenario are Cost ($\mathcal{C}1$), Material Recovery ($\mathcal{C}2$), Greenhouse Emission ($\mathcal{C}3$), Operational requirements ($\mathcal{C}4$), Worker requirement ($\mathcal{C}5$), Waste Reduction ($\mathcal{C}6$), Infrastructure ($\mathcal{C}7$), Occupational Hazard ($\mathcal{C}8$), Delamination Time ($\mathcal{C}9$), Cell Damage ($\mathcal{C}10$), Energy Requirement ($\mathcal{C}11$), Ecological Effect ($\mathcal{C}12$) [17].

Step 1. A panel of three DnMs have been considered to provide linguistic assessments of criteria significance and criterion-alternative interpretative performances. Then the linguistic data have been transformed into QFNs using the scale in Table 2.

Step 2. The decision matrix of DnMs over criteria significance has been aggregated by QFWA operator in Eq. (1). Here, three DnMs were prioritized with nearly equal weights $\langle 0.33,\ 0.33,\ 0.34 \rangle$.

Step 3. The aggregated criteria significance has been scored using the NSF defined in Eq. (2) and then the NSF-based SWARA method has been used to find criteria weight. The resultant criteria weights have been showcased in Table 3.

Step 4. Three decision matrices of criteria-alternative interpretative performances have been aggregated using Eq. (1) and then the 10×12 decision matrix has been considered for alternative ranking using the NSF-based COPRAS method. Table 4 and Fig. 1 represent the ranking with utility degree and the relative performance index of the alternatives, respectively.

Table 1 Alternatives

OSD	Organic solvent dissolution	Allows selective dissolution of encapsulant layer to segregate Si, Ag, glass, etc., keeping the other layers mostly unharmed
OSUI	Organic solvent and ultrasonic irradiation	Used to segregate glass and metal, ultrasonic high-amplitude irradiation uses waves to dismantle the compounds in dissolved EVA layer
ETH	Electro-thermal heating	Sends electric current to create immense heat by putting resistance, induction or arcs to segregate glass from crystallized solar plates
MSHC	Mechanical separation by hot-wire cutting	Cut the panels into smaller portions for further recovery of materials using a hot-wire, which is a thin nichrome alloy made of metal wire
PY	Pyrolysis	It is a thermal debonding process that dependents on the thermochemical and hydrodynamic functioning of module particles
SD	Solvent dissolution (HNO_3)	Use nitric acid (HNO_3) solution to dissolve the EVA layer to collect silicon
PD	Physical disintegration	Aim to recover glass and polyvinyl fluoride film from PV module by cutting into small pieces, then use shredding followed by crushing or grinding
DWMP	Dry and wet mechanical process	Wet mechanical erosion process by conducting friction is used to separate silicon, glass materials, followed by dry air recycling to remove water from recovered glass and metals
TT	Thermal treatment (two-step heating)	To separate Tedlar from module, it used to be heated at 330 °C for half an hour, then Tedlar and EVA layers were burnt at 400 °C for two hours
CE	Chemical etching	The procedure necessitates applying chemical solutions to the superficial portion of solar modules, dissolving semiconductor materials, and separating the glass substrates. The materials dissolved within can then be refined and recollected

Table 2 Quartic fuzzy scale of significance

Extremely high	Very high	High	Moderated high	Moderated	Moderated low	Low	Very low	Extremely low
⟨0.95, 0.08⟩	⟨0.85, 0.15⟩	⟨0.75, 0.3⟩	⟨0.65, 0.4⟩	⟨0.5, 0.5⟩	⟨0.4, 0.65⟩	⟨0.3, 0.75⟩	⟨0.15, 0.85⟩	⟨0.08, 0.95⟩

5 Result and Discussion

CE. MSHC and OSD came out as the best recycling technologies due to their high recovery rate, low energy requirements and less economic inclination and environmentally friendly way-outs. CE can precisely dissolve selective layers of EoL modules without damaging silicon wafers can recover high amount of silver, aluminum, and indium. As a result, high-value virgin silicon layers are recovered from EoL modules which can further be used for regeneration of new modules. Being a controlled closed-loop process and due to very less waste generation, CE is a sustainable choice for industrial level recovery and recycling plants.

Now, to examine the stability of the proposed integrated framework and to scrutinize the reliability of final results, the integrated MCGDM model has been analogously compared with subsisting MCDM methods and existing SFs.

5.1 Comparison with Subsisting MCDM Methods

Table 5 shows a little change in comparative ranking corresponding to existing MCDM methods and the relative performances of those alternatives are depicted in Fig. 2. Hence, Spearman's correlation coefficient (SRCC) [2] has been calculated and determined as $(1,\ \ 0.976,\ \ 0.988,\ \ 0.976,\ \ 0.998)^T$. SRCC runs between $[-1, 1]$ where -1 represents non-correlation between two quantitative variables and $+1$ represents fully correlation between them. As all the correlation values of existing MCDM methods with the proposed integrated MCGDM method are very close to $+1$, advising the proposed MCGDM method is well-correlated with subsisting ones, hence results related to SPVMRT selection are reliable and the proposed integrated method is well-derived and pertinent to deal with any real-life decision-making scenario.

5.2 Comparison with Subsisting SFs under QFE and Q-ROF Environment (Q-ROFE)

The comparative relative index of performance of alternatives corresponding to SF by Arora and Naithani [4] and q-ROF SFs by Li et al. [18], Mandal and Seikh

Table 3 Criteria weights

*C*1	*C*2	*C*3	*C*4	*C*5	*C*6	*C*7	*C*8	*C*9	*C*10	*C*11	*C*12
0.0916	0.0992	0.0848	0.0731	0.0730	0.0694	0.0802	0.0818	0.0916	0.0708	0.0850	0.0992

Table 4 Alternative ranking

	Utility degree	Rank
OSD	99.193	3
OSUI	98.517	6
ETH	96.128	9
MSHC	99.974	2
PY	95.077	10
SD	98.723	4
PD	98.115	7
DWMP	98.715	5
TT	96.476	8
CE	100	1

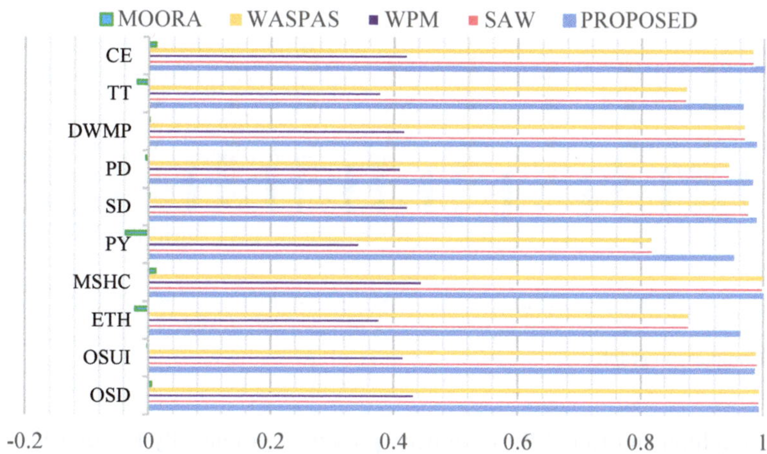

Fig. 1 Alternative relative performance index

Table 5 Ranking corresponding to existing MCDM methods

	OSD	OSUI	ETH	MSHC	PY	SD	PD	DWMP	TT	CE
Proposed	3	6	9	2	10	4	7	5	8	1
SAW	2	3	8	1	10	5	7	6	9	4
WPM	2	6	9	1	10	3	7	5	8	4
WASPAS	2	3	8	1	10	5	7	6	9	4
MOORA	3	6	9	1	10	4	7	5	8	2

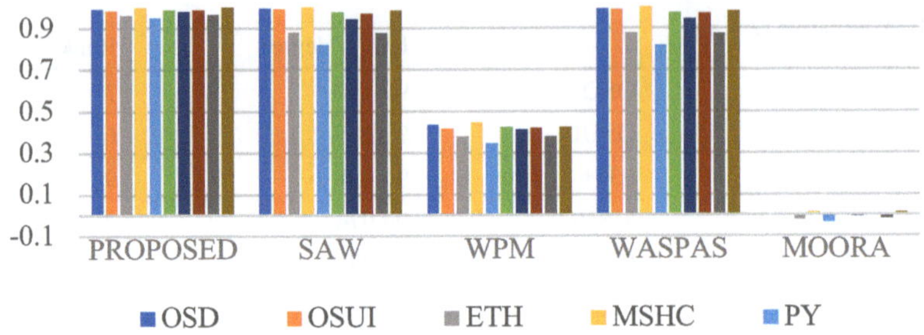

Fig. 2 Comparison graph

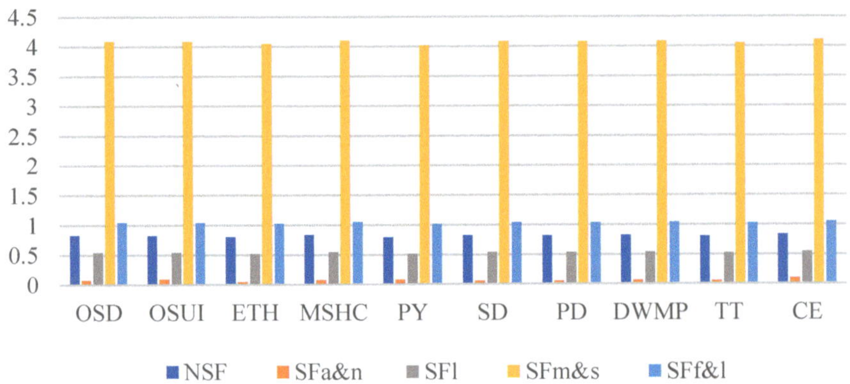

Fig. 3 Comparison graph

[19], Farhadinia and Liao [20] have been presented graphically through Fig. 3 and numerically through Table 6. Table 6 shows the ranking of alternatives using $SF_{a\&n}$ differs a little because of its drawbacks discussed in Sect. 3.1. While the ranking results for other SFs are nearly same with the proposed NSF-based framework. The SRCC are $(1,\ 0.9,\ 0.998,\ 0.998,\ 0.996)^T$, respectively. This indicates the Quartic fuzzy NSF is stable and the NSF-based integrated framework is capable enough to provide feasible solution dealing with real-life data. The superiority of NSF-based algorithm over the existing one defined by Arora and Naithani [4] has also been established through the variation of ranking results in Table 6, first and second row.

5.3 *Sensitivity Analysis*

Assorting the CWs concerning BC and CC, the sensitivity of the proposed model has been checked [21]. For all the cases, the alternative's rankings have been shown in Table 7. Table 7 shows that as the total percentage of weights for BC gets higher and the total percentage of weights of CC gets lower, the ranking differs from the

Table 6 Ranking corresponding to existing SFs

SFs	Fuzzy Environment	OSD	OSUI	ETH	MSHC	PY	SD	PD	DWMP	TT	CE
NSF	QFE	3	6	9	2	10	4	7	5	8	1
$SF_{a\&n}$ [4]	QFE	*5*	*1*	*10*	*4*	*3*	*6*	*9*	*7*	*8*	*2*
SF_{l} [18]	q-ROFE	3	6	9	1	10	4	7	5	8	2
$SF_{m\&s}$ [19]	q-ROFE	3	6	9	2	10	5	7	4	8	1
$SF_{f\&l}$ [20]	q-ROFE	3	6	9	1	10	5	7	4	8	2

The variation in the ranking order wirh respect to score function defined by Arora and Naithani [4] has been presented in Table 6 in terms of italics.

Table 7 Criteria weight variation

	OSD	OSUI	ETH	MSHC	PY	SD	PD	DWMP	TT	CE
90BC-10CC	6	10	5	2	1	9	7	8	3	4
80BC-20CC	6	10	5	2	1	8	9	7	4	3
70BC-30CC	4	10	9	2	1	7	8	6	5	3
60BC-40CC	4	9	10	1	3	5	7	6	8	2
50BC-50CC	3	7	10	1	8	4	6	5	9	2
40BC-60CC	3	6	9	1	10	4	7	5	8	2
Proposed	3	6	9	2	10	4	7	5	8	1
30BC-70CC	3	4	9	2	10	5	7	6	8	1
20BC-80CC	2	4	9	3	10	5	7	6	8	1
10BC-90CC	3	1	9	4	10	5	7	6	8	2

proposed one. The change of ranking for "OSD," "OSUI," "ETH," "PY," "SD," and "TT" are very significant. This indicates that the CCs play a crucial role in identifying the most suitable SPVMRT. Figure 4 also shows the sensitivity in degree of utility of alternatives for the same. Hence, the proposed integrated method is criteria sensitive.

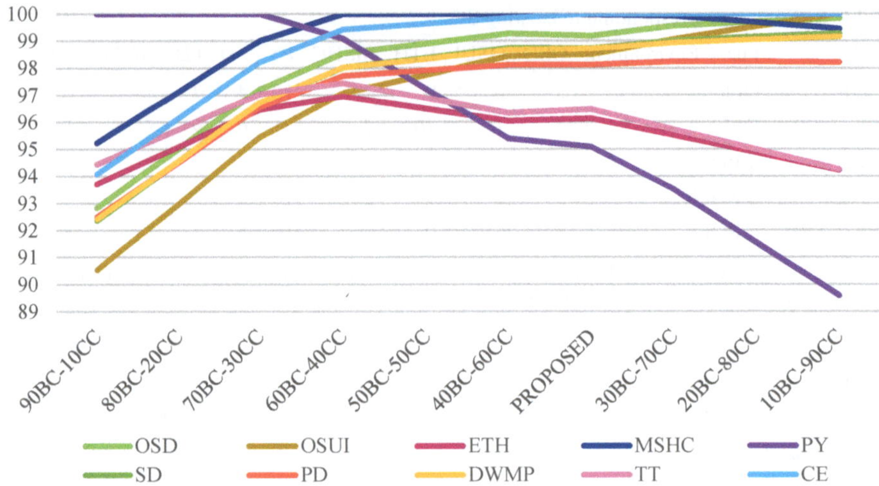

Fig. 4 Utility degree

6 Conclusion

Recycling EoL SPVMs and recovering high-value materials from it not only contribute in cost saving, also stabilize circular economy and material recycling and reuse. By this, the manufacturing costs of new SPVMs decrease, and the sustainability and longevity of earth metals increase. In India, the policies related to SPVM recycling are strict. The government authorities and private organizations have advanced facilities for SPVMRT in all the cities. In rural areas, government is incorporating many action plans to increase public awareness about the effectiveness and long-term benefits of solar power and initiating to install recycling plants with advanced technologies.

This research focuses on selecting the most suitable SPVMRT in those rural areas with high efficiency, cost effectiveness, minimal environmental repercussions and high recovery rates. The proposed NSF-based SWARA-COPRAS under QFE has successfully established its ability to find suitable SPVMRT as per the requirement. Other MCGDM problems (like food waste recycling, upcycling, medical diagnosis, etc.) can also be solved by this method by inventing new SFs, with new aggregation operators. The same problem can be encountered in different orthopair fuzzy environments which are special cases of q-ROFE (Intuitionistic fuzzy environment, Pythagorean fuzzy environment, Fermatean fuzzy environment, etc.) and other fuzzy environments like, Interval-valued fuzzy environment, Picture fuzzy environment, Spherical fuzzy environment, etc. A new problem focusing on SPVM treatment technologies considering a set of influential factors may bring a new direction to this proposed decision-making framework.

AI Use Disclosure Statement The authors used generative AI technologies only for proofreading. This includes minor corrections to grammar, spelling, and syntax. No AI was used for ideation, data analysis, or writing original content.

References

1. Rahman MM, Chowdhury FN (2023) Recycling of solar PV panels in Bangladesh along with south and South East Asia. Curr Opin Green Sustain Chem 44:100862
2. Basu R, Chakraborty S, Saha AK (2025) Novel q-rung Orthopair fuzzy distance based similarity measure and score function in real life decision making. Eng Appl Artif Intell 147:110348
3. Yager RR (2016) Generalized orthopair fuzzy sets. IEEE Trans Fuzzy Syst 25(5):1222–1230
4. Arora HD, Naithani A (2023) A new definition for quartic fuzzy sets with hesitation grade applied to multi-criteria decision-making problems under uncertainty. Decis Anal J 7:100239
5. Tao M, Chen T, Click N, Adcock R (2023) Recent progress and future prospects of silicon solar module recycling. Curr Opin Green Sustain Chem 44:100863
6. Lin D, Liu Z, Li X, Cao Z, Xiong R (2023) Development of metal-recycling technology in waste crystalline-silicon solar cells. Clean Energy 7(3):532–546
7. Pereira MB, de Souza GBM, Espinosa DCR, Pavao LV, Alonso CG, Cabral VF, Cardozo-Filho L (2023) Simultaneous recycling of waste solar panels and treatment of persistent organic compounds via supercritical water technology. Environ Pollut 335:122331
8. Li M, Widijatmoko SD, Wang Z, Hall P (2023) A methodology to liberate critical metals in waste solar panel. Appl Energy 337:120900
9. Martínez M, Barrueto Y, Jimenez YP, Vega-Garcia D, Jamett I (2024) Technological advancement in solar photovoltaic recycling: a review. Minerals 14(7):638
10. Ghahremani A, Adams SD, Norton M, Khoo SY, Kouzani AZ (2024) Delamination techniques of waste solar panels: a review. Clean Technol 6(1):280–298
11. Ndalloka ZN, Nair HV, Alpert S, Schmid C (2024) Solar photovoltaic recycling strategies. Sol Energy 270:112379. https://doi.org/10.1016/j.solener.2024.112379
12. Preet S, Smith ST (2024) A comprehensive review on the recycling technology of silicon based photovoltaic solar panels: challenges and future outlook. J Clean Prod:141661
13. Iakovou E, Pistikopoulos EN, Walzberg J, Iseri F, Iseri H, Chrisandina NJ, Nkoutche C (2024) Next-generation reverse logistics networks of photovoltaic recycling: perspectives and challenges. Sol Energy 271:112329. https://doi.org/10.1016/j.solener.2024.112329
14. Ali A, Shahid M, Qadir SA, Islam MT, Khan MW, Ahmed S (2024) Solar PV end-of-life waste recycling: an assessment of mechanical recycling methods and proposed hybrid laser and high voltage pulse crushing method. Resources 13(12):169
15. Lee J, Duffy N, Allen J (2025) A review of end-of-life silicon solar photovoltaic modules and the potential for electrochemical recycling. Adv Energy Sustainability Res 6(2):2400254
16. Sim Y, Ankit, Tay YB, Ravikumar D, Mathews N (2025) Open challenges and opportunities in photovoltaic recycling. Nat Rev Electr Eng 2(2):96–109
17. Basu R, Chakraborty S, Saha AK (2025) Assessment of end-of-life solar photovoltaic module recycling techniques using q-rung orthopair fuzzy information measure induced group decision-making. Int J Hydrog Energy 161:150610
18. Li J, Chen M, Pei S (2023) Generalized q-rung orthopair fuzzy interactive Hamacher power average and Heronian means for MADM. Artif Intell Rev 56(8):8955–9008
19. Mandal U, Seikh MR (2023) A novel score function-based EDAS method for the selection of a vacant post of a company with q-rung orthopair fuzzy data. Math Comput Sci 1:231–250
20. Farhadinia B, Liao H (2021) Score-based multiple-criteria decision making process by using p-rung orthopair fuzzy sets. Informatica 32(4):709–739
21. Basu R, Chakraborty S, Saha AK (2024) Healthcare waste treatment technology selection under Fermatean fuzzy environment. In: International conference on emerging trends in mathematical sciences & computing. Springer Nature Switzerland, Cham, pp 423–434

Quantification of the GSI Classification System Based on the Rock Mass Waves Velocity Utilizing the Artificial Intelligence Algorithms

Mohammad Rezaei and Pouya Koureh Davoodi

Abstract The geological strength index (GSI) is a fundamental parameter in rock mechanics and geotechnical engineering that reflects the structural integrity and mechanical behavior of rock masses. Accurate evaluation of this parameter is essential for ensuring the stability, safety, and economic efficiency of underground constructions, mining excavations, and slope stability analyses. However, conventional techniques for determining GSI mainly depend on field observations and qualitative assessments, which are inherently subjective and prone to human bias. These limitations often lead to inconsistencies in classification and may affect the reliability of engineering design decisions. To overcome these shortcomings, the present research employs artificial intelligence (AI) and metaheuristic optimization techniques to establish robust, data-driven predictive models for GSI estimation. A comprehensive dataset was compiled from the Beheshtabad Water Transfer Tunnel in southwestern Iran, comprising paired measurements of compressional (V_p) and shear (V_s) wave velocities, together with field-evaluated GSI values. Because seismic velocities are relatively inexpensive and easy to measure, they serve as practical predictor variables to replace or complement traditional classification systems such as Rock Mass Rating (RMR) and Q-system, which require extensive field and laboratory investigations. Three modeling strategies, i.e., the trust region reflective (TRR) algorithm, the genetic algorithm (GA), and a hybrid TRR–GA approach were explored, to model the nonlinear dependency between GSI and the seismic parameters. Five mathematical formulations, including exponential, logarithmic, power, quadratic polynomial, and linear functions, were systematically tested to determine the most suitable representation of the relationship. Among these, the quadratic polynomial model consistently achieved the highest predictive accuracy across all algorithms. Model performance was evaluated using a suite of statistical metrics,

M. Rezaei (✉)
Department of Mining Engineering, Faculty of Engineering, University of Kurdistan, Sanandaj, Iran
e-mail: m.rezaei@uok.ac.ir

P. K. Davoodi
Department of Mining Engineering, Faculty of Engineering, Urmia University, Urmia, Iran
e-mail: p.kourehdavoodi@urmia.ac.ir

A. Azizi et al. (eds.), *Intelligence at the Interface*, Emerging Trends in Mechatronics,
https://doi.org/10.1007/978-981-95-7584-8_4

including R^2, RMSE, MARE, MAPE, and MSE, along with additional indicators such as Index of Scatter (IOS), Index of Agreement (IOA), and the A20 accuracy index. Comparative analyses revealed that the hybrid TRR–GA model produced the most accurate and stable predictions, outperforming the standalone TRR and GA models. Although both V_p- and V_s-based models yielded satisfactory results, the V_p-based formulations demonstrated slightly superior predictive capability. These findings confirm the advantages of integrating the global optimization strength of GA with the local refinement capability of TRR, resulting in a more robust and efficient predictive model. The study ultimately demonstrates that GSI can be estimated reliably using non-destructive, low-cost seismic measurements, offering a viable alternative to traditional, subjective field-based methods. The proposed hybrid TRR–GA framework provides valuable insights for rock mechanics, tunneling, and geotechnical design, promoting safer and more cost-effective decision-making in engineering practice.

Keywords Geological strength index · Compressional wave velocity · Shear wave velocity · Trust region reflective · Genetic algorithm · Hybrid optimization

1 Introduction

The geological strength index (GSI) is a fundamental empirical parameter widely used in geotechnical and rock engineering to evaluate the structural quality, integrity, and mechanical behavior of rock masses. Since its introduction, the GSI system has become an essential component in the design and analysis of engineering structures such as tunnels, slopes, foundations, and underground caverns. Its accurate determination plays a crucial role in evaluating rock stability, selecting support systems, and optimizing the cost and safety of engineering projects. An underestimated GSI can lead to overly conservative designs and unnecessary expenses, while an overestimated GSI may result in structural instability and safety hazards [1–3].

Although the GSI system has proven valuable in engineering practice, its traditional determination methods are largely qualitative, relying heavily on field observations, visual assessments, and empirical charts. These methods often depend on the experience and judgment of field engineers, which introduces subjectivity and limited reproducibility. In heterogeneous geological environments, where lithological variations, discontinuities, and weathering conditions differ significantly, such subjectivity may cause inconsistent classification results among different observers. Therefore, there is an urgent need for quantitative, data-driven, and reproducible approaches that can enhance the objectivity and accuracy of GSI estimation [4].

During the last two decades, significant progress has been made toward developing predictive frameworks that link GSI to other measurable geotechnical and geomechanical properties. Among these, seismic wave velocities, particularly the compressional (V_p) and shear (V_s) wave velocities, have emerged as promising indicators. These parameters can be measured rapidly, non-destructively, and at relatively low cost, while providing meaningful insights into rock density, porosity, fracturing intensity, and stiffness. Consequently, they are well-suited for establishing empirical or semi-empirical correlations with GSI. However, the relationship between seismic velocities and GSI is highly nonlinear and complex, posing significant challenges for traditional regression or deterministic modeling methods [5–7].

To overcome these limitations, the application of artificial intelligence (AI) and machine learning (ML) techniques in geotechnical modeling has expanded considerably [8–16]. Regression-based methods, while simple and interpretable, often struggle to capture nonlinearities inherent in geotechnical data. In contrast, evolutionary and metaheuristic algorithms, such as the genetic algorithm (GA), offer powerful optimization capabilities capable of handling multidimensional and non-differentiable problems. Nevertheless, GA may occasionally exhibit slow convergence or premature stagnation in local minima. On the other hand, deterministic algorithms such as the trust region reflective (TRR) method are well-suited for fine-tuning and local optimization but may lack the global search capacity needed to explore complex solution spaces. In response to these challenges, hybrid algorithms that combine the complementary advantages of global and local optimization methods have gained attention. Integrating GA's global exploration ability with TRR's local refinement precision can produce more robust, accurate, and computationally efficient predictive models. These hybrid frameworks are particularly effective in rock mechanics and geotechnical engineering, where the relationships between parameters are nonlinear, noisy, and data-limited [15, 16].

Within this research context, the present study systematically evaluates the performance of TRR, GA, and their hybrid TRR–GA combination for predicting GSI using seismic velocity parameters (V_p and V_s) as inputs. The modeling process explores five mathematical functional forms comprising exponential, logarithmic, power, quadratic polynomial, and linear equations to identify the most appropriate relationship for reliable estimation. Model evaluation and comparison are conducted using multiple statistical indices to ensure a rigorous and comprehensive performance assessment.

The principal objectives of this study are:

1. To evaluate the accuracy and reliability of TRR, GA, and TRR–GA models in predicting GSI from seismic velocity data.
2. To investigate the advantages of hybrid metaheuristic modeling in improving prediction accuracy and model stability.
3. To provide practical recommendations for applying AI-based predictive methods in rock mass classification and tunnel engineering.

By addressing the above objectives, this research contributes to the advancement of quantitative and intelligent approaches in rock mechanics, offering a practical and reliable framework for estimating GSI based on measurable geophysical parameters. The outcomes of this study are expected to assist engineers and researchers in reducing subjectivity, enhancing design reliability, and optimizing cost efficiency in underground and surface geotechnical projects.

2 Background and Literature Review

2.1 The Geological Strength Index: Definition and Applications

The geological strength index (GSI) is one of the most widely applied empirical systems for assessing rock mass quality and mechanical behavior in geotechnical engineering. Developed by Hoek and Brown, the GSI method provides a quantitative framework for evaluating the degree of fracturing, block interlocking, and surface condition of discontinuities in rock masses. By integrating these characteristics, GSI helps to define the overall structural integrity of rock formations and serves as a cornerstone for determining design parameters in various engineering applications [6].

The applications of GSI are broad and well-established. It plays a pivotal role in the design and stability analysis of tunnels, underground caverns, foundations, and slopes, where accurate estimation of rock mass behavior is essential for safety and cost efficiency. The GSI not only assists engineers in evaluating stress–strain responses, failure modes, and deformation characteristics, but also helps in optimizing excavation and support systems for underground projects. Furthermore, by linking field observations to rock mechanics parameters such as uniaxial compressive strength and deformation modulus, the GSI system serves as a practical bridge between qualitative geological assessments and quantitative engineering analyses [17].

Despite its usefulness, traditional approaches to GSI estimation are primarily subjective, relying heavily on visual assessment, expert judgment, and empirical charts. This reliance can lead to inconsistencies, especially in projects where field conditions are heterogeneous or the available data are limited. To overcome these challenges, researchers have increasingly turned to quantitative, data-driven, and computational methods capable of improving the reproducibility and accuracy of GSI estimation. Among these, statistical regression and optimization-based modeling approaches have gained particular prominence. When combined with easily measurable geophysical variables such as the compressional (V_p) and shear (V_s) wave velocities, these methods can significantly enhance prediction precision and reduce uncertainty.

Previous studies have shown that integrating empirical relationships with metaheuristic algorithms such as genetic algorithms (GA) and trust region reflective (TRR) can substantially improve model generalizability and robustness. These hybrid approaches capture the inherent complexity and nonlinearity in rock mass data, providing better predictive performance compared to traditional empirical methods [18].

Consequently, GSI is now not only a conventional field classification parameter but also a central element in intelligent predictive modeling frameworks used in rock mechanics and tunneling. In this context, GSI serves a dual function:

1. As a diagnostic index that reflects the in-situ mechanical and structural condition of the rock mass.
2. As a target variable for computational intelligence models aiming to simulate and predict rock mass quality from physical and measurable indicators.

This dual perspective reinforces GSI's role as both a practical engineering parameter and a scientific benchmark for developing more advanced, objective, and data-driven geotechnical methodologies.

2.2 Developed Relationships between GSI and Other Rock Mass Parameters

Over the past decades, extensive research has focused on establishing empirical and semi-empirical relationships between GSI and other rock mass classification systems. These efforts have aimed to improve the accuracy and practicality of rock mass characterization in engineering design. Table 1 summarizes several widely cited correlations proposed by previous researchers between GSI and indices such as the Rock Mass Rating (RMR) and the Q-system.

While these equations have provided useful empirical insights, most of them are indirect relationships, linking GSI with other classification indices rather than with fundamental physical or mechanical rock properties. As a result, they remain limited in their generalizability and require field-specific calibration to achieve satisfactory accuracy. In contrast, direct relationships between GSI and physical parameters such as seismic wave velocities (V_p and V_s) have received comparatively less attention, despite their strong theoretical and practical relevance. Seismic wave velocities are

Table 1 Summary of previously developed relationships between GSI and other rock mass classification parameters

No.	Equation	R^2	References
1	$Q = 0.08e^{0.049GSI}$	–	[3]
2	RMR = 0.42GSI + 33	–	
3	GSI = RMR-5 (FOR RMR > 23)	–	[19]
4	GSI = RMR ± 5	–	[20]
5	GSI = 0.42RMR + 23.08	0.44	[21]
6	GSI = 0.739RMR + 12.097	0.759	[22]
7	GSI = 0.7394RMR + 0.57	0.57	[23]
8	GSI = 0.9934RMR – 4.913	0.84	[24]
9	GSI = 1.2092RMR – 18.6143	–	[25]
10	GSI = 1.265RMR – 21.49	–	[26]
11	$GSI = 0.5RQD + 1.5J_C$	–	[27]
12	GSI = 0.793RMR + 2.001	0.736	[28]
13	GSI = 0.876RMR + 0.935	0.876	

inherently related to the elastic and structural characteristics of rocks, including fracture intensity, density, porosity, and stiffness. Consequently, they can serve as powerful, non-invasive proxies for assessing rock mass quality. Moreover, they offer considerable advantages in terms of speed, cost, and accessibility compared to the traditional field and laboratory-based approaches [29, 30].

The underutilization of seismic parameters in GSI modeling represents a significant gap in the literature. Relying solely on visual or empirical classification methods often leads to higher costs, longer evaluation times, and increased uncertainty. In contrast, seismic-based approaches can provide a quantitative, reproducible, and scalable alternative for real-world engineering applications.

To address this research gap, the present study emphasizes the integration of V_p and V_s as core predictors in developing reliable GSI estimation models. This approach not only enhances the precision and efficiency of rock mass characterization but also lays the groundwork for broader application of geophysical–geomechanical integration in rock engineering. By leveraging the correlation between wave propagation properties and rock mass integrity, this work proposes a more practical, data-driven, and transferable framework for predicting GSI, with direct implications for tunneling, mining, and slope stability assessments.

3 Study Area Description

The present study was conducted using data collected from the Beheshtabad Water Transfer Tunnel, located in southwestern Iran, one of the country's most significant water conveyance projects. The tunnel is designed to transfer water from the Karun River Basin toward central Iran, addressing the chronic water scarcity issues in that region. Given its considerable length and the diversity of rock formations encountered along the alignment, the Beheshtabad tunnel provides an ideal case study for evaluating the relationship between seismic wave velocities and the geological strength index (GSI).

3.1 Geographical Location

The Beheshtabad tunnel alignment begins near the downstream section of the Beheshtabad River at coordinates 32°02' north and 50°38' east. Running for around 65 km in a northeastward orientation, the route terminates upstream of the Cham Asman Dam on the Zayandeh-Rood River, positioned at 51°12' E and 32°22' N. Throughout its trajectory, the tunnel passes through a diverse range of lithological formations, encompassing igneous, metamorphic, and sedimentary rock types. The variability in rock composition, combined with the occurrence of fractures, faults, and other structural discontinuities, results in significant differences in the mechanical and physical behavior of the rock mass. To account for these variations, comprehensive

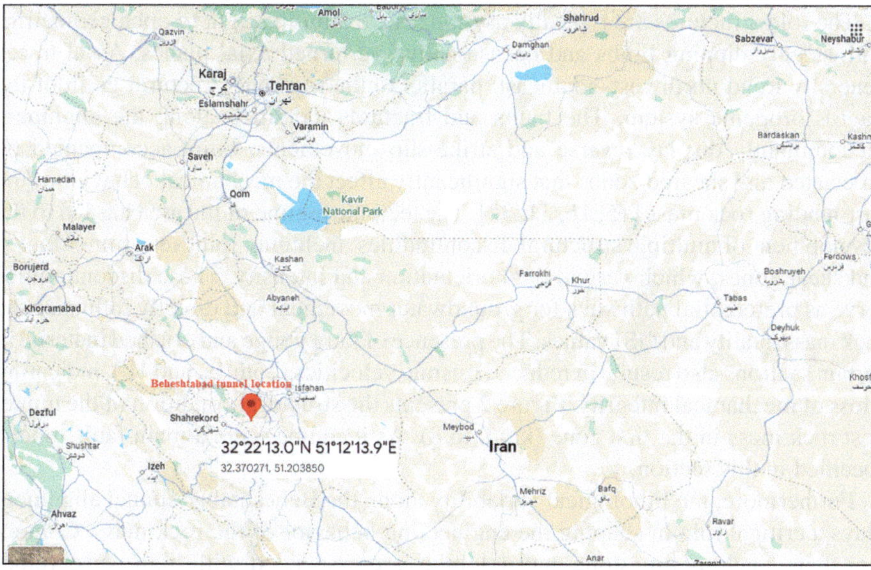

Fig. 1 Beheshtabad tunnel's location on the Iran's map between the Shahrekord and Isfahan cities

geotechnical investigations have been carried out along the tunnel alignment, with particular attention to key parameters such as compressional and shear wave velocities (V_p and V_s), and various rock mass classification schemes. These parameters are essential for evaluating excavation stability, designing effective support systems, and predicting how the rock mass will respond to stresses induced by construction activities [15, 16, 31, 32]. The Beheshtabad tunnel's location on the Iran's map are represented in Fig. 1, which is located between Shahrekord and Isfahan cities.

3.2 *Geological and Tectonic Setting Along the Beheshtabad Tunnel Alignment*

The Beheshtabad tunnel alignment traverses a geologically intricate segment of the Zagros Fold–Thrust Belt, one of the most tectonically dynamic and structurally complex regions of Iran. This belt, formed primarily as a result of the collision between the Arabian and Eurasian plates, extends northwest–southeast and is characterized by intense folding, faulting, and thrusting. The geological sequence encountered along the tunnel includes sedimentary formations of Mesozoic and Cenozoic ages, consisting mainly of limestone, marl, dolomite, shale, conglomerate, and cherty interbeds. Throughout the tunnel's route, significant variations in stratigraphy and structural features are observed, reflecting the diversity and complexity of the region's geodynamic evolution [15, 16, 31, 32].

The fold structures are typically asymmetric, exhibiting tight to open geometries with steeply dipping limbs and occasionally overturned axial planes. Axial traces generally trend northwest–southeast, parallel to the regional structural grain of the Zagros orogenic system. The faults and fractures identified along the alignment predominantly display reverse and strike-slip movements, often accompanied by brecciated and sheared zones that significantly affect the mechanical behavior of the surrounding rock mass [15, 16, 31, 32]. The tectonic regime of the area has led to the development of multiple structural discontinuities, including joint sets, minor faults, and shear zones, which vary in both orientation and intensity. These discontinuities serve as preferential pathways for groundwater movement and directly influence the rock mass quality and GSI values. The presence of fault gouge and crushed material in certain sections also results in reduced seismic velocities (both V_p and V_s), indicating a loss of mechanical integrity. Figure 2 presents the structural condition of the tunnel host rock mass in the first zone (km 0 to 8). As it can be seen, a main fault zone is specified in this section.

Furthermore, the lithological variability along the Beheshtabad tunnel alignment plays a critical role in shaping the engineering behavior of the rock mass. Competent units, such as limestone and dolomite, are associated with high wave velocities and elevated GSI values, whereas weaker layers, including marl, shale, and clay-rich interbeds, exhibit lower velocities, greater deformability, and increased

Fig. 2 Structural condition of the tunnel host rock mass in the first section of the tunnel route (km 0 to 8)

vulnerability to weathering. In certain sections, alternating sequences of marl and limestone generate a complex pattern of strong and weak strata, which poses additional challenges for both excavation and support design. One of the major challenges encountered during excavation was the presence of numerous fault zones distributed throughout the project area (an example is shown in Fig. 2). For structural assessment, the tunnel has been divided into eight distinct segments, each approximately eight kilometers long. Geological and geophysical investigations reveal that each segment possesses unique structural characteristics, reflecting the pronounced heterogeneity of the subsurface formations [15, 16, 31, 32]. A representative sample of the lithology and solubility of Asmari limestones in the seventh section of the tunnel route (a segment between kilometers 48 and 56) is presented in Fig. 3.

To illustrate the geological and structural complexity of the tunnel route, Fig. 4 presents the tunnel entrance. As it can be seen from this figure, tunnel entrance is particularly influenced by extensive faulting and unstable lakes deposits, making it one of the most geologically and structurally challenging portions of the tunnel.

Hydrogeologically, the region is influenced by the karstic nature of the carbonate formations, where subsurface water circulation occurs primarily through interconnected fractures and bedding planes. This results in localized zones of water inflow during tunneling, particularly within fractured limestone and dolomitic units. The presence of water-bearing fractures not only affects the excavation process but also modifies the dynamic response of the rock mass by reducing effective stress and increasing saturation levels [15, 16, 31, 32].

Fig. 3 Lithology and solubility of Asmari limestones in the seventh section of the tunnel route (kilometers 48 and 56)

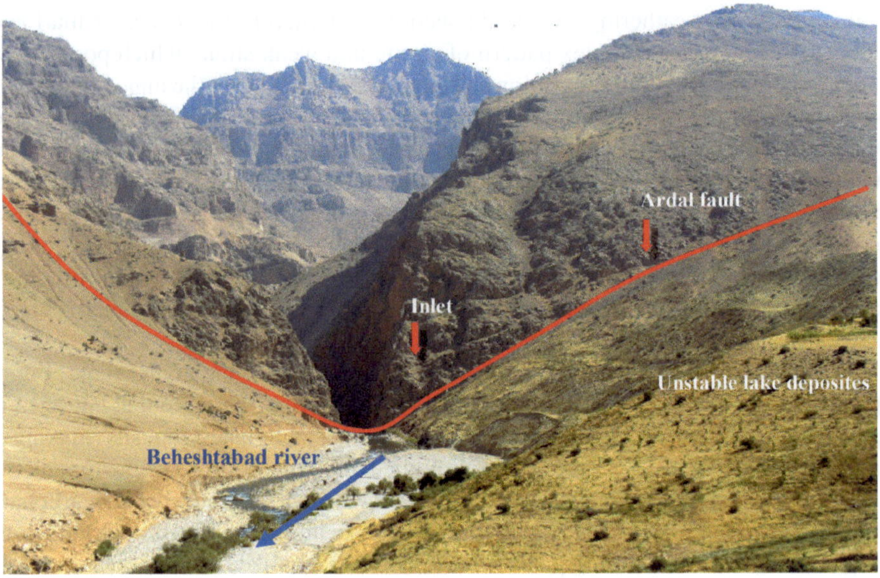

Fig. 4 Ardel fault and unstable lakes marls at the tunnel entrance

Overall, the Beheshtabad tunnel alignment provides a natural laboratory for examining the interplay between geological structure, lithology, and rock mass properties. Along this tunnel, a wide range of lithological and structural units is encountered, each exhibiting distinct mechanical and geophysical characteristics. Table 2 presents a summary of the main geological formations and their engineering significance. The combination of intense tectonic deformation, lithological heterogeneity, and varying hydrogeological conditions offers an exceptional opportunity to evaluate how seismic wave velocity parameters (V_p and V_s) correlate with rock mass quality indices such as GSI.

Table 2 Petrological, structural, and hydrological characteristics of the rocks encountered along the tunnel alignment

Parameter	Description
Rock type	Limestone, marl, shale, dolomite, chert, conglomerate, sandstone, marlstone, mudstone, andesite, serpentine, schist, tuff
Bedding thickness	Ranges from thin to massive; transitions between thinly bedded marl–limestone alternations to thick, competent limestone blocks
Tectonic features	Numerous faults, fractures, and shear zones; fault gouge and breccia observed in several sections; intense jointing in weak marl layers
Hydrogeological conditions	Groundwater inflow concentrated in fractured carbonate zones; marl and shale units act as aquicludes; water saturation varies significantly along the route

3.3 Seismic Testing and GSI-Index Evaluation Methodology

3.3.1 Data Acquisition: Downhole Seismic Testing

In this study, the fundamental geophysical parameters, namely the compressional (V_P) and shear (V_S) wave velocities of the rock mass, were obtained through the downhole seismic testing (DHT) method. This in-situ technique is widely regarded as one of the most dependable approaches for evaluating the dynamic elastic behavior of rock materials under natural field conditions.

The downhole testing process was implemented through the following systematic stages:

- Borehole preparation: A total of thirteen boreholes were drilled at selected points along the Beheshtabad water conveyance tunnel route, each reaching the required investigation depth. To preserve the borehole's structural integrity during testing, steel casing was installed along its wall.
- Source and receiver setup: A three-component (3-C) geophone receiver along with a clamped geophone or a hydrophone was carefully positioned and firmly fixed at the borehole bottom. The seismic source consisted of surface impacts generated by a heavy hammer striking a steel baseplate placed close to the borehole collar. This configuration ensured a controlled and repeatable generation of seismic energy.
- Wave generation and recording: When the impact source was activated, compressional (P) and shear (S) seismic waves propagated through the surrounding rock mass and were captured by the downhole receiver. To distinctly register the shear waves, a polarized horizontal S-wave source was employed by striking a wooden plank in a fixed horizontal orientation. This arrangement effectively separated S-wave signals from P-wave arrivals, minimizing signal interference [33].
- Data acquisition: The first-arrival times of P-wave and S-wave traveling from the surface source to the downhole receiver were accurately recorded by a high-sensitivity seismograph. Measurements were repeated at several discrete depths by progressively repositioning the receiver upward in the borehole. The corresponding travel times and known source–receiver offsets allowed the construction of a detailed travel-time–depth profile for both wave types.
- Velocity computation: The compressional and shear wave velocities (V_{p} and V_{s}) at each test interval were determined by dividing the measured travel distance between source and receiver by the respective first-arrival time. Altogether, 150 high-quality datasets containing paired V_{p} and V_{s} measurements were collected from various geological units along the tunnel alignment, providing a comprehensive representation of the spatial variability in rock mass properties [34].

3.3.2 The GSI Determination

The geological strength index (GSI) at each investigation point was quantitatively determined through systematic field mapping and detailed face logging, carried out in accordance with the procedure proposed by Hoek et al. (2013). The GSI framework provides a well-established means of estimating the extent to which geological factors, such as fracturing, block size, and joint surface conditions, reduce the overall strength of a rock mass [16].

The evaluation relied primarily on the visual interpretation of two key geological features:

- Geological configuration of the rock mass: This aspect describes the overall structure and degree of fracturing within the rock mass, reflecting its transition from intact, massive blocks to severely crushed or disintegrated material.
- Surface condition of discontinuities: This parameter characterizes the quality and appearance of discontinuity surfaces, considering factors such as joint roughness, degree of weathering, and the nature of any infilling materials present.

At each measurement site, field engineers compared the observed characteristics of the rock mass with the standardized GSI charts to assign an appropriate GSI value. This process involved integrating multiple field observations, including the spacing and number of joint sets, surface roughness, and the extent of weathering. By reaching a consensus based on expert judgment, a reliable and consistent GSI estimate was obtained for each location.

In total, 150 GSI values were established by an experienced geological team, ensuring both internal consistency and representativeness of the data across the varied geological settings encountered along the tunnel alignment.

3.3.3 Integration for Analysis

The compiled dataset, consisting of 150 paired observations of V_p, V_s, and GSI, served as the fundamental input for the subsequent analysis. The principal aim was to investigate the quantitative relationships between the measured seismic velocities, which physically represent the stiffness and integrity of the rock mass, and the GSI, which captures its structural quality in empirical terms.

3.4 Statistical Characteristics of the Utilized Data

To achieve this, advanced hybrid artificial intelligence algorithms, namely the trust region reflective (TRR), genetic algorithm (GA), and their combined TRR–GA model, were employed to establish, refine, and optimize the predictive correlations. The detailed methodology and computational framework of these hybrid approaches are discussed in the following sections.

For constructing a dependable predictive model, it is first necessary to develop a comprehensive understanding of the statistical behavior of the employed dataset. This section presents a detailed statistical overview of the 150 recorded data points corresponding to each of the key parameters, i.e., P-wave velocity (V_p), S-wave velocity (V_s), and the geological strength index (GSI). The main statistical descriptors considered include the minimum, maximum, mean, and standard deviation values, which collectively illustrate the range, central tendency, and variability within the dataset.

The analysis of P-wave velocity (V_p) data reveals pronounced variability in the stiffness and continuity of the rock mass along the tunnel alignment. The maximum recorded velocity, 4655 m/s, corresponds to zones of highly competent, intact rock with strong elastic properties. In contrast, the minimum measured value of 1738 m/s indicates the presence of considerably fractured or weathered rock masses with reduced seismic transmission efficiency. Across all measurements, the mean V_p was calculated to be 3225 m/s, representing the overall average condition of the tested sections. The corresponding standard deviation of 892.9 m/s reflects a notably wide dispersion, confirming the heterogeneous nature of the subsurface materials.

Similarly, the shear wave velocity (V_s) data exhibit considerable variation, highlighting the differences in shear stiffness and mechanical integrity of the rock mass. The maximum V_s value of 2195 m/s was obtained in dense, unweathered rock layers, while the minimum value of 1212 m/s represents weaker or highly jointed zones. The average V_s across the dataset was 1710 m/s, with a standard deviation of 280 m/s, again pointing to meaningful variability in the mechanical response of the rock units encountered.

The distribution of GSI values further emphasizes the geological diversity of the study area. The highest recorded GSI of 73 represents a very good quality rock mass with well-interlocked blocks and limited joint alteration. Conversely, the lowest GSI value of 22 corresponds to extremely poor-quality rock, often fragmented or resembling soil-like material in behavior. The mean GSI across all samples was 56.12, while the standard deviation of 12.34 demonstrates a wide range of rock mass conditions spanning from fair to very good categories.

This extensive range of V_p, V_s, and GSI values confirms the pronounced heterogeneity of the geological formations encountered along the Beheshtabad tunnel alignment. Such variability is, in fact, advantageous for the development of a robust and generalizable artificial intelligence model, as it ensures that the trained system can accurately predict GSI values under a broad spectrum of geomechanical conditions. For an enhanced visual comparison of these statistical characteristics, all parameters are collectively illustrated in Fig. 5.

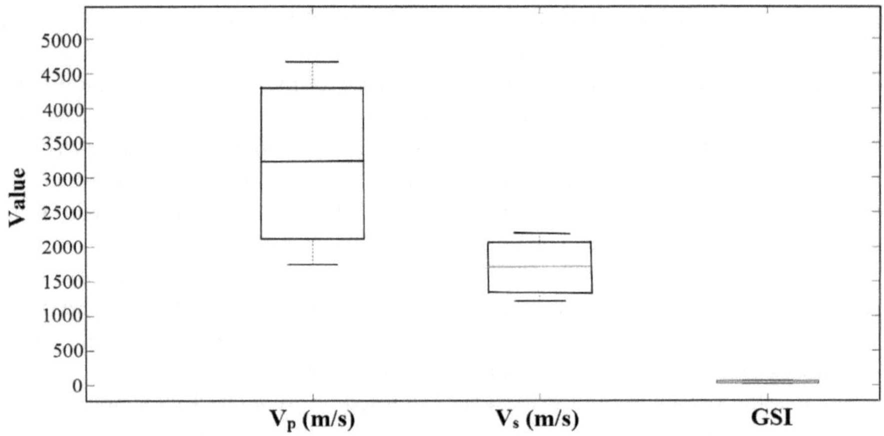

Fig. 5 Statistical spread of measured V_p, V_s, and GSI values

4 Methodology

This section presents a detailed explanation of the methodological framework employed in the present research. The principal aim of this study is to investigate how the seismic wave velocities of a rock mass influence its GSI classification by applying advanced artificial intelligence.

In the first stage, the metaheuristic algorithms adopted in this study, namely the trust region reflective (TRR) algorithm [15], the genetic algorithm (GA) [35], and the hybrid TRR-GA algorithm [15] are introduced. Each of these algorithms offers distinct strengths in dealing with nonlinear optimization problems and is specifically employed to capture the complex and nonlinear interactions between seismic velocity parameters and the GSI index.

The second stage involves the actual modeling and prediction of GSI through two separate computational approaches:

- Modeling based on V_p.
- Modeling based on V_s.

Both approaches make use of the aforementioned algorithms to develop predictive models that can accurately estimate GSI from the measured V_p and V_s. The organization of this section is designed to provide a clear and comprehensive understanding of the computational methods utilized, as well as their respective contributions to the predictive analysis process.

4.1 Description of the Applied Algorithms

To achieve an accurate prediction of GSI from the measured rock mass wave velocities, this research employs three distinct optimization algorithms including the trust region reflective (TRR) algorithm, the genetic algorithm (GA), and a hybrid TRR–GA approach. The choice of these algorithms stems from their demonstrated success in addressing complex nonlinear regression problems, their adaptability to irregular and non-uniform datasets, and their ability to balance global exploration with local refinement in the search for optimal solutions.

The following subsections describe the theoretical basis, computational mechanism, and the specific rationale for employing each algorithm in the context of this study.

4.1.1 The TRR Algorithm Methodology

The trust region reflective (TRR) algorithm is a deterministic optimization technique widely recognized for its efficiency in solving large-scale, nonlinear least-squares problems that include boundary constraints. The algorithm operates by iteratively constructing and solving simplified local approximations of the objective function within a defined "trust region," which represents the neighborhood where the model is assumed to be a reliable reflection of the true function.

The operational framework of the TRR algorithm can be summarized in four main steps:

1. Local model construction: At each iteration, a quadratic approximation of the objective function is formulated using the available gradient and Hessian information. This approximation serves as a simplified model of the local behavior of the target function.
2. Definition of the trust region: A trust region radius is established, specifying the maximum allowable step size from the current estimate. This constraint stabilizes the optimization process and prevents sudden or unstable movements in regions where the function curvature is high.
3. Trial step computation: A candidate step is computed by minimizing the local quadratic model under the imposed trust region constraint. This step represents a tentative movement toward the potential optimum.
4. Acceptance and update: If the predicted improvement from the model closely matches the actual reduction in the objective function, the step is accepted, and the trust region radius may be increased. Conversely, if the reduction is smaller than anticipated, the region is contracted to maintain model reliability.

The "reflective" property of the TRR algorithm allows it to effectively handle bound constraints by reflecting steps that would otherwise extend beyond permissible parameter limits back into the feasible solution space. This feature is particularly important in geomechanical modeling, where physical parameters such as velocities or GSI values must remain within realistic and meaningful ranges [36].

In this research, the TRR algorithm is applied to determine the optimal coefficients in regression-based models that establish the relationship between V_p, V_s, and GSI. Owing to its rapid convergence near the optimal solution, the TRR approach is especially advantageous for fine-tuning parameter values after an initial estimation phase, ensuring both computational efficiency and high predictive precision.

4.1.2 The GA Methodology

The genetic algorithm (GA) is a stochastic, population-based optimization approach inspired by the principles of Darwinian natural selection and Mendelian genetics. It evolves a population of potential solutions over successive generations, progressively improving their performance with respect to a defined fitness measure. The algorithm's evolutionary process is governed by three main operators: selection, crossover, and mutation.

- Parent selection (selection): In this stage, the most successful individuals within the current population, those achieving higher fitness values, are preferentially chosen as parents for the next generation. The likelihood of selection is directly proportional to the fitness score, which, in the present study, corresponds to the predictive accuracy of GSI estimation.
- Genetic recombination (crossover): The selected parents exchange segments of their encoded parameters to produce new offspring. This recombination process allows advantageous traits from different parent solutions to combine, fostering the generation of improved candidate solutions in subsequent iterations.
- Genetic alteration (mutation): To maintain population diversity and prevent premature convergence to suboptimal solutions, random modifications are occasionally applied to certain parameters. These small, random perturbations help the algorithm explore new areas of the solution space that might otherwise remain undiscovered.

The key strength of GA lies in its global search capability in which it does not require derivative information and can effectively explore complex and multidimensional spaces. Consequently, it is particularly suitable for optimization problems where the objective function is nonlinear, discontinuous, or non-differentiable [35].

Within the context of this study, the GA is employed to identify the regions of the parameter space that are most likely to contain optimal or near-optimal solutions for GSI prediction. Its ability to explore a wide range of possible solutions minimizes the likelihood of being trapped in local minima, an issue often encountered in wave velocity–GSI correlation modeling due to the inherent complexity and noise within geotechnical data.

4.1.3 Methodology of the Hybrid TRR-GA Algorithm

The hybrid TRR–GA algorithm integrates the complementary advantages of the GA and TRR methods by combining GA's global exploration capability with the precise local refinement potential of TRR. This combined approach allows for both a broad search across the solution domain and fine-tuning of the most promising candidates, thereby enhancing predictive performance and computational efficiency.

The hybridization process is carried out in two main phases:

1. Global search using GA: In the initial phase, the genetic algorithm is executed over multiple generations to perform an extensive exploration of the parameter space. This step identifies a diverse set of potential solutions that exhibit relatively high fitness scores while avoiding premature convergence to local optima.
2. Local refinement using TRR: The best-performing solutions derived from the GA are subsequently used as the initial input for the TRR algorithm. TRR then conducts localized optimization, adjusting model parameters to minimize the residual error between the predicted and observed GSI values with high precision.

The principal advantage of this hybrid optimization strategy is its ability to significantly reduce computational time compared to running GA independently for a large number of generations, while simultaneously improving the likelihood of achieving the true global optimum, something that pure TRR optimization alone might fail to accomplish [15].

In the present research, the TRR–GA hybrid model proved highly effective in capturing the nonlinear and interdependent relationships between seismic wave velocities (V_p and V_s) and GSI. The combined algorithm achieved both high predictive accuracy and remarkable computational efficiency, making it particularly valuable for engineering geology applications where precise characterization of rock mass quality is vital for ensuring structural safety and optimizing project costs. It should be noted that the performance of hybrid intelligent models is commonly better than that of the base models [37–39].

4.2 *Modeling of GSI*

In this section, the central aim is to establish the most appropriate mathematical model that can describe how the output variable depends on the input parameters by employing three different metaheuristic optimization methods, i.e., TRR, GA, and the hybrid TRR-GA algorithm. In this research, the input parameters consist of the compressional and shear wave velocities (V_p and V_s), while the output parameter is the geological strength index (GSI). Each modeling process is conducted independently for V_p and V_s so that the influence of each velocity type on the GSI can be distinctly analyzed and compared.

To achieve this goal, every algorithm is designed to construct five alternative functional relationships between the input and output variables, as follows:

1. Linear relationship.
2. Power relationship.
3. Exponential relationship.
4. Quadratic polynomial relationship.
5. Logarithmic relationship.

Once these five relationships are developed, their predictive capabilities are examined using several statistical performance measures. The evaluation criteria include the coefficient of determination (R^2), which reflects model accuracy, and multiple error-based indicators such as mean square error (MSE), mean absolute relative error (MARE), root mean square error (RMSE), and mean absolute percentage error (MAPE). These indicators are calculated using the following standard formulations [40–47]:

$$R^2 = 1 - \frac{\sum_{i=1}^{n} (y_i - \hat{y}_i)^2}{\sum_{i=1}^{n} (y_i - \overline{y}_i)^2} \tag{1}$$

$$\text{MAPE} = \frac{100\%}{n} - \sum_{i=1}^{n} \left| \frac{y_i - \hat{y}_i}{y_i} \right| \tag{2}$$

$$\text{RMSE} = \sqrt{\frac{1}{n} \sum_{i=1}^{n} (y_i - \hat{y}_i)^2} \tag{3}$$

$$\text{MARE} = \frac{1}{n} \sum_{i=1}^{n} \frac{|y_i - \hat{y}_i|}{|y_i|} \tag{4}$$

$$\text{MAE} = \frac{1}{n} \sum_{i=1}^{n} |y_i - \hat{y}_i| \tag{5}$$

$$\text{MSE} = \frac{1}{n} \sum_{i=1}^{n} (y_i - \hat{y}_i)^2 \tag{6}$$

In Eqs. (1) to (6), y_i and ? signify the i-th amount of actual data (observed) and their related average amount, respectively, y_i is related to the i-th amount of forecasted data (predicted), and n shows the number of datasets,

The optimal model for each algorithm is determined based on two conditions: it should yield the highest R^2 value while simultaneously minimizing the associated error measures (MSE, MARE, RMSE, and MAPE). The model that best satisfies these criteria is considered the most reliable predictive function for estimating GSI within the context of that particular algorithm.

This modeling process is conducted in two main stages:

- Stage 1: Developing the relationship between V_p and GSI using the TRR, GA models along with the TRR-GA hybrid algorithm.

- Stage 2: Developing the relationship between V_s and GSI using the same set of algorithms.

This two-step framework allows for a clear comparison of how each algorithm performs when applied to different types of wave velocity data, ultimately identifying the most accurate and practically useful relationship for GSI prediction in geotechnical and engineering geological applications.

4.2.1 Modeling of GSI Based on V_p

In this part of the study, the relationship between compressional wave velocity (V_p) and the geological strength index (GSI) was explored using three metaheuristic modeling strategies comprising the genetic algorithm (GA), the trust region reflective (TRR) approach, and a combined hybrid model integrating both techniques (TRR-GA). The main goal was to formulate a reliable mathematical relationship that effectively characterizes the nonlinear correlation between the GSI and V_p.

For this purpose, five types of mathematical functions were examined within each modeling approach, i.e., linear, power, exponential, logarithmic, and quadratic polynomial forms. Among these, the quadratic polynomial function consistently produced the most accurate and stable results across all three algorithms, closely matching the observed GSI values. The extracted mathematical relations between GSI and V_p from the TRR, GA, and TRR-GA models are given in Eqs. (7) to (9), respectively.

$$\text{GSI} = 0.0000003294V_p^2 + 0.0126305305V_p + 4.088 \tag{7}$$

$$\text{GSI} = 0.0000003321V_p^2 + 0.01266170939V_p + 4.0956 \tag{8}$$

$$\text{GSI} = 0.0000001457V_p^2 + 0.014216558V_p + 0.8805 \tag{9}$$

The performance of each model was quantitatively validated using several statistical indicators, including R^2, MSE, MARE, RMSE, and MAPE. In the case of the TRR model, the quadratic polynomial function achieved an R^2 value of 0.9443, meaning that approximately 94.43% of the observed variation in GSI could be explained by V_p. The associated error measures were RMSE = 3.2008, MARE = 0.0540, MAPE = 5.58%, and MSE = 10.2449, all of which indicate high predictive precision and minimal deviation from the measured data. The fitted model curve along with the relevant statistical indices is presented in Fig. 6, highlighting the model's strong alignment with experimental findings.

Similarly, the GA-based model produced results that were closely aligned with those of the TRR algorithm, yielding an R^2 value of 0.9427, RMSE = 3.2215, MARE = 0.0541, MAPE = 5.60%, and MSE = 10.3780, as illustrated in Fig. 7. Although these results were marginally less optimal than those obtained using the TRR model,

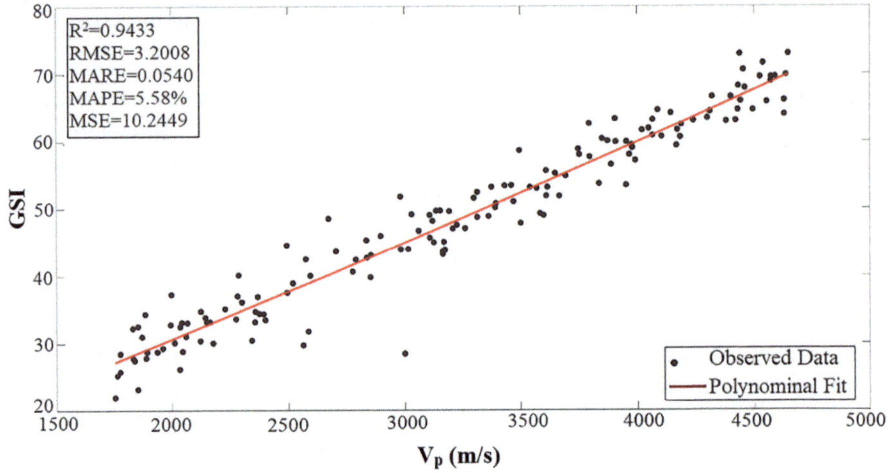

Fig. 6 TRR model of V_p versus GSI with statistics

they still demonstrate a high level of predictive accuracy and overall consistency. It is worth noting that the inherently stochastic nature of GA may lead to minor fluctuations in the quality of the fitted model, depending on population diversity and the random initialization process.

In contrast, the hybrid TRR-GA model achieved the best overall performance among all approaches. This integrated TRR-GA algorithm recorded an R^2 value of 0.9509, RMSE = 3.0813, MARE = 0.0551, MAPE = 5.61%, and MSE = 9.4944, outperforming the individual methods in both accuracy and stability. The enhanced

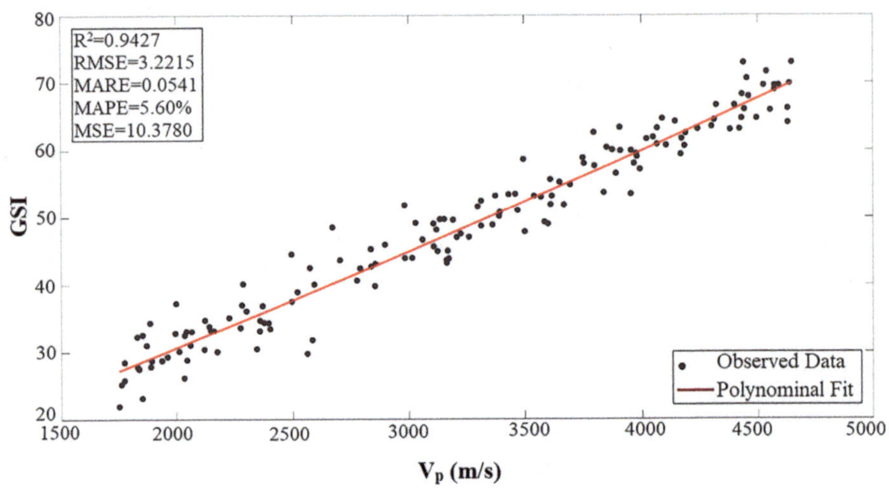

Fig. 7 GA model of V_p versus GSI with statistics

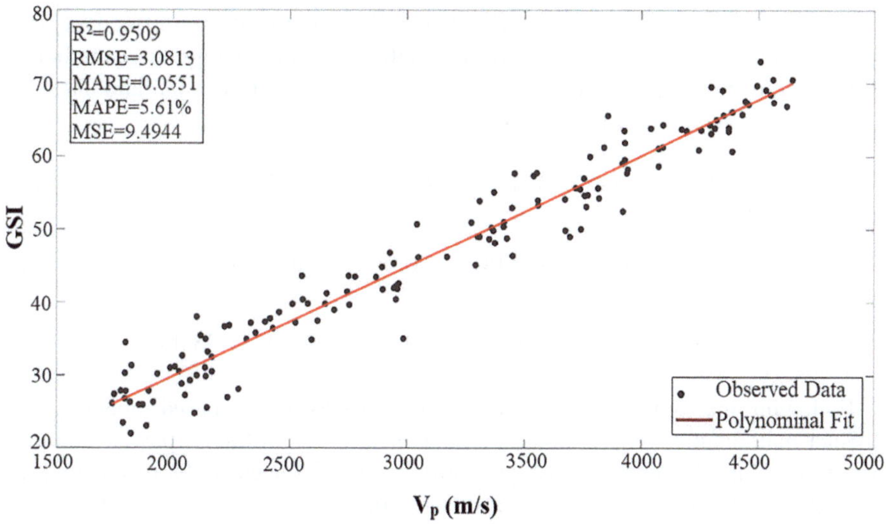

Fig. 8 TRR-GA model of V_p versus GSI with statistics

performance reflects the hybrid model's ability to combine the global exploration capability of GA with the local refinement efficiency of TRR. Through this synergy, the algorithm can effectively navigate the search space, escape local optima, and converge toward the true global solution. The corresponding prediction curve and related statistical indices are presented in Fig. 8.

The comparative analysis clearly indicates that integrating GA and TRR substantially improves the precision of GSI estimation based on V_p measurements. While both standalone TRR and GA models independently provide reliable and robust predictions, their combination enhances model stability and accuracy through complementary strengths of TRR's deterministic convergence and GA's adaptive global search.

Overall, these findings confirm that the quadratic polynomial function is the most suitable mathematical form for representing the intricate nonlinear relationship between compressional wave velocity and rock mass quality (GSI). Moreover, the success of the hybrid TRR-GA model underscores the growing potential of hybrid metaheuristic techniques as powerful tools for improving predictive modeling and parameter estimation in geotechnical and rock engineering applications.

4.2.2 Modeling of GSI Based on V_s

The relationship between the geological strength index (GSI) and shear wave velocity (V_s) was analyzed using the same three optimization algorithms applied earlier trust region reflective (TRR), genetic algorithm (GA), and their hybrid TRR-GA model. The main purpose of this stage was to determine a mathematical formulation that best characterizes the dependency of GSI on V_s and to evaluate how effectively each algorithm captures this correlation. Based on the current modeling, three mathematical relations between GSI and V_s were developed from TRR, GA, and TRR-GA models, as presented in Eqs. (10) to (12), respectively.

$$GSI = -0.0000015741 V_s^2 + 0.0469465825 V_s - 25.3481 \tag{10}$$

$$GSI = -0.0000015548 V_s^2 + 0.046922781 V_s - 25.354 \tag{11}$$

$$GSI = -0.00000113031 V_s^2 + 0.046120974 V_s - 25.4307 \tag{12}$$

Following the same methodological framework as the V_p-based modeling, five distinct mathematical expressions, i.e., exponential, logarithmic, power, quadratic polynomial, and linear relationships, were examined for each algorithm. In every case, the quadratic polynomial function provided the best fit, offering the highest determination coefficients (R^2) and the lowest associated error indices among the tested forms.

For the TRR model, the quadratic polynomial function produced an R^2 of 0.8413, indicating that approximately 84.13% of the variability in GSI can be explained by V_s. The accompanying error indicators were RMSE = 4.9829, MARE = 0.0825, MAPE = 8.60%, and MSE = 24.8296. These statistics suggest a satisfactory predictive capability, although with somewhat higher deviations compared to the V_p-based models. The fitted regression curve and related numerical evaluations are presented in Fig. 9.

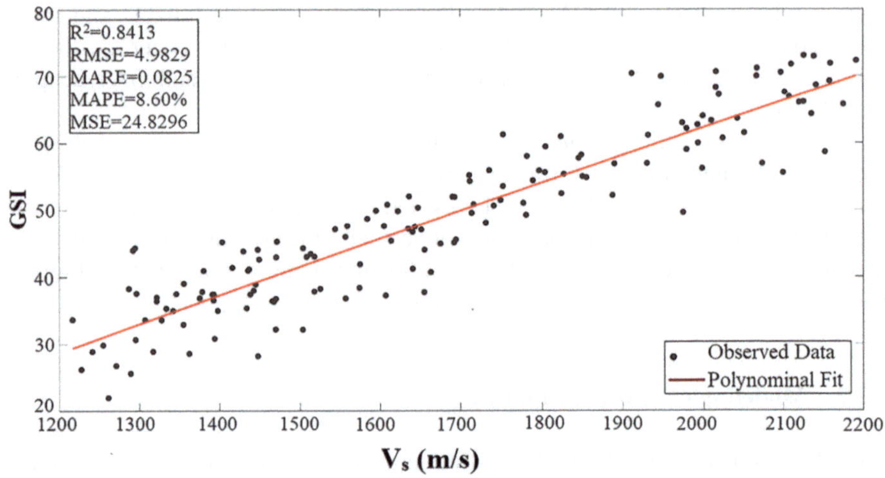

Fig. 9 TRR model of V_s versus GSI with statistics

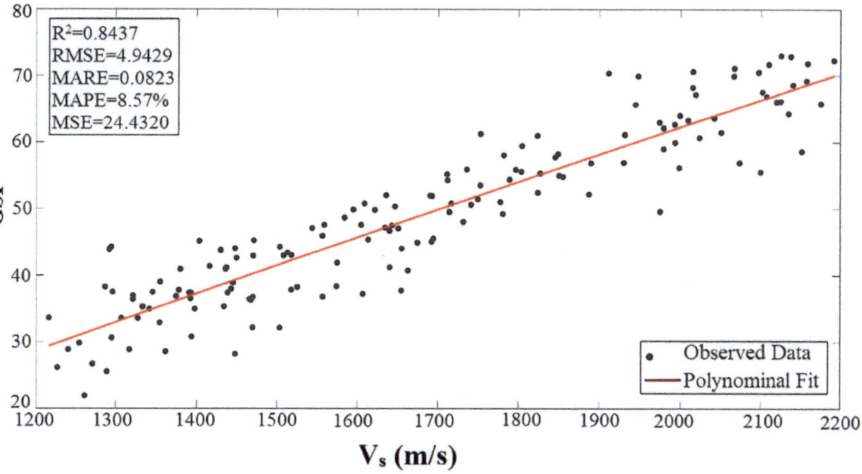

Fig. 10 GA model of V_s versus GSI with statistics

Similarly, the GA model achieved comparable performance, yielding an R^2 of 0.8437, RMSE = 4.9429, MARE = 0.0823, MAPE = 8.57%, and MSE = 24.4320, as illustrated in Fig. 10. These outcomes demonstrate that the GA algorithm successfully captures the overall behavioral trend of the dataset, achieving nearly the same accuracy level as the TRR model.

The integrated TRR-GA framework exhibited superior performance compared to the individual algorithms, achieving an R^2 value of 0.8595, an RMSE of 4.6988, a MARE of 0.0804, a MAPE of 8.31%, and an MSE of 22.079. This improved accuracy highlights the effectiveness of the hybrid strategy in enhancing the predictive model by leveraging the global optimization capabilities of the Genetic Algorithm alongside the local refinement strengths of the TRR method. Figure 11 presents the predicted curve together with the corresponding statistical indicators for this model.

Although both TRR and GA algorithms provided reasonable predictions, the overall predictive strength of V_s-based models was slightly lower than that of V_p-based models, as reflected in the reduced R^2 values and increased error measures. This difference likely arises from the more complex and less direct relationship between shear wave velocity and the mechanical behavior of rock masses, compared to compressional wave velocity.

Nevertheless, the hybrid TRR-GA algorithm again demonstrated superior performance, outperforming both individual methods. By combining GA's global exploration capability with TRR's efficient local optimization, the hybrid approach produced more stable and accurate estimates, mitigating the weaknesses inherent to each single algorithm. This improvement underscores the importance of using integrated optimization frameworks when dealing with nonlinear, multivariable geotechnical relationships.

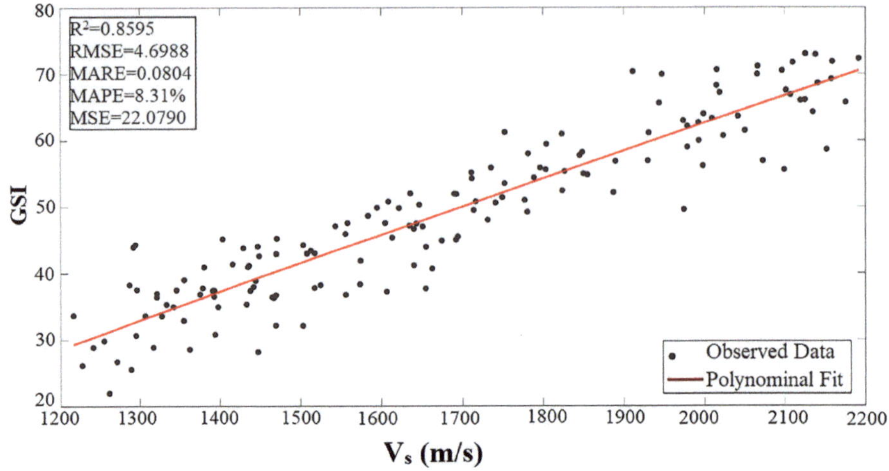

Fig. 11 TRR-GA model of V_s versus GSI with statistics

In summary, the results highlight that while the dependency of GSI on V_s is more intricate and exhibits greater variability, the hybrid TRR-GA model significantly enhances predictive reliability. This finding emphasizes the necessity of employing advanced hybrid metaheuristic algorithms to improve accuracy and robustness in geotechnical assessments involving seismic wave parameters and rock mass classification.

5 Evaluation and Validation of the Derived Models

5.1 Models Comparison Using Statistical Metrics

To evaluate the predictive efficiency of the applied metaheuristic algorithms, i.e., trust region reflective (TRR), genetic algorithm (GA), and the hybrid TRR-GA, a comprehensive statistical analysis was conducted. The performance of each model was assessed through five widely accepted statistical metrics: the coefficient of determination (R^2), mean square error (MSE), root mean square error (RMSE), mean absolute relative error (MARE), and mean absolute percentage error (MAPE). The comparative results for the V_p-based and V_s-based GSI predictions are presented in Figs. 12 and 13, respectively. These indicators collectively provide a clear understanding of both the accuracy and stability of each modeling approach.

According to the obtained results, the hybrid TRR-GA model consistently outperformed both the standalone TRR and GA algorithms across all statistical indices. This superior performance demonstrates that combining the global search strength of GA with the local convergence precision of TRR significantly enhances the model's

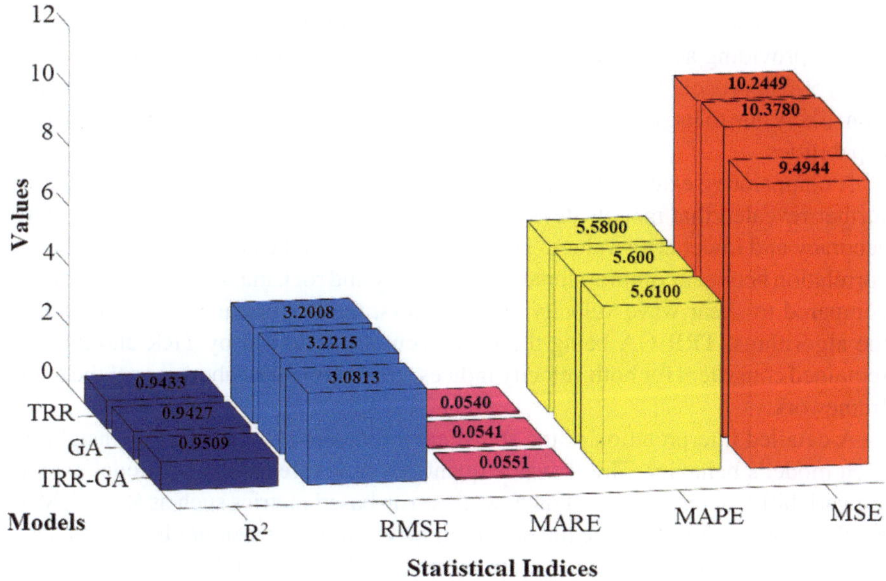

Fig. 12 Comparison of statistical indices (R^2, RMSE, MARE, MAPE, MSE) for TRR, GA, and combined TRR-GA approaches in GSI prediction based on V_p

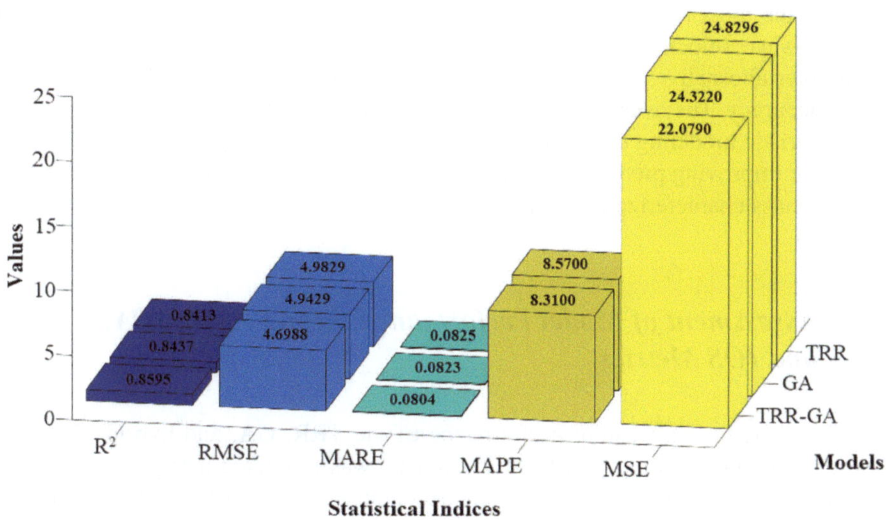

Fig. 13 Comparison of statistical indices (R^2, RMSE, MARE, MAPE, MSE) for TRR, GA, and combined TRR-GA approaches in GSI prediction based on V_S

prediction accuracy and robustness. The TRR algorithm ranked second in performance, providing accurate and stable predictions but with slightly higher errors than the hybrid approach. In contrast, the GA model exhibited comparatively lower precision, likely due to its stochastic search mechanism and the absence of local refinement capabilities.

A comparative examination of the results for V_p-based and V_s-based predictions further revealed that models developed using V_p generally produced slightly better accuracy and lower error values. This improvement can be attributed to the stronger correlation between compressional wave velocity and rock mass mechanical behavior compared to shear wave velocity. Nevertheless, the performance hierarchy among the algorithms, TRR-GA being the most accurate, followed by TRR and then GA remained consistent for both velocity indices, confirming the robustness of the hybrid framework.

A detailed interpretation of the statistical indicators provides further insight into each model's behavior. The R^2 value quantifies how effectively the model explains the variability in observed GSI data, while error-based metrics such as RMSE, MSE, MARE, and MAPE describe the deviation magnitude between predicted and actual values. The lower values of these errors for the TRR-GA model confirm its ability to minimize prediction discrepancies and maintain stable performance even in the presence of data variability.

In summary, the analysis conclusively demonstrates that the TRR-GA hybrid algorithm is the most robust and efficient modeling approach for predicting GSI from seismic wave velocity data. While the standalone TRR and GA models can serve as simpler alternatives for preliminary or rapid estimations, they exhibit relatively lower accuracy. The overall findings emphasize the substantial advantages of hybrid metaheuristic modeling techniques in geotechnical engineering applications, particularly for improving prediction reliability and supporting informed decision-making in rock mass characterization and design.

5.2 Assessment of Model Performance through A20, IOA, and IOS Metrics

In this study, the forecasting performance of the TRR, GA, and hybrid TRR-GA approaches in estimating GSI from V_p and V_s was evaluated using three complementary quantitative indicators, i.e., the acceptable 20% accuracy index (A20), the index of agreement (IOA), and the index of scatter (IOS).

The above-mentioned metrics were chosen to capture both the precision of predictions and the consistency between observed and estimated values. The A20 index measures the proportion of predictions whose relative errors are within $\pm$ 20%, providing a practical sense of the models' accuracy. The IOA quantifies the overall agreement between observed and predicted data, ranging from 0 (no agreement) to 1 (perfect agreement), with values closer to 1 indicating stronger concordance.

Meanwhile, the IOS complements these metrics by reflecting the scatter of predicted values relative to actual observations, offering an additional perspective on model reliability [15, 16, 48–51].

The results of these evaluations are illustrated in radar charts for both V_p- and V_s-based predictions (Figs. 14 and 15). The hybrid TRR-GA model consistently outperformed the standalone TRR and GA approaches across all three metrics. Notably, the A20 value for the combined approach reached a perfect score of 1.0000, indicating that all GSI predictions fell within ± 20% of the measured data. Similarly, the IOA and IOS values for this model were very close to 1, demonstrating its high predictive accuracy and minimal scatter. Although the GA and TRR models also provided reasonably accurate predictions, their performance lagged behind that of the hybrid TRR-GA approach.

This comparison clearly highlights the superior ability of the combined TRR-GA methodology to produce more accurate and reliable estimates of GSI, underlining its advantage for practical applications.

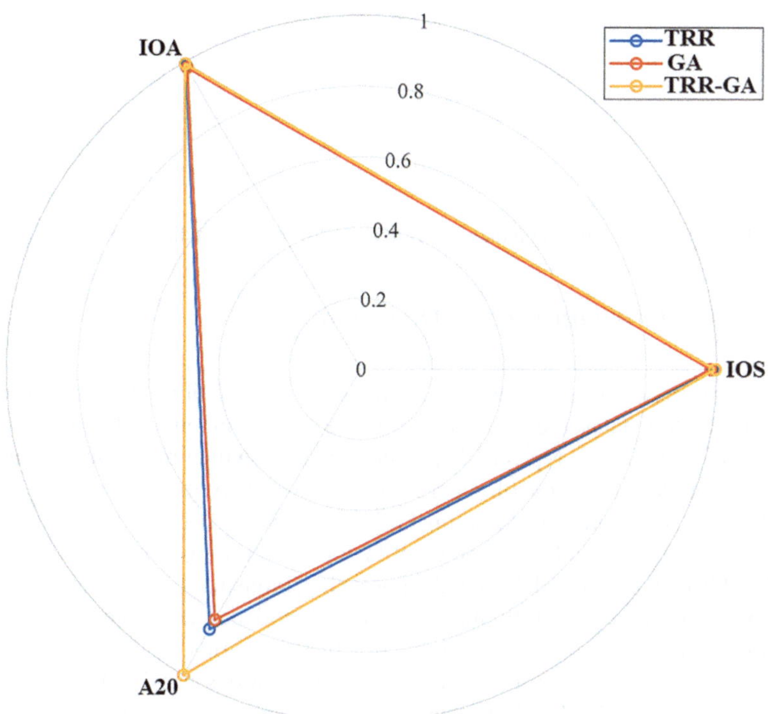

Fig. 14 Radar chart of statistical indices (IOS, IOA, A20) for GSI prediction based on V_p

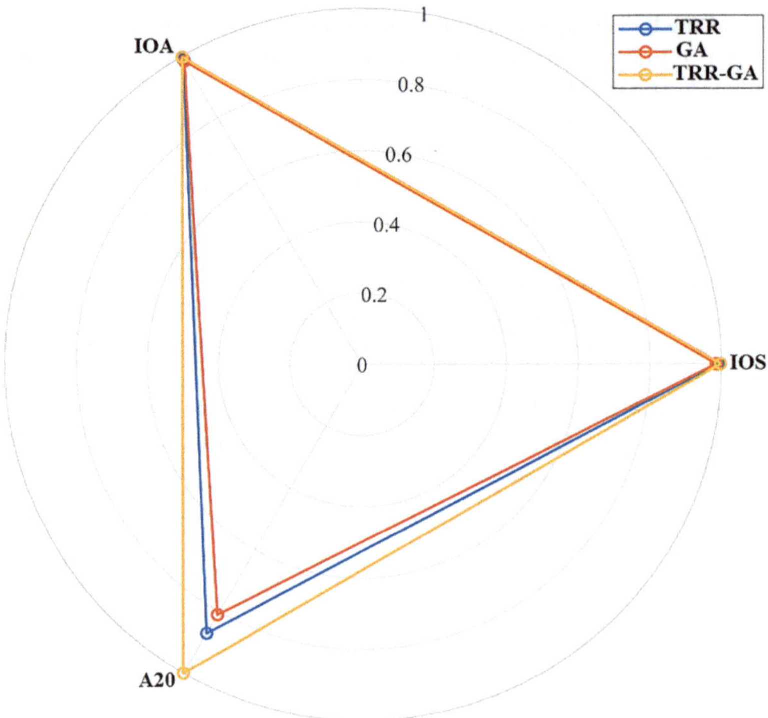

Fig. 15 Radar chart of statistical indices (IOS, IOA, A20) for GSI prediction based on V_s

5.3 Constraints and Future Direction

While the results of this study are promising, several limitations should be acknowledged. First, the dataset employed for model development and validation was restricted to certain rock types and specific locations, which may limit the applicability of the models to other geological contexts. Second, the models depend solely on V_p and V_s measurements for estimating GSI, which may not fully capture the complex geological, structural, and environmental factors affecting rock mass quality. Additionally, although the hybrid TRR-GA approach demonstrated the highest predictive performance, its increased computational demands and the need for careful parameter calibration may constrain its use in some practical scenarios.

In light of these limitations, several recommendations can be made for future research. Expanding the dataset to include a wider range of rock types and geological conditions would enhance the robustness and generalizability of the models. Incorporating additional rock mass properties or environmental variables could further improve the accuracy of GSI predictions. Moreover, investigating alternative or more sophisticated hybrid algorithms may help achieve a better balance between predictive performance and computational efficiency. Finally, conducting field validations of the developed models across diverse mining or tunneling projects is crucial to confirm their practical reliability and applicability.

6 Conclusion

The geological strength index (GSI) is widely acknowledged as a key parameter in rock mechanics and geotechnical engineering, serving as a measure of rock mass quality and a basis for evaluating its mechanical behavior. Accurate estimation of GSI is crucial for the safe and cost-effective design of underground structures, including tunnels, mines, and slopes. Despite its importance, traditional approaches for determining GSI largely rely on field observations and expert judgment, which can introduce errors, inconsistencies, and limited reproducibility. Consequently, recent research has increasingly turned to data-driven and artificial intelligence techniques to provide more quantitative, reliable, and precise predictions.

In this study, we sought to address these limitations by using simple, practical, and cost-effective rock mass parameters, namely seismic wave velocities (V_p and V_s), as predictors of GSI. Three modeling strategies were employed and compared: trust region reflective (TRR), genetic algorithm (GA), and a hybrid TRR-GA approach. In addition, five functional forms, including exponential, logarithmic, power, quadratic polynomial, and linear equations, were tested to identify the best mathematical representation of the relationship between seismic velocities and GSI. The analysis revealed that the quadratic polynomial form achieved the most accurate fit, effectively capturing the complex nonlinear relationships between the input parameters and GSI.

A comprehensive statistical assessment using R^2, RMSE, MARE, MAPE, and MSE, supplemented by IOS, IOA, and A20 indices, showed that the hybrid TRR-GA model consistently outperformed the standalone TRR and GA methods. By combining the global search capability of GA with the local refinement strength of regression, the hybrid model effectively reduced prediction errors while maintaining high stability. Among the individual models, TRR demonstrated superior performance compared with GA, ranking second overall. This suggests that while GA is powerful for global exploration, its standalone application may not consistently deliver stable and highly accurate predictions.

Another notable finding was that GSI predictions based on V_p were slightly more precise than those derived from V_s. This can be attributed to V_p's higher sensitivity to variations in rock density, porosity, and mass integrity factors closely linked to GSI. Nonetheless, the ranking of model performance remained consistent for both seismic parameters: TRR-GA first, TRR second, and GA third, highlighting the robustness and generalizability of the hybrid approach across different geological conditions.

From a practical standpoint, these findings offer valuable insights for rock engineering applications. First, easily measurable and cost-effective parameters such as V_p and V_s can reliably support GSI prediction, reducing dependence on time-consuming and expensive laboratory tests. Second, hybrid approaches like TRR-GA provide more accurate and dependable tools for decision-making in large-scale geotechnical projects. Third, selecting an appropriate functional form is critical, and as demonstrated, quadratic polynomials appear to offer the most reliable predictive outcomes.

Nevertheless, this study has certain limitations. The dataset was primarily obtained from the Beheshtabad Water Transfer Tunnel, which may not represent the full geological diversity of other regions. Moreover, only two seismic parameters (V_p and V_s) were considered, while other mechanical and geological properties, such as uniaxial compressive strength, elastic modulus, and joint conditions, were not included. Finally, despite its superior predictive capability, the TRR-GA hybrid model's higher computational demand and the need for careful parameter tuning may restrict its practical application in some contexts.

Despite these limitations, the present research opens promising directions for future studies. Expanding the dataset to include more diverse rock types and geological conditions would enhance model generalizability. Incorporating additional geomechanical parameters alongside seismic velocities could further improve prediction accuracy and stability. Exploring more advanced hybrid algorithms may provide an optimal balance between computational efficiency and predictive performance. Finally, field validation of the proposed models across various mining and tunneling projects is essential to assess their reliability and practical applicability.

AI Use Disclosure Statement

The authors confirm that the scientific content, data analysis, and interpretations presented in this book chapter are entirely their original work. No artificial intelligence (AI) tools were used in generating the research findings or conclusions. Minor language and grammatical corrections were applied solely to improve clarity and readability, and these were performed with the assistance of AI-based proofreading tools. The use of such tools did not influence the intellectual content, data interpretation, or scientific insights of this chapter.

References

1. Gokceoglu C, Sonmez H, Kayabasi A (2003) Predicting the deformation moduli of rock masses. Int J Rock Mech Min Sci 40(5):701–710
2. Narimani S, Davarpanah SM, Bar N, Török Á, Vásárhelyi B (2023) Geological strength index relationships with the Q-system and Q-slope. Sustainability 15(14):11233
3. Wijaya BJ, Warmada IW, Indrawan IGB (2023) Correlation between GSI, RMR, and Q systems of rock masses at the construction site of spillway tunnel at Meninting dam. AIP Conf Proc 2629:030007
4. Abbas N, Li KG, Emad MZ, Qin Q, Li M, Sultan Shah K, Yue R, Qiu S (2024) Empirical evaluation of RMR, GSI, and Q for underground excavations. Iran J Sci Technol Trans Civ Eng 48(4):2583–2594

5. Laderian A, Abaspoor MA (2012) The correlation between RMR and Q systems in parts of Iran. Tunn Undergr Sp Technol 27(1):149–158
6. Bieniawski ZT (1989) Engineering rock mass classifications. John Wiley & Sons, p 272
7. Morelli GL (2018) Correction to: variability of the GSI index estimated from different quantitative methods. Geotech Geol Eng 36:1417
8. Khoshjavan S, Mazlumi M, Rezai B, Rezai M, (2010) Estimation of hardgrove grindability index (HGI) based on the coal chemical properties using artifical neural networks. Orient J Chem 26(4):1271–1280
9. Majdi A, Rezaei M (2013) Application of artificial neural networks for predicting the height of destressed zone above the mined panel in longwall coal mining. In: 47th U.S. rock mechanics/ geomechanics symposium, 23–26 June, San Francisco, California, USA, pp 1665–1673
10. Rezaei M (2020) Feasibility of novel techniques to predict the elastic modulus of rocks based on the laboratory data. Int J Geotech Eng 14(1):25–34
11. Asadizadeh M, Rezaei M (2021) Surveying the mechanical response of non-persistent jointed slabs subjected to compressive axial loading utilising GEP approach. Int J Geotech Eng 15(10):1312–1324
12. Rezaei M, Rajabi M (2021) Assessment of plastic zones surrounding the power station cavern using numerical, fuzzy and statistical models. Eng Comput 37:1499–1518
13. Wang Y, Rezaei M, Abdullah RA, Hasanipanah M (2021) Developing two hybrid algorithms for predicting the elastic modulus of intact rocks. Sustainability 15(5):4230
14. Ding X, Hasanipanah M, Rezaei M (2025) Assessment of mechanical properties of rock using deep learning approaches. Measurement 250:117180
15. Koureh Davoodi P, Hajizadeh F, Rezaei M (2025) Evaluating the RMR correlation with the rock mass wave velocity using the meta-heuristics algorithms. S Sci Rep 15(1):17716
16. Koureh Davoodi P, Hajizadeh F, Rezaei M (2025) Application of the metaheuristic algorithms to quantify the GSI based on the RMR classification. Sci Rep 15:28950
17. Mirabedini ME, Haghshenas E, Ganjian N (2022) Estimating the geological strength index (GSI) in regional seismic-landslide zonation using the empirical regression model. Adv Civ Eng 1(6):4798523
18. Perez Ruiz DA (2016) Modern optimization with R. J Stat Softw 70 (Book Review 3)
19. Ceballos F, Olalla C, Jiménez R (2014) Relationship between RMRb and GSI based on in situ data. The ISRM regional symposium—EUROCK 2014, ISRM-EUROCK-2014-058, Vigo, Spain
20. Hoek E, Brown ET (2019) The Hoek–Brown failure criterion and GSI—2018 edition. J Rock Mech Geotech Eng 11(3):445–463
21. Cosar S (2004) Application of rock mass classification systems for future support design of the dim tunnel near Alanya. Middle East Technical University, Ankara, p 217
22. Irvani I, Wilopo W, Karnawati D (2015) Determination of nuclear power plant site in West Bangka based on rock mass rating and geological strength index. J SE Asian Appl Geol 7(1):42–50
23. Singh JL, Tamrakar NK (2013) Rock mass rating and geological strength index of rock masses of Thopal-Malekhu River areas, Central Nepal lesser Himalaya. Bull Dep Geol 16:29–42
24. Ali W, Mohammad N, Tahir M (2014) Rock mass characterization for diversion tunnels at Diamer Basha dam, Pakistan—a design perspective. Int J Sci Eng Technol 3(10):1292–1296
25. Zhang Q, Huang X, Zhu H, Li J (2019) Quantitative assessments of the correlations between rock mass rating (RMR) and geological strength index (GSI). Tunn Undergr Sp Technol 83:73–81
26. Siddique T, Khan EA (2019) Stability appraisal of road cut slopes along a strategic transportation route in the Himalayas, Uttarakhand. India SN Appl Sci 1:409
27. Hoek E, Carter TG, Diederichs MS (2013) Quantification of the geological strength index chart. In: The 47th US rock mechanics/Geomechanics symposium held in, San Francisco, CA, USA, pp 1757–1764
28. Somodi G, Bar N, Kovács L, Arrieta M, Török Á, Vásárhelyi B (2021) Study of rock mass rating (RMR) and geological strength index (GSI) correlations in granite, siltstone, sandstone and quartzite rock masses. Appl Sci 11(8):3351

29. Rezaei M, Hosseini SP, Jahed Armaghani D, Khandelwal M (2025) Improving the P-wave velocity determination by considering the effects of joint properties in artificial rock samples. J Min Environ 16(3):1009–1025
30. Rezaei M, Ahmadi SR, Nguyen H, Jahed Armaghani D (2025) Improved determination of the S-wave velocity of rocks in dry and saturated conditions: application of machine-learning algorithms. Transp Geotech 49:101371
31. Zayandab Engineering Consulting Company (2010) Geotechnical report of Beheshtabad tunnel (In Persian)
32. Rezaei M, Habibi H (2023) Designing of the Beheshtabad water transmission tunnel based on the hybrid empirical method. Struct Eng Mech 86(5):621–633
33. Rezaei M, Koureh Davoodi P (2021) Determining the relationship between shear wave velocity and physicomechanical properties of rocks. Int J Min Geo-Eng 55(1):65–72
34. Rezaei M, Koureh Davoodi P, Najmoddini I (2019) Studying the correlation of rock properties with P-wave velocity index in dry and saturated conditions. J Appl Geophys 169:49–57
35. Sivanandam SN, Deepa SN, Sivanandam SN, Deepa SN (2008) Genetic algorithm optimization problems. In: Introduction to genetic algorithms. Springer, Heidelberg, pp 165–209
36. Golanbari B, Mardani A, Hosainpour A, Taghavifar H (2025) Predicting terrain deformation patterns in off-road vehicle-soil interactions using TRR algorithm. J Terramechanics 117:101021
37. Armaghani DJ, Mohamad ET, Narayanasamy MS, Narita N, Yagiz S (2017) Development of hybrid intelligent models for predicting TBM penetration rate in hard rock condition. Tunn Undergr Space Technol 63:29–43
38. Zhou J, Guo H, Koopialipoor M, Jahed Armaghani D, Tahir MM (2021) Investigating the effective parameters on the risk levels of rockburst phenomena by developing a hybrid heuristic algorithm. Eng Comp 37(3):1679–1694
39. Jahed Armaghani D, Kumar D, Samui P, Hasanipanah M, Roy B (2021) A novel approach for forecasting of ground vibrations resulting from blasting: modified particle swarm optimization coupled extreme learning machine. Eng Comp 37(4):3221–3235.
40. Basahel H, Mitri H (2017) Application of rock mass classification systems to rock slope stability assessment: a case study. J Rock Mech Geotech Eng 9(6):993–1009
41. Alejano LR (2024) Rock mass classification systems: a useful rock mechanics tool, often misused. Rock Mech Rock Eng. https://doi.org/10.1007/s00603-024-04087-y
42. Santos V, Da Silva PF, Brito MG (2018) Estimating RMR values for underground excavations in a rock mass. Minerals 8(3):78
43. Sheng D, Yu J, Tan F, Tong D, Yan T, Lv J (2023) Rock mass quality classification based on deep learning: a feasibility study for stacked autoencoders. J Rock Mech Geotech Eng 15(7):1749–1758
44. Hussain S, Mohammad N, Khan M, Rehman ZU, Tahir M (2016) Comparative analysis of rock mass rating prediction using different inductive modeling techniques. Int J Min Eng Miner Process 5:9–15
45. Sahu A, Sinha S, Banka H (2024) Fuzzy inference system using genetic algorithm and pattern search for predicting roof fall rate in underground coal mines. Int J Coal Sci Technol 11:1. https://doi.org/10.1007/s40789-023-00630-4
46. Huang F, Xiong H, Chen S, Lv Z, Huang J, Chang Z, Catani F (2023) Slope stability prediction based on a long short-term memory neural network: comparisons with convolutional neural networks, support vector machines and random forest models. Int J Coal Sci Technol 10:18. https://doi.org/10.1007/s40789-023-00579-4
47. Qi Q, Yue X, Duo X, Xu Z, Li Z (2023) Spatial prediction of soil organic carbon in coal mining subsidence areas based on RBF neural network. Int J Coal Sci Technol 10:30. https://doi.org/10.1007/s40789-023-00588-3
48. Khatti J, Grover KS (2025) Estimation of uniaxial strength of rock: a comparison between Bayesian-optimized machine learning models. Mining. Metall Explor 42:133–154
49. Khatti J, Grover KS (2024) Assessment of uniaxial strength of rocks: a critical comparison between evolutionary and swarm optimized relevance vector machine models. Transp Infrastruct Geotechnol 11:4098–4141

50. Fissha Y, Khatti J, Ikeda H, Singh GK, Owada N, Toriya H, Adachi T, Kawamura Y (2024) Predicting ground vibration during rock blasting using relevance vector machine improved with dual kernels and metaheuristic algorithms. Sci Rep 14(1):20026
51. Khatti J, Polat BY (2024) Assessment of short and long-term pozzolanic activity of natural pozzolans using machine learning approaches. Structure 68:107159

Improving Balanced Scorecard Implementation Through Machine Learning: The Conditional Impact of Environmental Uncertainty in SMEs

Mahshid Lonbani

Abstract Environmental uncertainty requires firms to respond rapidly to unforeseen changes in order to survive, especially small and medium enterprises (SMEs) which are more adaptive to market changes and are unable to directly control the external environment. A successful strategy implementation starts in the formulation stage and a failure to find the linkage between strategy formulation and strategy implementation is a step toward strategy failure. Hence, this study investigates the roles of environmental uncertainty as moderation between strategy and Balanced Scorecard (BSC) as a way of strategy implementation system among Malaysian SMEs. Using Partial Least Square, this study demonstrated that environmental uncertainty acts as moderation between BSC and all Miles and Snow's strategy typology except for Defender strategy. Furthermore, this paper introduces a Machine Learning (ML)—supported framework that leverages SMEs' performance and market data to predict BSC outcomes under uncertain environments. The integration of ML provides data-driven insights, improves adaptability, and complements traditional statistical approaches.

Keywords Balanced scorecard · Small and medium enterprise · Machine learning · Strategy implementation · Environmental uncertainty · Moderation effect

1 Introduction

The Balanced Scorecard (BSC) has evolved from performance measurement tool to a system of strategy implementation [52, 53]. Strategy implementation is an important concern in any organization. Nowadays, BSC is being popular among both large and small companies ([37], SMES). Generally, BSC can help managers to use the boundaries of control in their organizations which are not strict (reference). Hence, in this study, the BSC was considered as part of management control system specifically

M. Lonbani (✉)
Crown Institute of Higher Education, Sydney, NSW, Australia
e-mail: lonbanimahshid@gmail.com

A. Azizi et al. (eds.), *Intelligence at the Interface*, Emerging Trends in Mechatronics,
https://doi.org/10.1007/978-981-95-7584-8_5

through its strategic implementation. This is based on the suggestion of Norreklit [70] that the BSC is not only considered as a strategic measurement system, but also as a strategic control system which could align organizational goals and objectives to the strategy.

There are some certain problems found by the SMEs implementing BSC. SMEs experience a very dynamic process in strategy formulation and its linkage to operation [38]. Environments have been viewed as key contingency variables for relationships between strategy and strategy implementation. To the best of authors' knowledge, the contingent effect of business strategy and environmental uncertainty on BSC implementation among SMEs has not been addressed yet. This is important to evaluate the continent effect of market uncertainty on the effectiveness of BSC implementation among SMEs. Therefore, it is important to fill this gap and evaluate these relationships.

2 Literature Review

During the time (between 1992 and 1996), BSC became a strategic management device which emphasized the causality. This means that the cause and effect between measures has been focused enormously [44, 69]. In the mid-1990s, the issue of causality encompassed strategic objectives rather than measures [57]. Papers of Kaplan and Norton in 1996 can significantly show this transition. Therefore, as highlighted before, the changes enabled BSC to change from measurement system to a management system [53]. Regarding emphasis of BSC on supporting the management of implementing strategy, Kaplan and Norton considered this development of BSC as a core element of strategic management system. This type of BSC has been changed to a new strategic control system in the next generation. The second generation was refined through the integration of destination statement. This development was based on the idea of strategic objective validation and setting targets [57]. Additionally, a broad activity and outcome perspectives have been introduced as opposed to the traditional four perspectives.

Regarding different generations, the BSC has become a tool for strategic management which leads managers to be involved in the process of managing the strategy. The BSC's popular implementation has motivated studies on it. However, in the years following the emergence of the concept of the BSC there has practically been no empirical studies on the application of this management tool for SMEs [82]. Studies conducted by Kaplan and Norton [52] emphasized on organizations or companies with a large size. This statement is approved mostly over the cases examined in early BSC studies. Users or intended users expressed in the studies are banks and technology firms of substantial size, while SMEs are not usually focused [50]. In this regard, Tanyi [94] stated that the smaller companies have the least number of BSC users, but when compared to the mid-sized companies, it becomes difficult to draw a fine line.

There are a large body of studies which addressed the use of BSC in large organizations; however, the available literature regarding BSC implementation in small firms is very few. In spite of this scarcity, a review of literature related to this specific group of organization will be presented in this section.

Although there are numerous studies on the general restrictions of BSC concept (e.g., [2, 23, 26, 70, 72]), there are only limited investigations which focused on the boundaries of BSC usage among SMEs. This may be the result of rare application of BSC in these environment compared to their larger counterparts [81].

Strategy formulation and linkage to operations is a very dynamic process in SMEs [39]. This dynamism is a reflection of the ever increasing market turbulence that SMEs find themselves in [16]. Regarding this issue, Rompho [81] referred to frequent changes of SMEs' strategy as the most important cause of BSC failure, since starting to apply the BSC, different measures were added or reviewed in surveyed companies. This was not in line with the ideas presented in the previous studies. It was mentioned that in the context of emergent strategy and confronting new situation measures are mainly unchanged and the only changes are related to the presentation of new initiatives [54]. This concept does not support Rompho's results, where the only measures unaffected were related to financial and learning perspective, with most measures referring to the internal business process and customers needing to be adjusted.

According to Rompho [81], the BSC is limited only due to the nature of the SMEs in which response to market changes is given more frequently compared to that in a large organization that operates in much more stable market. He stated that this problem was not a common reason for the failure of BSC in large organizations. This means that during two years, many revisions were done on the measures in BSC due to inevitable strategy changes in a quickly-changing business environment. This varying environment caused confusion among employees, even among the manager/owners. The numerous revisions on BSC brought about the need for new data for new measures. As a result, it was impossible to track the cause and effect relationship between measures in BSC, which led to a waste of effort and time.

Meanwhile, Kaplan and Norton [53] believed that using BSC, companies are capable of modifying their strategies. As mentioned earlier, in an extremely-dynamic context, companies must change their strategies constantly. Consequently, measures in BSC have to be changed frequently. Norreklit [70] analyzed BSC critically and asserted that in dynamic context, companies change their measures frequently, which results in an uncertainty margin in regard to the utility of defined indicators. Generally, for an organization, it is not easy to establish measures for performances with which it has no or very little experience. As a result, as measuring effects are difficult in those companies that continually have to adapt themselves to new situations, BSC cannot be applied to companies in extremely-dynamic contexts.

Therefore, it can be concluded that the formalized nature of BSC which needs a well-developed strategy is not compatible with characteristics of SMEs' business strategy. As mentioned above SMEs should have a flexible strategy in response to their market changes. These frequent changes of strategy together with their dynamic environment cause problem while BSC implementation. Therefore, this problem is going to be addressed in this study.

2.1 Strategic Planning in SMEs

Strategic planning in SMEs does not reflect exhaustive strategic analysis, but rather, a personality driven, opportunistic or instinctive approach, channeled through an emergent planning process [7]. In defining strategy, small business decision makers assimilate information and advice from a wide array of professional, business and personal sources.

In numerous SMEs, there are fewer senior managers compared to larger firms; this means that the capacities of only one person have a deep influence. It can be referred to the lack of management expertise [35] that in SMEs, the strategic business planning is generally confined and it is short term in focus, with a prevalent philosophy of fire-fighting 'react and adapt' [28, 60, 71, 102]. The general influence of the intensive competitive environment where SMEs operate is that strategic planning often becomes a useless exercise.

Considering the difficulties that are accompanied with aligning the strategy to external competitive context, Argument et al. [6] concluded that most of the SMEs operating in the automotive sector are not worried about the future strategic developments. To survive in the supply chain, they merely have to maintain a reactive strategy. Furthermore, according to Brouthers et al. [14], planning in SMEs is generally less political, less rational, less controlled, and more intuitive in comparison with planning in large firms, and a lot of existing evidence suggest that numerous established SMEs rely only on internal or financial planning as the most important approach to getting prepared for the future [74, 99].

Adobi [1] carried out a research in Nigeria to investigate existing strategies of SMEs in order to explore whether the Nigerian SMEs adopted unconscious actions or deliberately-planned patterned behaviors (PPB) that guided their actions. His study demonstrated that 75% of SMEs adopted more of emergent and preferred their choices over the other, whereas 25% used more of planned even though spontaneously acted to issues. The 25% that adopted PPB were from manufacturing sector. The obtained results showed that the strategic adoption process of the Nigerian SMEs is informal and this is dependent on marketing forces of demand and supply, with other environmental factors. Essentially, their strategies originated from their attempt to overcome the environmental factors and, spontaneously, taking advantage of opportunities appropriately. This finding was in line with that of Westhead et al. [100] who claimed that the attitude of SMEs in the UK is not planned, but reactive.

Harvesting from the above review, it can be concluded that the attitude of companies toward adopting a well-formulated strategy are different. It is said that large organizations/companies have a well-developed strategy but small to medium sized enterprises mostly lack such an essential parameter. Apart from the above discussion, the following are the scholars who have similar opinion:

- According to Atkins and Lowe [9] and Mintzberg [66], SMEs have a limited planning and control system.
- Greatbanks and Boaden [36] believed that SMEs have poor strategic planning, and they are not completely aware of the critical success factors. SMEs should be supported by performance measurement systems (PMS) in managing uncertainty, sustaining the evolution and change processes, and innovating their services and products.
- Singh et al. [89] stated that SMEs are not following any comprehensive framework for developing their strategies and quantifying their competitiveness.
- Jennings and Beaver [49] indicated that strategic management in small organizations is almost unconscious process which is very much a function of the personality, experience and ability of the owner-manager.
- According to Russo and Martins [82], most of the SMEs have no formalized vision, a long-term strategy, and mission; instead, they prefer to informally define their objectives for a short-term (1, 2 or 3) years.
- Saindis et al. [83] confirmed that smaller companies rarely record their business plan or strategies. This condition is not due to the impossibility of providing defined strategies; rather, this is because they prefer short-term plans and unwritten, flexible, informal strategies. In this situation, the leaders can determine more adequate strategies that are appropriate for the continuously shifting environment of the markets.

2.2 SME's Reaction to External Environment

Literature contains studies which indicate that SMEs are flexible to the market changes [28, 102]. This is considered as a positive feature suggesting the existence of some correlation between SMEs and the market culture model [90]. On the other hand, according to Pelham [75]; this attribute is rooted in the fact that SMEs generally have a confined overview of markets wherein they operate; which leads to having no control over their competitive positions. Instead of actively driving the market, they have to passively react and adapt themselves to the market changes [46, 93].

SMEs consider no value for formal market research; thus, it is rarely carried out. Managers of small manufacturing firms dedicate the least amount of time to spend for the marketing analysis. According to Mohan-Neill [67], fewer than 50% of managers of small companies collect continuously the information regarding the market segments or market growth. In small firms, the lack of data collection, confined planning, and the confined resources, can amplify both indirect and direct effects of the industry environment on their performance.

Smaller companies possess more constrained resources compared to larger ones [11]. When they face external pressures for environmental considerations, they often cannot invest their limited resources in political resistance. Rather, they normally invest their resources to address the immediate environmental concerns [13]. In addition, Man et al. [58] stated that SMEs are typically more susceptible to external effects in comparison with larger companies. This is because their competitive advantages are less sustainable when the environment changes.

Porter [77] claimed that the strategy selection depends on how well a firm is aligned with the environment. For instance, in situations where there is a high uncertainty in customer, technology, or competitive environments, firms should be capable of accommodating to the environmental changes. According to DeSarbo et al. [20], the environmental uncertainty demands a firm to be capable of responding more rapidly to unanticipated changes to survive and do extremely well in the industry. In general, the management is about responding to the environmental factors, including the increasing competition, uncertainty in economic growth, improvements of technology as well as planning necessary activities for the functioning and structure of the entire company [43].

Managers should take into account factors of environmental uncertainty, including the extent of predictability of capital and financial markets, governmental intervention and regulation, the competitors' actions, the suppliers' actions, and general conditions the organization faces with [92]. According to Atherton [8], small firms are not able to directly control their external environment and they are often uncertain of the impact and effects on the enterprise. Alpkan et al. [3] stated that planning has long been recognized as an influential factor for SMEs to address uncertainty produced by quick changes occurring in the environment. Therefore, it can be found that SMEs need to adapt to external environment in order to survive.

3 Underlying Theories

This study explored different theories which provided the basis for developing hypotheses and the conceptual framework. They are mainly related to theory of strategic management in SMEs, and contingency theory. Before going further the research model which is a summaries of every step done for this study is illustrated in Fig. 1.

3.1 Strategic Management Based on Miles and Snow's Typology

One of the most critical tasks for managers is constantly searching for the best strategy that can enhance performance quality. There are studies that have classified the business strategies in order to effectively explore the relationship that exist between strategy and other variables. Generally, typologies are employed for some objectives, including the contextualization of the business level strategies in the strategic management, identifying the ways organizations should behave in particular situations, determining various decision-making categories, or describing various

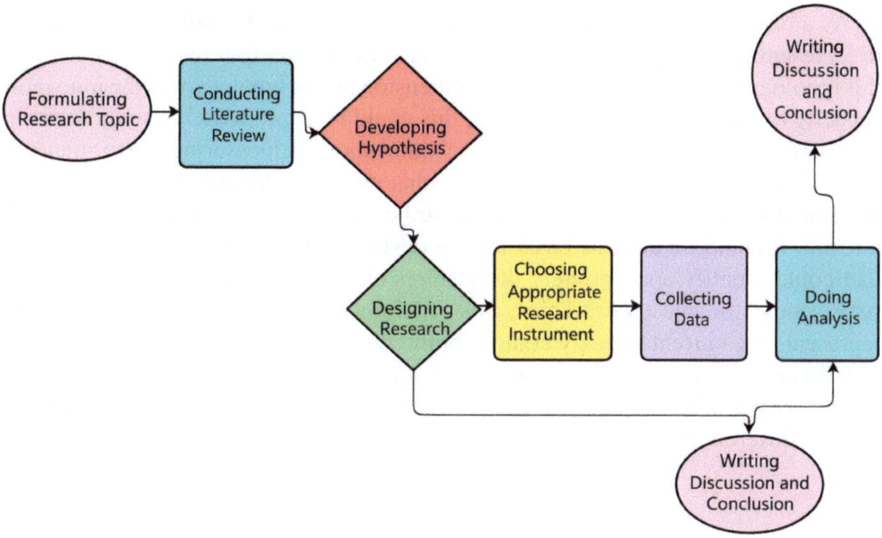

Fig. 1 Research model

views of strategy. Miles and Snow [62] and Porter [76] developed the most popular and effective typologies. The two typologies have distinct characteristics. Kald and Nilsson [51] suggested that, primarily, the typology developed by Miles and Snow [62] focuses on the strategic orientations, while that of Porter [76] focuses on the strategic positioning.

The strategic typology proposed by Miles and Snow [62] has received a great deal of attention among marketing and management authors (e.g., [18, 20]). This typology is still one of the most commonly adopted widespread models of strategy [20, 21]. This is employed in various industry settings because of its consistency with the real strategic positions of firms in many industries and numerous countries as well as its industry-independent nature. Originally, the strategic typology of Miles and Snow considered the strategic behavior of large-scale companies; although, several studies have been carried out for the validation of typology within the SME's context as well (e.g., [5]).

In the alternative framework of Miles and Snow [62], four strategic types are identified, namely Prospector, Defender, Analyzer, and Reactor. The Prospectors consider the environment as an uncertain and dynamic entity. They preserve the flexibility and innovate in order to overcome the environmental changes. Generally, they become industry designers. On the contrary, for the Defenders, the environment is certain and stable. They look for control and stability in operations in order to obtain the highest degree of efficiency. On the other hand, Analyzers take into account both flexibility and stability; they attempt to make profits from the best of both of them. Finally, Reactors have no consistency in their strategic choice and their performance is poor. Different empirical studies have proved the validation of this typology [24, 87], which is considered consistent internally and acceptable academically [25].

As a result, this study adopted the typology developed by Miles and Snow [62] to represent the strategy variables. Specifically, there were four reasons for selecting this typology. First, the most important dimension of this typology was the innovation capacity of an organization [32]. Thus, this could be appropriately used to examine its relationship with BSC since it is an innovation in measuring performance as well as strategy implementation. Second, this was considered to be holistic once it integrated the relationships between the strategy, processes, and structure [97]. As a result, it could be used as basis for the development of a theoretical framework that could identify appropriately the features of the BSC, which were appropriate for various strategic types. Third, this typology was suitably documented and it represents the current strategy content of the literature, and it was empirically well tested and internally consistent and well accepted [32]. Fourth, it shared similar characteristics with the other strategy archetypes, e.g., build and harvest strategies [34], or differentiation and low-cost strategies [76].

3.2 Use of BSC as Part of Management Control System (MCS)

Gomes and Salas [31] reviewed the literature regarding management control system from the late 1980s and provided a comprehensive approach to control. According to their systemic approach, the company is aimed to meet different stakeholders' expectations within the organization as far as possible. According to the scholars, the aim of management control system is to promote the companies' success by widely engaging people in the process of management in a way to achieve both business and individual objectives. Changes that have occurred within the business environment from the 1980s onwards have led the companies to a more open, dynamic, international context. Therefore, their management control systems are assumed to add novel indices to the conventional economic, and quality, financial, and productivity indicators. Some new indices are related to the organizational contexts, environmental factors, culture, interests of managerial stakeholders, competitive position, and in particular the customer satisfaction with processes, services, and products.

Recently-proposed approaches have suggested methodologies for evaluating the performance of companies and a boundary between informal control mechanisms and formal ones in order to enhance the potential of the processes of decision-making. In addition, some of the new management control systems have complemented formal control systems with informal mechanisms in a way to encourage individuals to behave according to the managerial expectations. Accordingly, the model introduced by BSC can be considered as a part of management control system through the integration of these perspectives for both formal and informal control and because of its significant role in the strategic implementation process [22].

There are very explicit connections between the organizational goals and management control systems, as they mainly measure the progress toward the achievement of the desired organizational aims. This is a helpful exercise while assessing the features of management control systems that are employed for reporting goals to make a judgment about the degree to which they accommodate the followings: multiple stakeholders, measure efficiency, effectiveness and equity, capture non-financial and financial outcomes; horizontal links across the value chain and vertical links between operations and strategy; information regarding the way the company is related to its external environment, and the capability of adaptation. Most likely, BSC or the performance hierarchies offer a methodology for tackling many of the above-mentioned issues [15]. The MCS mechanism is designed to mediate the implementation of strategies of a company. This is considered as one of the major components of management process [4] since it provides a basis for implementing the plans in a way to achieve both the stakeholders' and business's objectives. As argued by Montgomery and Porter [68], the implementation of strategies is a bridge from identification of desirable objectives to achieving them successfully.

Dias Jordão and Casas Novas [22] conducted the study on the employment of BSC as a part of the management control system for the implementation of strategies in a large Brazilian Mixed Economy Company (MEC). The results obtained from this study showed that:

1. BSC helps to translate the adopted strategy of MEC into goals and objectives aligned to Financial, learning and growth, improvement of internal processes and customer perspectives as suggested by the BSC model.
2. The model can be applied as an MCS, which improves the process of corporate strategy implementation and its diffusion at various organization levels.
3. In the case of organizations, the evaluation of performance through integrating the quantitative and qualitative measures is a significant mechanism in implementing such strategies.

3.3 Relationship Between Business Strategy and Strategy Implementation

In the area of strategic management, both strategy formulation and strategy implementation are important. Hrebiniak [45] recommend that strategy implementation was worthy of more attention than strategy formulation in some extent. The relationship between strategy and strategy implementation has been evaluated by different scholars. Among available studies, there were numerous tools which were considered as a ways of strategy implementation. Following paragraphs refer to some methods of strategy implementation including BSC and elaborate their relationship with strategy.

The BSC is used to communicate and coordinate the translation of the business strategy [53]. The BSC is directed top down, both in its contents and in its development as a management system. Therefore, to be able to clarify and translate the strategy, top management must agree on the strategy. The goal is to create a common and comprehensible strategic basis in the form of a formulated BSC [52]. Because of this, the verbally formulated strategy should be translated and causally linked in terms of objectives and measures.

Kaplan and Norton [53] highlighted the importance of aligning the BSC with organizational strategy. Yu et al. [101] found the importance of strategy-linked BSC in determining the effectiveness of a BSC. The scholars believed that with strategy-linked BSC, the BSC becomes a strategy implementation instrument which helps company to communicate its strategy through all levels.

Baroto et al. [10] also emphasized the effect of strategy formulation and strategy implementation. They asserted that strategy formulation has a significant positive relationship with the success of strategy implementation. The improvements in strategy formulation will increase the effectiveness of strategy implementation.

Therefore, based on the above discussion the following hypotheses were developed:

H1: Business strategy has a strong influence on BSC as a strategy implementation system among SMEs.
H1a: Defender business strategy has a strong influence on BSC implementation.
H1b: Prospector business strategy has a strong influence on BSC implementation.
H1c: Analyzer business strategy has a strong influence on BSC implementation.
H1d: Reactor business strategy has a strong influence on BSC implementation.

3.4 Contingency Theory Regarding Management Control System (MCS)

Contingency-based study has a long tradition in the investigation of the management control systems. A number of studies have been carried out to explain the MCS effectiveness through the examination of designs that are best suit to the nature of technology, environment, structure, size, strategy, and national culture. Recently, contingency-based research has kept its popularity with the studies that include these variables but redefined them in contemporary terms [15].

In order to respond to uncertainty, the contingency theory framework predicts that control systems would lead to improved performance if they act in concurrence with other factors of the firm's environment [34, 47]. For example, Govindarajan and Gupta [33] suggested that a good fit between the elements of the firm's external environment, the firm's strategy, and its choice of control system would result in better corporate performance. Based on contingency theory framework, Chenhall [15] predicted that positive perceptions about the ease of use and usefulness of MCS would lead to better usage of the system, which would accordingly improve performance.

3.5 Organizational Effectiveness and the Concept of Fit

Venkatraman [97] developed a conceptual framework and six different perspectives have been identified, namely fit as mediation, fit as moderation, fit as covariation, fit as matching, fit as profile-deviation, fit as gestalts; each perspective implies distinctive theoretical meanings and requires the utilization of specific analytical schemes. It emphasizes the isomorphic nature of correspondence between a given conceptualization and the testing scheme(s); however, scholars have employed interchangeably these perspectives, they often invoke one perspective in theoretical discussion and employ another in the empirical research [97]. This study mainly focused on one of these aspects of fit which is called moderator.

3.6 Fit as Moderator

Venkatraman [98] and Fairchild and MacKinnon [27] stated that the moderation perspective specifies that the influence of predictor variable on a criterion variable depends fundamentally upon the level of a third variable (i.e., moderator). As a result, the fit between moderator and predictor significantly determines the performance. Generally, a moderator can be considered as either a categorical variable (e.g., organizational types, stages of product life cycle, or different types of environment) or a quantitative internal-level variable (e.g., competitive intensity or the degree of business- relatedness), which has effect on the direction and/or the strength of the relation between a dependent variable (e.g., performance) and a predictor variable (e.g., a strategy characteristic).

3.7 The Moderating Role of Environmental Uncertainty

In general, organizations should be adapted completely to the organizational environment that is associated with the organizational performance. The dynamicity of the environment brings about uncertainty for the organizations. The intensity of competition and uncertainty of the organizational environment are considered as threat to the organizational performance. To be rational, a firm should attempt to decrease the uncertainties. In addition, the contingency theory indicates that a firm is capable of reducing the uncertainties through changing its activities and how it operates within uncertain environments. Those firms that face relatively static environments face relatively a less degree of uncertainty. Similarly, those firms that operate within a less dynamic environment experience less competition. Examining environmental scanning activities as potential moderators of the strategy/performance relationship is firmly grounded in the information processing perspective advanced by Daft et al. [19] and allows an exploration of the detailed linkage among strategy, environmental scanning, and organizational performance advanced by contingency theorists (e.g., [96]).

Apart from the above scholars there are many other researchers who have highlighted the moderating role of environment on strategy-performance relationship (e.g., [12, 30, 61, 73]). However, there are very few studies on the relationship between strategy and strategy implementation.

In this regard, Hassan et al. [42] conducted a study to investigate the direct impact of marketing strategy creativity (MSC) and marketing strategy implementation effectiveness (MSIE) on performance and mediating role of marketing strategy implementation effectiveness on the relationship between strategy creativity and firm performance. The study also focused on examining the moderating role of environmental uncertainty (EU) on the relationship between MSC and MSIE and the impact of MSC and MSIE on the performance across different business strategies. Results of the study showed that both marketing strategy creativity and marketing strategy implementation effectiveness have a positive and strong link with performance, MSIE was considered as mediator between MSC and performance; however, environmental uncertainty had not moderated the relationship between marketing strategy creativity and marketing strategy implementation effectiveness.

Combe et al. [17] investigated the development of strategic options and the complex relationship to strategic flexibility. Findings of the study pointed to two major variables moderating the relationship between management cognition and the implementation of strategic options, and thereby strategic flexibility. According to the scholars, environmental turbulence (competition) is considered as a moderator between management cognition and strategy formulation and implementation.

Regarding the moderating role of external environment (environmental uncertainty) on strategy formulations and strategy implementation relationship, Qi and Zhao [79] also highlighted that the environmental uncertainty is an important moderator between supply chain strategy formulation and implementation.

As discussed earlier, environmental uncertainty can be defined as the function of change with regard to customer preferences, competitors move or technology [48]. According to contingency theorists, the firm must try to match its strategic direction and management control system to the ever changing environment, for example, ever changing preferences of customers or technological uncertainty [33]. The nature of today's business involves high uncertainty and frequent changes to almost everything. Even organizations that are mature and have an outstanding track record of success face challenges with time when comes to strategy implementation [80].

Based on the above discussion, it can be concluded that most of the studies ascertain the moderating role of external environment on strategy-performance relationship. However, the number of studies which examined the moderating role of environment on strategy and strategy implementation was very few. Specifically, there was no study to evaluate the moderating effect of environmental uncertainty on the relationship between business strategy and BSC. Hence, the relationship between SMEs' strategy and BSC implementation regarding environmental uncertainty can be hypothesized as follows:

H2: Environmental uncertainty moderates the relationship between business strategy and BSC implementation.
H2a: Environmental uncertainty moderates the relationship between Defender business strategy and BSC implementation.
H2b: Environmental uncertainty moderates the relationship between Prospector business strategy and BSC implementation.
H2c: Environmental uncertainty moderates the relationship between Analyzer business strategy and BSC implementation.
H2d: Environmental uncertainty moderates the relationship between Reactor business strategy and BSC implementation.

4 Research Model

Based on the literature related to strategic management in SMEs, SMEs' contextual variables, basic concepts regarding BSC, the use of this system as a component of management control system and specifically contingency theory the conceptual framework of this study has been developed. The relationship between strategy, environmental uncertainty and BSC as strategy implantation tools is shown in Fig. 2. Based on this figure, in the relationship between strategy with four mentioned factors and BSC implementation, environmental uncertainty can act as a moderator.

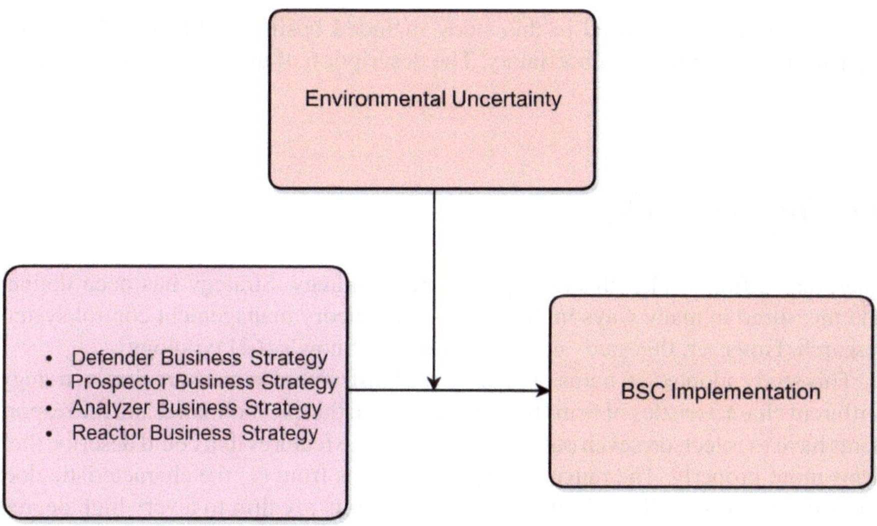

Fig. 2 Research framework

5 Research Design

This study evaluated the use of BSC as a technique to implement strategy among SMEs. Therefore, a judgmental/purposive sampling method was adopted to collect data. The basis for selecting judgment sampling over random sampling was to ensure accurate and representative information was gathered for the research area [59]. It allowed the researcher to draw the sample that would best fit the research intended objectives [95]. The sample for this study was SMEs which had the experience of using BSC in their firms. Therefore, respondents were recruited using the purposive sampling technique. In a purposive sample, the sample elements were selected because it was expected that they could serve the research purpose. The study utilized the cross-sectional survey method to achieve the objectives of this research. The survey method is usually associated with the positivistic philosophy. It is a popular and common method in business and management research [84]. Most of the past studies employed the survey method (e.g., [65, 91, 41]). The researcher was able to personally distribute questionnaire among different SMEs with the experience of using BSC. Three hundred questionnaires were distributed among SMEs and a total of 257 were returned.

6 Description of Measures

The conceptual model used in this study included business strategy, BSC implementation, environmental uncertainty. The description of measures are explained as follows.

6.1 Business Strategy

There are different classifications of business strategy. Strategy has been defined and measured in many ways in the contingency theory management control system research. However, this study adopted Miles and Snow's [62] typology.

This study adopted 26 items of Segev's [85] instrument to measure firm's strategy. Different characteristics of firms have been listed in the questionnaire, and the respondents have to select, on seven points Likert scale, the features that could describe their firms more properly. The range of respondents was from (1) the characteristic does not suit my firms at all, to (7) the characteristic suits my firm to a very high degree. According to the description of four types of Miles and Snow's strategy, nine items were developed to evaluate Defender type of strategy, eight items for Prospector, five and four items were used to measure Analyzer and Reactor strategy respectively.

6.2 *BSC Implementation*

The second part of the questionnaire was related to respondents' opinion regarding BSC implementation in their firms. In order to measure the opinion of the respondents regarding the implementation of BSC, eight questions based on a seven point Likert rating scale was adapted from Giannopoulos et al. [29]. This section asked the respondents about the effectiveness of BSC as strategy implementation system, effect of using BSC in strategy evaluation, etc.

6.3 *Environmental Uncertainty*

This study found technology and customer to be an important part of external environment for SMEs which could lead to environmental uncertainty. For these two elements, the measures have been adopted from the study of DeSarbo et al. [20] which involve the technological change evaluation, the difficulty of technological forecasting, the extent of technical opportunity, etc. The evaluation of the market environment uncertainty regarding customers was conducted based on changes occurred in customers' preferences, price sensitivity of customers, product needs of customers, the change of customer base, and ease of predicting changes in the marketplace.

7 Machine Learning–Enhanced BSC Implementation

7.1 *Machine Learning as a Complement to BSC*

Machine Learning (ML) has recently been highlighted as a promising tool to support SMEs in dealing with uncertainty and complexity in decision-making [56]. While the BSC provides a structured framework to link strategy and performance, SMEs often face difficulties in translating uncertain market signals into timely strategic adjustments. ML can complement BSC implementation by identifying hidden patterns in financial, customer, process, and learning and growth data [86]. Unlike traditional approaches, ML models can adapt to nonlinear dynamics and provide predictive insights on how environmental uncertainty will affect strategy–performance alignment.

Table 1 Machine learning approaches for BSC in SMEs

ML technique	Application in SMEs' BSC implementation	Benefit under environmental uncertainty
Regression and time-series models	Predict key performance indicators across BSC perspectives	Forecast shifts in financial or customer outcomes
Decision trees and random forests	Identify drivers of BSC success/failure by strategy type	Improve interpretability of uncertainty impacts
Clustering (e.g., K-means)	Segment SMEs by strategy typology and performance patterns	Tailor strategy to group-specific uncertainty
Neural networks (Deep learning)	Model complex nonlinear interactions between strategy and outcomes	Capture hidden relationships under volatile environments
Anomaly detection (Isolation forest, autoencoders)	Detect unexpected deviations in BSC performance	Early warning for strategic misalignment

7.2 Potential Applications in SMEs

1. Predictive Modeling—Forecast BSC performance indicators (financial returns, customer satisfaction) under varying levels of environmental uncertainty.
2. Clustering and Segmentation—Group SMEs with similar strategic profiles (Defender, Analyzer, Prospector, Reactor) and assess how uncertainty impacts them differently.
3. Anomaly Detection—Identify early warning signals when strategies deviate from expected performance.
4. Optimization—Recommend the best resource allocation strategy to maintain performance under uncertainty.

To illustrate how ML can complement the Balanced Scorecard framework in SMEs, Table 1 summarizes common ML techniques, their potential applications in BSC implementation, and the benefits they provide under environmental uncertainty. These approaches range from predictive models that forecast key performance indicators to clustering methods that segment SMEs by strategy type, offering managers practical tools for data-driven decision-making.

Figure 3 illustrates how ML can be integrated with Miles and Snow's typology, BSC perspectives, and environmental uncertainty. SMEs' data feeds into ML models, which generate predictive outputs that strengthen strategy—BSC linkage under uncertainty.

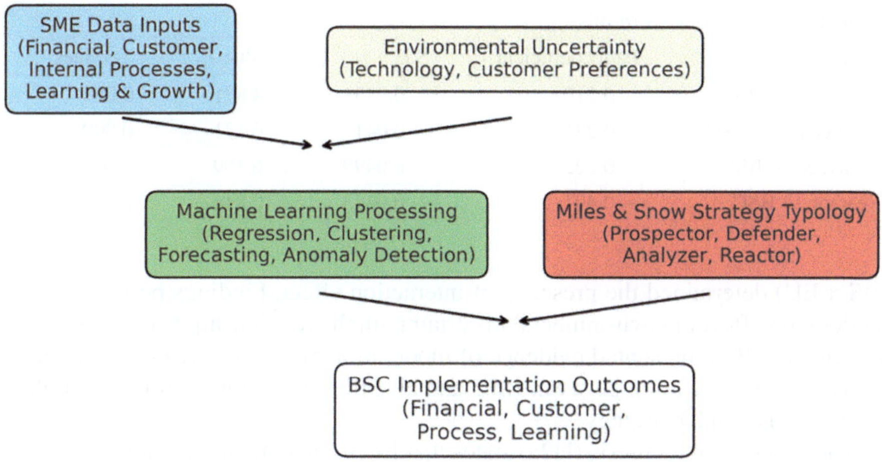

Fig. 3 Proposed ML–enhanced BSC framework

8 Hypothesis Testing and Results

SEM statistical techniques were used in the current study to specify the measurement models and the structure model quality criteria for practical work. Then, after fixing the indicators of the measurement, the research model was developed and transferred to an SEM, and then the PLS approach was applied. At first stage of the PLS, goodness of measurement model was tested. This was done through testing of convergent and discriminant validity of the constructs. Once convergent and discriminant validity has been established, the next stage was to test structural model to confirm the hypothesized relationship in term of predictive inference. In addition, the predictive relevance and coefficient of determination ($R2$) were computed. The structural model results can be used to test the research hypotheses. The results of the hypotheses can be tested on the basis of path coefficients, p-value and t-values, with a significance level of 0.05.

Findings related to the effect of business strategy on BSC implementation suggested that there is a relationship between these two variables. This means that this relationship is positive in Defender, Prospector, and Analyzer; however, the relationship is negative regarding Reactor strategy and BSC implementation. Table 2 shows the results of the direct hypothesis.

The moderation hypothesis is related to environmental uncertainty and the relationship between business strategy and BSC implementation which was tested based on the guidelines prescribed by Preacher et al. [78] using SEM. In this approach, the items for environmental uncertainty and business strategy were first mean centered. Next, every item of each dimension of business strategy was cross multiplied with every item of the environmental uncertainty. This moderator constructs along with two other constructs (business strategy and BSC Implimentation) were included in the structural model. The significance of the direct effect of the moderator constructs

Table 2 Direct hypothesis results

Paths	Path coefficient	S.E	Value-*P*	*T*-values
Defender → BSC	0.210	0.0431	4.871	0.000
Prospector → BSC	0.281	0.0415	6.771	0.000
Analyzer → BSC	0.322	0.0499	6.459	0.000
Reactor → BSC	− 0.491	− 0.0453	10.844	0.000

(BS x EU) determined the presence of interaction effect. Findings pertained to the moderation effect of environmental uncertainty in the relationship between business strategy and BSC presented evidence of moderation effect of environmental uncertainty on the association between BSC and all Miles and Snow's strategy typology except for Defender strategy.

The proposed hypothesis (H2a) states that Environmental uncertainty moderates the relationship between Defender business strategy and BSC implementation. The interaction term was insignificant since coefficient value is 0.058, and *T*-value is 0.974 which is less than 1.96, indicating the non-presence of moderation effects. So H6a was not supported.

H2b states that Environmental uncertainty moderates the relationship between Prospector business strategy and BSC implementation. The interaction term was significant since coefficient value is − 0.156, and *T*-value is 4.06 which is above 1.96, indicating the presence of moderation effects. So H6b was supported. Figure 4 shows the interaction effect of environmental uncertainty and Prospector.

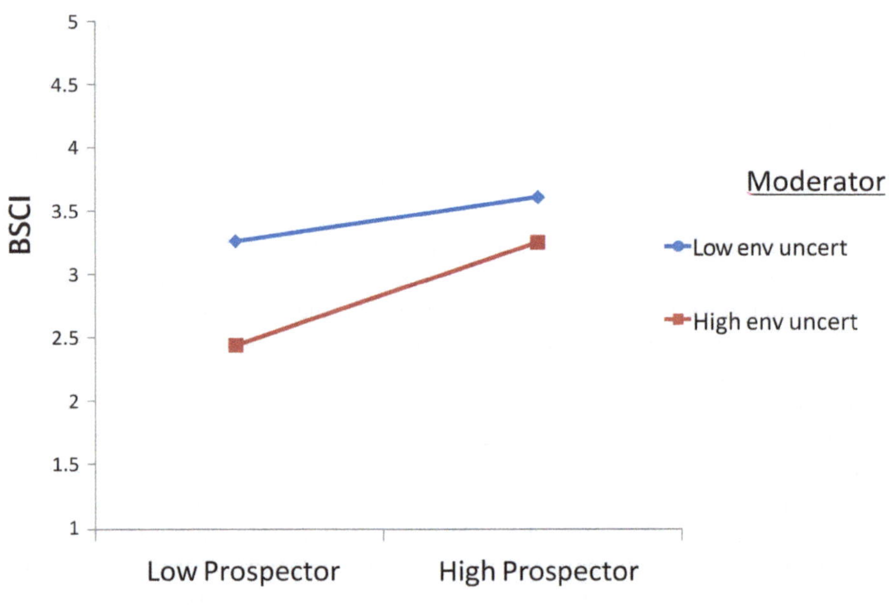

Fig. 4 Interaction effect of environmental uncertainty and prospector

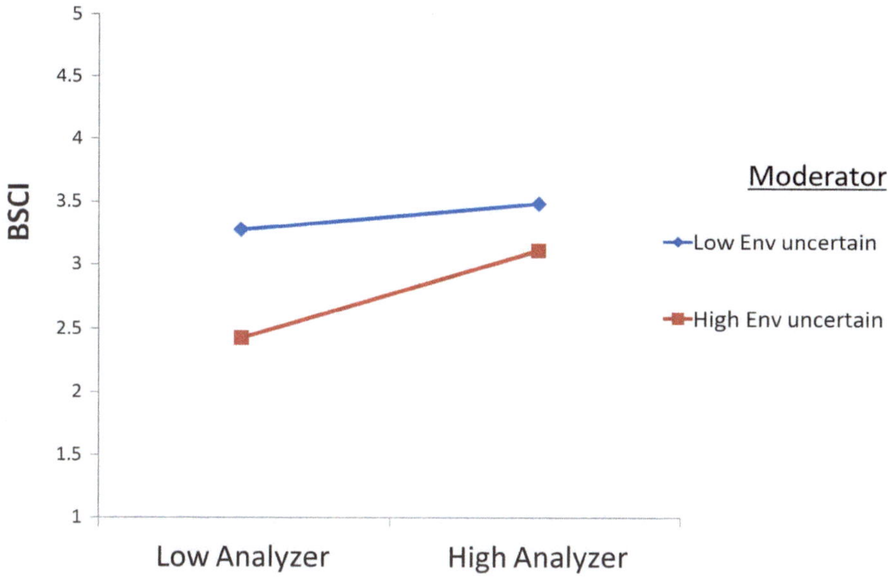

Fig. 5 Interaction effect of environmental uncertainty and analyzer

Figure 4 shows that the value of BSCI increases with the value of Prospector strategy. However, this rate of increase becomes lower (due to negative coefficient value of moderation effect) when the environment is highly uncertain for this type of business strategy.

Hypothesis H2c states that Environmental uncertainty moderates the relationship between Analyzer business strategy and BSC implementation. The interaction term was significant since coefficient value is − 0.104, and *T*-value is 2.479 which is above 1.96, indicating the presence of moderation effects. So H2c was supported. Figure 5 shows the interaction effect of environmental uncertainty and Analyzer.

Figure 5 shows that the value of BSCI increases with the value of Analyzer strategy. However, this rate of increase becomes lower (due to negative coefficient value of moderation effect) when the environment is highly uncertain for this type of business strategy.

Hypothesis H2d suggested that environmental uncertainty moderates the relationship between Reactor business strategy and BSC implementation. The interaction term was significant since coefficient value is − 0.148, and *T*-value is 3.444 which is above 1.96, indicating the presence of moderation effects. So H2d was supported. Figure 6 shows the interaction effect of environmental uncertainty and Reactor.

Figure 6 shows that the value of BSCI decreases with the value of Reactor strategy. Additionally, this rate of decrease is higher (due to negative coefficient value of moderation effect) when the environment is highly uncertain for this type of business strategy.

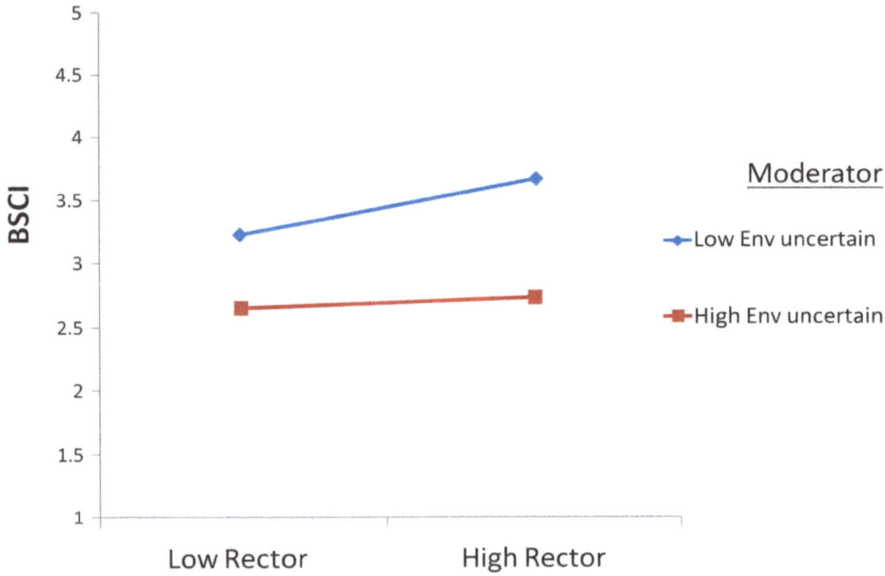

Fig. 6 Interaction effect of environmental uncertainty and reactor

9 Discussion and Conclusion

The related results of this study on the relationship between Malaysian business strategy based on Miles and Snow's typology and BSC implementation showed that the relationship is significant in all four types of strategy. However, this relationship is negative regarding Reactors and positive regarding Prospectors, Analyzers, and Defenders. According to Kaplan and Norton [52, 53], BSC is potentially capable of serving various purposes, including clarification and achievement of consensus on strategy, alignment of the organizational and individual goals to the strategy, formation of a link between strategic objectives and long-term targets and annual budgets, identification and alignment of strategic initiatives, and gaining necessary feedback for learning about and improving the strategy. In addition, they pointed out that BSC helps organizations in being focused on all strategic areas in the organization, taking into consideration the short-term results from four perspectives of BSC, and communicating and implementing the strategies in the organization. As a result, the BSC effectiveness is dependent mostly on the strategy link within BSC [52, 55].

The findings on the effect of SMEs' business strategy on BSC support the results of study conducted by Rompho [81]. The author claimed that BSC failed to be implemented in SMEs due to frequent strategy changes. The study highlighted the effect of strategy on BSC and referred to the importance of strategy stability while BSC implementation. In current study, the negative effect of Reactor type of strategy and positive influence of Analyzer, Prospector, and Defender strategy on BSC can be explained by the above reasoning. The main difference between firms which adopt Reactor strategy and those which implement Defender, Prospector, or Analyzer strategy is related to their approaches. All Miles and Snow's strategic types have some specific approach toward environment except Reactor type of strategy. It means that the Reactor strategy seems the least aggressive, and companies often have no predictive strategy. These companies are not proactive; when events occur, they react to any changes. In such context, strategies are enforced when the companies have to provide a response to the changes occurred within the market, or other environmental factors. This reactiveness and lack of clear strategy can influence BSC negatively. Conversely, results showed that the effect of Defenders, Prospectors, and Analyzers is positive on BSC implementation because they have more stable strategy compared to Reactor type of strategy. Analyzers, Prospectors, and Defenders all make commitment to particular strategic approaches, while Reactors recipes are more haphazard.

The heart of the BSC is a framework of four major categories or perspectives for strategy implementation—financial, customer, internal business, and innovation and learning. In order to ensure an effective BSC implementation, Malaysian SMEs should follow a more specific approach toward formulating their business strategy. BSC can be an effective way of translating an overall strategy to the many parts of an organization. However, the findings highlighted that organizations must have a clear idea of what they want to accomplish in order to successfully implement the BSC. When the decided strategy is not clear and well-defined (Reactors), the implementation of BSC would be declined. One possible reason for this situation is related to nature of strategy-centered BSC. Therefore, based on the findings, SME managers are strongly advised to clearly develop their strategy in order to implement BSC properly. In other words, Malaysian SMEs should avoid being unclear about their strategic intention if they seek to use management control system such as BSC successfully.

Finally, findings pertained to the moderation effect of environmental uncertainty in the relationship between business strategy and BSC presented evidence of moderation effect of environmental uncertainty on the association between BSC and all Miles and Snow's strategy typology except for Defender strategy.

Different types of Miles and Snow's business strategy perceive market environment differently. As suggested by some researchers [88, 40], an organization that pursues a Prospector strategy is loaded with a high level of perceived environmental uncertainty (PEU). On the other hand, a Defender-type organization is generally associated with a low level of PEU [62, 63, 64]. When a company selects a Defender strategy, it will operate in a relatively stable environment. Analyzers operate in an environment with changes moderately. On the other hand, Reactors do not have clear organizational approach as they depend on current needs and react to any occurred conditions of the market.

The results of this study suggested that environmental uncertainty which includes customer and technological uncertainty significantly interact with strategies to affect BSC implementation regarding Analyzer, Prospector, and Reactor. However, there is no any moderation effect for Defenders. One possible explanation for the nonexistence of moderation for Defenders is that they perceive environment more stable rather than dynamic and uncertain. Therefore, environmental uncertainty cannot affect Defenders.

In the face of high levels of uncertainty and change, there is a need for a dynamic approach in which strategy formulation are carried out. It is difficult to implement strategy via BSC as the environment becomes increasingly uncertain. It means that, environmental uncertainty significantly decrease the effect of SMEs' business strategy on BSC implementation. In this study as highlighted before, only one type of business strategy (Defender) is not influenced by the market uncertainty. The effect of other types of business strategies (Prospector, Analyzer, and Reactor) on BSC implementation decreased under environmental uncertainty. It is because of negative coefficient of the moderation effect. Previously, it was found that Defender, Prospector and Analyzer types of SMEs' business strategy could significantly and positively influence BSC implementation; however, this significant relationship was negative regarding Reactor strategy. When considering the interaction effect of market environment as a moderator, the implementation of BSC for companies adopting Prospector, Analyzer, and Reactor business strategy becomes less successful. These types of strategies find environment uncertain and need to be adjusted to those changes occurred in the market. The journey from business strategy to BSC could be highly influenced by market environment when it is highly uncertain. The higher perceived environmental uncertainty by SME leaders, the lesser the business strategy has effect on BSC implantation. For instance, while companies are attempting to implement BSC based on firm's adopted strategy for lowering their prices on specific product, they unexpectedly face customers' new taste and preferences. This is how environmental uncertainty can decrease the effect of business strategy on BSC implementation. Therefore, environmental uncertainties threaten the business strategy as it hampers the ability to develop long-range plans, which can consequently affect the BSC implementation. The findings implied that the market in Malaysia can interact with the relationship between three types of strategy and BSC implementation. Therefore, the weakening effect of environmental uncertainty should be considered enormously in Malaysian SMEs.

Beyond the statistical findings, emerging technologies such as Machine Learning can extend these insights. By integrating ML, SMEs can enhance the predictive power of BSC. ML tools can anticipate the weakening effect of uncertainty, allowing managers to simulate outcomes before executing strategies. This hybrid approach combines the theoretical rigor of BSC with the adaptive, data-driven insights of ML, leading to more resilient SME strategies.

The SME managers are recommended not to put emphasis on the formulation of strategy over its implementation; this is because the strategy formulation cannot guarantee necessarily a successful implementation of the strategy. In comparison, the strategy formulation is a more intellectual process in a relatively higher hierarchical level; whereas, the strategy implementation is more operational and requires involvement and coordination of the whole organization. Managers should carefully develop proper strategies; although, they must allocate a large part of available resources to those activities that are linked to the strategy implementation. This study has showed that BSC plays an important role in SMEs' context and warrants the special attention of managers toward a clearer strategy approach and applying BSC in their firm's environment in order to enhance performance.

Acknowledgements This research paper is made possible through the support of the Universiti Teknologi Malaysia (UTM) and the author wish to appreciate their help and support.

Conflict of Interest The author states that there is no conflict of interest.

AI Use Disclosure Statement The authors used generative AI technologies only for proofreading. This includes minor corrections to grammar, spelling, and syntax. No AI was used for ideation, data analysis, or writing original content.

References

1. Adobi TJ (2012) Strategic focus of Nigerian SMSEs: more of emergent or planned. J Emerg Trends Econ Manag Sci 3(3):229–232
2. Aidemark LG (2001) The Meaning of balanced scorecards in the health care organisation. Financ Account Manag 17(1):23–40
3. Alpkan L, Yilmaz C, Kaya N (2007) Market orientation and planning flexibility in SMEs performance implications and an empirical investigation. Int Small Bus J 25(2):152–172
4. Anthony RN, Govindarajan V (2004) Management control systems. McGraw-Hill, New York
5. Aragón-Sánchez A, Sánchez-Marín G (2005) Strategic orientation, management characteristics, and performance: a study of Spanish SMEs. J Small Bus Manag 43(3):287–308
6. Argument L, Harrison D, Wainwright C (1997) Manufacturing strategy within the SME sector. In: Advances in manufacturing technology XI (Conference proceedings). Glasgow Caledonian University, Glasgo
7. Ates A (2008) Strategy process in manufacturing SME. Ph.D. thesis, Department of Design, Manufacture and Engineering Management, University of Strathclyde, Glasgow, UK
8. Atherton A (2003) The uncertainty of knowing: an analysis of the nature of knowledge in a small business context. Hum Relat 56(11):1379–1398
9. Atkins MH, Lowe JF (1997) Sizing up the small firm: UK and Australian experience. Int Small Bus J 15(3):42–55
10. Baroto MB, Arvand N, Ahmad FS (2014) Effective strategy implementation. J Adv Manag Sci 2(1):50–54
11. Bianchi R, Noci G (1998) Greening SMEs' competitiveness. Small Bus Econ 11(3):269–281
12. Bierly PE, Daly PS (2007) Alternative knowledge strategies, competitive environment, and organizational performance in small manufacturing firms. Entrep Theory Pract 31(4):493–516
13. Bowen FE (2002) Organizational slack and corporate greening: broadening the debate. Br J Manag 13:305–316
14. Brouthers K, Andriessen F, Nicolaes I (1998) Driving blind: strategic decision making in small companies. Long Range Plan 31(1):130–138
15. Chenhall RH (2003) Management control systems design within its organizational context: findings from contingency-based research and directions for the future. Acc Organ Soc 28(2):127–168
16. Chin TA, Hamid ABA, Rasli A, Tat HH (2014) A literature analysis on the relationship between external integration, environmental uncertainty and firm performance in Malaysian SMEs. Procedia Soc Behav Sci 130:75–84
17. Combe IA, Rudd JM, Leeflang PS, Greenley GE (2012) Antecedents to strategic flexibility: management cognition, firm resources and strategic options. Eur J Mark 46(10):1320–1339

18. Conant JS, Mokwa MP, Varadarajan PR (1990) Strategic types, distinctive marketing competencies and organizational performance: a multiplemeasures-based study. Strateg Manag J 11:365–383
19. Daft RL, Sormunen J, Parks D (1988) Chief executive scanning, environmental characteristics, and company performance: an empirical study. Strateg Manag J 9(2):123–139
20. DeSarbo WS, Anthony Di Benedetto C, Song M, Sinha I (2005) Revisiting the Miles and Snow strategic framework: uncovering interrelationships between strategic types, capabilities, environmental uncertainty, and firm performance. Strateg Manag J 26(1):4774
21. DeSarbo WS, Di Benedetto A, Jedidi K, Song M (2006) A constrained latent structure multivariate regression methodology for empirically deriving strategic types. Manag Sci 52(6):909–924
22. Dias Jordão RV, Casas Novas JL (2013) A study on the use of the balanced scorecard for strategy implementation in a large brazilian mixed economy company. J Technol Manag Innov 8(3):98–107
23. Dinesh D, Palmer E (1998) Management by objectives and the balanced scorecard: will Rome fall again. Manag Decis 36(5):363–369
24. Doty DH, Glick WH, Huber GP (1993) Fit, equifinality, and organizational effectiveness: a test of two configurational theories. Acad Manag J 36(6):1196–1250
25. Dvir D, Segev E, Shenhar A (1993) Technology's varying impact on the success of strategic business units within the Miles and Snow typology. Strateg Manag J 14(2):155–161
26. Epstein M, Manzoni JF (1998) Implementing corporate strategy: from tableaux de bord to balanced scorecards. Eur Manag J 16(2):190–203
27. Fairchild AJ, MacKinnon DP (2014) Using mediation and moderation analyses to enhance prevention research. Springer, USA
28. Ghobadian A, Gallear D (1997) TQM and organization size. Int J Oper Prod Manag 17(2):121–163
29. Giannopoulos G, Holt A, Khansalar E, Cleanthous S (2013) The use of the balanced scorecard in small companies. Int J Bus Manag 8(14):1–22
30. Goll I, Rasheed AMA (1997) Rational decision making and firm performance: the moderating role of environment. Strateg Manag J 8(7):583–591
31. Gomes JS, Salas JA (1997) Controle de gestão: uma abordagem contextual e organizacional – Textos e casos. Atlas, São Paulo
32. Gosselin M (1997) The effect of strategy and organizational structure on the adoption and implementation of activity-based costing. Acc Organ Soc 22(2):105–122
33. Govindarajan V, Gupta A (1985) Linking control systems to business unit strategy: impact on performance. Acc Organ Soc 10(1):51–66
34. Govindarajan V (1984) Appropriateness of accounting data in performance evaluation: an empirical examination of environmental uncertainty as an intervening variable. Acc Organ Soc 9(2):125–135
35. Gray C, Mabey C (2005) Management development, key differences between small and large businesses in Europe. Int Small Bus J 23(5):467–485
36. Greatbanks R, Boaden R (1998) Can SMEs afford to measure performance. In: Proceedings of the performance measurement–theory and practice conference, 14–17 July, Cambridge, pp 120-132
37. Gumbus A, Lussier RN (2006) Entrepreneurs use a balanced scorecard to translate strategy into performance measures. J Small Bus Manag 44(3):407–425
38. Gunasekaran A, Marri HB, Mcgauahey R, Grieve RJ (2001) Implications of organization and human behavior on the implementation of CIM in SMEs: an empirical analysis. Int Jof CIM 14(2):175–185
39. Gunasekaran A, Okko P, Martikainen T, Yli-Olli P (1996) Improving productivity and quality in small and medium enterprises: cases and analysis. Int Small Bus J 15(1):59–72
40. Gupta AK (1987) Matching managers to strategies: point and counterpoint. Hum Resour Manag 25(2):15–234

41. Harrington RJ, Kendall KW (2006) Strategy implementation success: the moderating effects of size and environmental complexity and the mediating effects of involvement. J Hosp Tour Res 30(2):207–230
42. Hassan MU, Qureshi SU, Sharif S, Mukhtar A (2013) Impact of marketing strategy creativity on organizational performance via marketing strategy implementation effectiveness: empirical evidence from pakistani organizations. Middle-East J Sci Res 16(2):264–273
43. Hill CWL, Jones GR (2013) Strategic management; an integrated approach, 10th edn. Sounth Western Cengage Learning, Mason, USA
44. Hoque Z (2014) 20 years of studies on the balanced scorecard: trends, accomplishments, gaps and opportunities for future research. Br Account Rev 46(1):33–59
45. Hrebiniak LG (2006) Obstacles to effective strategy implementation. Organ Dyn 35(1):12–31
46. Hyvarinen L (1990) Innovativeness and its indicators in small and medium sized industrial enterprises. Int Small Bus J 9(1):64–79
47. James W, Elmezughi A (2010) The combined effect of costing and performance management systems on performance, moderated by strategy: Australian context. Account Account Perform 16(1):51–84
48. Jaworski BJ, Kohli AK (1993) Market orientation: antecedents and consequences. J Mark 53–70
49. Jennings P, Beaver G (1997) The performance and competitive advantage of small firms: a management perspective. Int Small Bus J 15(2):63–75
50. Johanson U, Skoog M, Backlund A, Almqvist R (2006) Balancing dilemmas of thebalanced scorecard. Account Audit Account J 19(6):842–857
51. Kald M, Nilsson F (2000) Performance measurement at Nordic companies. Eur Manag J 18(1):113–127
52. Kaplan RS, Norton DP (1996a) Using the balanced scorecard as a strategic management system. Harv Bus Rev 74(1):75–85
53. Kaplan RS, Norton DP (1996b) The balanced scorecard: translating strategy into action. Harvard Business Press, Chicago
54. Kaplan RS, Norton DP (2001a) Transforming the balanced scorecard from performance measurement to strategic management: part II. Account Horiz 15(2):147–160
55. Kaplan RS, Norton DP (2001b) The strategy-focused organization. Harvard Business School Press, Boston
56. Kim JS, Seo D (2023) Foresight and strategic decision-making framework from artificial intelligence technology development to utilization activities in small-and-medium-sized enterprises. Foresight 25(6):769–787
57. Lawrie G, Cobbold I (2004) Third-generation balanced scorecard: evolution of an effective strategic control tool. Int J Product Perform Manag 53(7):611–623
58. Man TWY, Lau T, Chan KF (2002) The competitiveness of small and medium enterprises: a conceptualization with focus on entrepreneurial competencies. J Bus Ventur 17(2):123–142
59. Marshall MN (1996) Sampling for qualitative research. Fam Pract 13(6):522–526
60. McAdam R (2000) Quality models in an SME context: a critical perspective using a grounded approach. Int J Qual Reliab Manag 17(3):305–323
61. McArthur AW, Nystrom PC (1991) Environmental dynamism, complexity, and munificence as moderators of strategy-performance relationships. J Bus Res 23(4):349–361
62. Miles RE, Snow CC (1978) Organizational strategy, structure and process. McGraw-Hill, New York
63. Miller D (1988) Relating Porter's business strategies to environment and structure: analysis and performance implications. Acad Manag J 31(2):280–308
64. Miller D, Dröge C, Toulouse JM (1988) Strategic process and content as mediators between organizational context and structure. Acad Manag J 31(3):544–569
65. Milovanovic BM, Wittine Z (2014) Analysis of external environment's moderating role on the entrepreneurial orientation and business performance relationship among Italian small enterprises. Int J Trade Econ Financ 5(3):224–229
66. Mintzberg H (1981) Organization design: fashion or fit. Harv Bus Rev:103–116

67. Mohan-Neill SI (1995) The influence of firm's age and size on its environmental scanning activities. J Small Bus Manag 33(4):10
68. Montgomery CA, Wernerfelt B (1991) Sources of superior performance: market share versus industry effects in the US brewing industry. Manag Sci 37(8):954–959
69. Newing R (1995) Wake up to the balanced scorecard: balanced scorecard as an aid to management. Manag Account 73(3):22–27
70. Norreklit H (2000) The balance on the balanced scorecard—a critical analysis of some of its assumptions. Manag Account Res 11(1):65–88
71. Oakes I, Lee G (1999) Between a rock and a hard place: some dilemmas for smaller component suppliers. Int J Qual Reliab Manag 16(3):252–262
72. Olson EM, Slater SF (2002) The balanced scorecard, competitive strategy, and performance. Bus Horiz 45(3):11–16
73. Oyewobi LO, Windapo AO, Cattell KS, Rotimi JOB (2013) The effects of business environments on corporate strategies and performance of construction organisations. In: Proceedings of 29th annual ARCOM conference, 2–4 Sept 2013, Reading, UK. Association of Researchers in Construction Management, pp 691–701
74. O'Regan N, Ghobadian A (1998) The need to re-think strategy in SMEs. In: First international conference on stimulating manufacturing excellence in SMEs. Sheffield University, Sheffield, pp 225–233
75. Pelham AM (1999) Influence of environment, strategy, and market orientation on performance in small manufacturing firms. J Bus Res 45(1):33–46
76. Porter ME (1980) Competitive strategy. Basic Books, New York
77. Porter ME (2004) Competitive advantage: creating and sustaining superior performance. Free Press, New York
78. Preacher KJ, Rucker DD, Hayes AF (2007) Assessing moderated mediation hypotheses: theory, methods, and prescriptions. Multivar Behav Res 42:185–227
79. Qi Y, Zhao X (2006) Moderating effect of environmental uncertainty on supply chain strategy formulation and implementation. In: Proceedings of the 11th annual conference of Asia Pacific decision sciences institute conference, June 14–18, Hong Kong, pp 882–884
80. Rajasekar J (2014) Factors affecting effective strategy implementation in a service industry: a study of electricity distribution companies in the Sultanate of Oman. Int J Bus Soc Sci 5(9):169–183
81. Rompho N (2011) Why the balanced scorecard fails in SMEs: a case study. Int J Bus Manag 6(11):39–46
82. Russo J, Martins A (2005) The balanced scorecard in SMEs. J Account 344:401–418
83. Saindis E, Gilli R, White A (2001) Emergent strategies in SMEs. In: SME international conference, May 14–16, Alborg, Denmark, pp 123–130
84. Saunders M, Lewis P, Thornhill A (2003) Research methods for business students, 3rd edn. Pearson Education, U.K
85. Segev E (1987) Strategy, strategy-making, and performance in a business game. Strateg Manag J 8(6):565–577
86. Shahid MK, Mahmood K, Altayyar RS, Abdullah H (2025) Role of AI capabilities in sustainable firm performance: mediating role of innovative work behavior and the moderating effect of environmental dynamism. Int J Product Perform Manag 1–21
87. Shortell SM, Zajac EJ (1990) Perceptual and archival measures of Miles and Snow's strategic types: a comprehensive assessment of reliability and validity. Acad Manag J 33(4):817–832
88. Simons R (1987) Accounting control systems and business strategy: an empirical analysis. Account Organ Soc 12:357–374
89. Singh KS, Garg SK, Deshmukh SG (2007) Strategy development for competitiveness: a study on Indian auto component sector. Int J Product Perform Manag 56(4):285–304
90. Singh H, Mahmood R (2014) Manufacturing strategy and export performance of small and medium enterprises in Malaysia: moderating role of external environment. Int J Bus Commer 3(5):37–52

91. Slater SF, Olson EM, Hult GTM (2006) The moderating influence of strategic orientation on the strategy formation capability–performance relationship. Strateg Manag J 27(12):1221–1231
92. Snow CC, Hrebiniak LG (1980) Strategy, distinctive competence, and organizational performance. Adm Sci Q 25:317–336
93. Storey D, Sykes N (1996) Uncertainty, innovation and management. In: Burns P, Dewhurst J (eds) Small business and entrepreneurship, 2nd edn. Macmillan Press, New York
94. Tanyi E (2011) Factors influencing the use of balanced scorecard. Master thesis. Henken School of Economics
95. Tongco MDC (2007) Purposive sampling as a tool for informant selection. Ethnobot Res Appl 5:147–158
96. Van de Ven A, Astley W (1981) Mapping the field to create a dynamic perspectives in organization design and behaviour. University of Pennsylvania, Center for the Study of Organizational Innovation, Philadelphia
97. Venkatraman N (1989) The concept of fit in strategy research: toward verbal and statistical correspondence. Acad Manag Rev 14(3):423–444
98. Venkatraman N, Prescott R (1990) Environment-strategy coalignment: an empirical test of its performance implications. Strateg Manag J 11(1):1–23
99. Waalewijn P, Segaar P (1993) Strategic management: the key to profitability in small companies. Long Range Plan 26(2):24–30
100. Westhead P, Wright M, Ucbasaran D (2002) International market selection strategies selected by 'micro' and 'small' firms. Omega 30(1):51–68
101. Yu L, Perera S, Crowe S (2008) Effectiveness of the balanced scorecard: the impact of strategy and causal links. J Appl Manag Account Res 6(2):37–56
102. Yusof S, Aspinwall E (2000) TQM implementation issues: review and case study. Int J Oper Prod Manag 20(6):634–655

An ML Framework for Equivalent Viscous Damping Ratio of RC Shear Walls

Atta Ullah Noor and Hao Wu

Abstract Reinforced concrete (RC) shear walls are commonly used in mid-to high-rise buildings to provide lateral strength and stiffness. Energy dissipation is primarily achieved through flexural yielding in well-confined boundary regions, provided the wall has sufficient shear capacity to prevent brittle shear failure. In this study, the equivalent viscous damping ratio (ξ_{eq}) across drift levels was predicted using 200 experimental test results. First, sensitivity analysis was performed using parameter weights to identify which inputs most strongly influence ξ_{eq}. The results indicate that ρ_l, ρ_{bl}, $P/\left(A_g f_c'\right)$, f_c', AR, l_w, s/d_b, b_0, and t_w, are the most important in predictors. Six machine learning algorithms, two single models (MLP and PS) and four ensembles (CatBoost, XGBoost, LightGBM, and RF), were used to develop ξ_{eq} models, with the ensemble CatBoost standing out for its high accuracy. The goodness-of-fit ($R^2 = 0.95$) shows predictions tightly clustered around experimental values. CatBoost achieved the lowest RMSE = 0.014 and mean error (ME) = 0.0002 with $\sigma_{ME} = 0.014$, indicating negligible error, the model is highly balanced and reliable. The Overall Balanced Grade (OBG) = 0.03 further suggests that CatBoost outperforms the other models. To provide a practical solution, PS was derived from a simplified MLP with one hidden neuron, achieving $R^2 = 0.63$ and ME = − 0.001, which is acceptable given its purpose as a compact, user-oriented equation for real-world scenarios.

Keywords RC shear wall · Machine learning · Equivalent viscous damping ratio · Practical solution

---

A. U. Noor (✉) · H. Wu
State Key Laboratory of Disaster Reduction in Civil Engineering, Tongji University, Shanghai, China
e-mail: noor27266@tongji.edu.cn

H. Wu
e-mail: 2009wuhao_tom@tongji.edu.cn

A. Azizi et al. (eds.), *Intelligence at the Interface*, Emerging Trends in Mechatronics,
https://doi.org/10.1007/978-981-95-7584-8_6

1 Introduction

Reinforced concrete (RC) shear walls are vertical structural elements that form part of a building's lateral force-resisting system. They provide lateral stiffness and strength to resist earthquake and wind loads, control story drift, and contribute to gravity load resistance. In seismic regions, they control story drift, and contribute to gravity load resistance. In seismic-prone areas, properly detailed shear walls are often the primary components that limit in-plane deformations and protect the structure from damage.

Although high-fidelity finite element (FE) models can evaluate multi-degree of freedom (MDOF) nonlinear response under seismic/impulsive loads, simpler, faster approximations are widely used. Peak response is commonly estimated using (i) inelastic response spectra or elastic single degree of freedom (SDOF) peaks scaled by a damping and stiffness-dependent factor, or (ii) an equivalent linear SDOF with effective damping and stiffness [1–3]. Accurate estimation of equivalent properties particularly the equivalent damping ratio underpins displacement-based analysis/design, which uses these properties to verify drift targets and is increasingly preferred over force-based design [3–6]. In addition, accurate estimation of the damping ratio is important because it governs the estimation of energy dissipation in design practice [7–10]. The energy-based design method aims to reduce the limitations of force and displacement-based design methods, which do not consider the balance among input energy from ground motion, a structure's capacity to dissipate it, and the duration of the ground motion [11–15]. The equivalent damping ratio concept was first introduced by Jacobsen [16] and was thereafter thoroughly examined by several researchers [1, 17–22]. Many researchers have investigated the damping characteristics of RC walls, infilled frames, masonry walls, and other individual structural components. Buttmann [23] subjected two reinforced concrete and masonry specimens to sinusoidal horizontal shaking.

In recent years, many studies have focused on predicting seismic parameters using machine learning (ML). Researchers have developed ML models for the shear strength of RC beams, RC beam-column connections, and the peak shear strength of RC walls [24–26]. Vu and Hoang [27] used a hybrid ML technique (LS-SVM) on concrete slabs reinforced with FRP bars to predict punching shear capacity. Siam and Ezzeldin [28] applied Projection to Latent Structures (PLS) and PCA to reinforced masonry walls to identify failure modes and peak drift ratio. Mangalathu et al. [29] applied different ML approaches to RC shear walls for classification and prediction of failure modes. Jafari et al. [30, 31] investigated the cyclic behavior of a hybrid self-centering wall. Xiong et al. [32] used PCA-based feature extraction to improve failure-mode prediction while using nine ML models [33].

Based on the literature review, existing ML-based evaluations of ξ_{eq} for RC shear walls have been limited to a 1.0% drift level [34]. In this research, an ML model was used to predict ξ_{eq} for RC shear walls across drift levels. In this context, widely used single and ensemble ML algorithms were utilized. Using the comprehensive dataset, feature sensitivity analysis was performed with feature weights from the trained NCA method. Features were ranked from highest to lowest weight to create subsets and evaluate each subset's performance to identify the most influential features for the model output. PS was developed using a simplified MLP architecture with a single neuron in the hidden layer. The purpose of the PS is to provide a practical mathematical equation for users. Six widely used algorithms were utilized to predict ξ_{eq} and to compare the performance of the developed models.

2 Characterization and Derivation of ξ_{eq} from Experimental Load-Displacement

The ξ_{eq} associated with each specimen in the dataset is calculated using Eq. (1), based on the equivalent damping equation for resonant steady state proposed by Jacobsen [16, 17].

$$\xi_{eq} = \frac{1}{4\pi} \frac{ED}{E_{sO}} \tag{1}$$

Where ξ_{eq} is the equivalent damping ratio and ED is the cyclic dissipated energy. E_{sO} is the elastic strain energy, given by $E_{sO} = \frac{1}{2} k u_0^2$, where k is the equivalent stiffness for the loop at lateral displacement u_0. Figure 1 shows these parameters for an assumed hysteresis loop and cycle number, where f_0 is the maximum lateral force for the maximum u_0. The ML algorithms require two intermediate output parameters (ED and f_0). These parameters are obtained from the experimental load-displacement curves of the database specimens. The extraction procedure is illustrated in Fig. 1 for specimen WSH2 [35] which was a cantilever wall with a height of 4560 mm.

As observed in Fig. 1, among the two cycles near the maximum drift ratio, the cycle with the maximum displacement is selected to avoid additional effects of cyclic degradation in the calculation. The damping ratio values at various displacements are computed using the two parameters above and u_0 (Fig. 2).

3 Experimental Dataset

Over the past few years, different databases of RC shear walls have been developed or collected by various researchers. Some of the datasets used in the literature have been used to derive empirical relationships for predicting and modeling various properties of RC shear walls, such as shear strength, deformation capacity, and hysteretic behavior [26, 36, 37], while others have been used to assess provisions in building codes [38, 39]. There are databases that were developed for very specific objectives, such as the one by Nie and Wang [40], which contains 15 specimens for the study of shear-critical RC walls under combined "tension-bending-shear" loading. Additionally, Mangalathu et al. [29] compiled a database of 393 specimens for developing ML models to predict failure modes of RC shear walls. This study uses data from

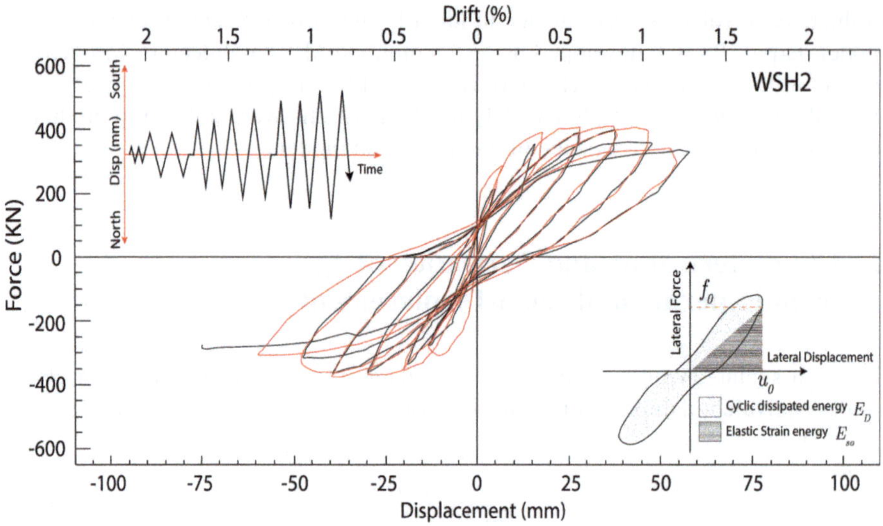

Fig. 1 Elastic strain and cyclic dissipated energy definitions and loading protocol and load-displacement curve for WSH2

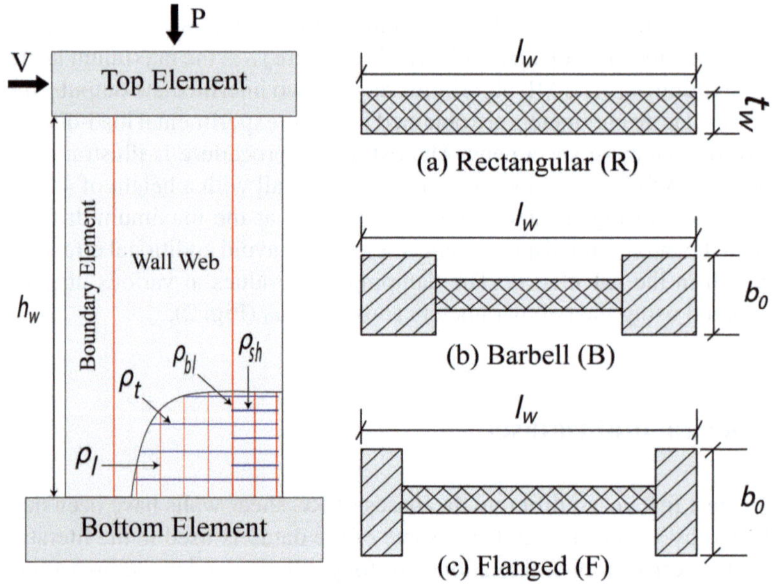

Fig. 2 RC shear wall cross-section types and geometric properties

the Seismic Engineering Research Infrastructures for European Synergies (SERIES) database, a comprehensive resource of experimental data on RC shear walls subjected to incremental reversed cyclic loading; specimens subjected to monotonic loading were not included. The details are given in Table 1.

The database includes information on 200 walls categorized by their cross-section: rectangular ($n = 132$), barbell ($n = 55$), flanged ($n = 8$) and T-shape (5) and by failure modes: flexural ($n = 90$), shear ($n = 32$), shear-tension ($n = 5$), shear-compression ($n = 28$), flexural-shear ($n = 21$), lap-splice failure ($n = 3$), sliding-shear ($n = 8$), and nonfailure ($n = 13$). Among these, 132 have rectangular cross-sections. Given the higher number of rectangular walls compared with barbell or other cross-sectional shapes, and their broader adoption since the late 1980s in many countries [41], rectangular walls dominates the dataset. The distribution of all parameters is given in Fig. 3.

Table 1 Design properties of wall specimens in the database

No.	Features	Min	Max	μ	σ	COV (%)
1	Length l_w (mm)	400	3500	1380	638	46.26
2	Height h_w (mm)	495	5486	2309	1132	49.02
3	Thickness t_w (mm)	26	305	111	46.57	41.45
4	Concrete compressive strength f_c' (MPa)	13.38	93.6	35.57	15.54	43.70
5	Transverse web reinforcement yield strength f_{yt} (MPa)	0	1187	453.6	164.92	36.35
6	Transverse boundary reinforcement yield strength f_{ysh} (MPa)	0	1375	454.9	214.85	47.22
7	Vertical web reinforcement yield strength f_{yl} (MPa)	160	1000	448.9	135.40	30.16
8	Vertical boundary reinforcement yield strength f_{ybl} (MPa)	0	900	456.2	127	27.86
9	Transverse web reinforcement ratio ρ_t (%)	0.001	0.025	0.006	0.0041	68.95
10	Transverse boundary reinforcement ratio ρ_{sh} (%)	0	0.042	0.007	0.0052	74.30
11	Vertical web reinforcement ratio ρ_l (%)	0	0.029	0.006	0.0061	94.00
12	Vertical boundary reinforcement ratio ρ_{bl} (%)	0	0.031	0.006	0.0051	82.30
13	Axial load ratio $P/\left(A_g f_c'\right)$	0	0.86	0.126	0.1336	105.3
14	Boundary element depth b_0 (mm)	45	3045	240.9	408.03	169.3
15	Boundary element length d_b (mm)	0	500	179.0	91.40	51.06
16	Hoop spacing/boundary element length s/d_b	0	47.66	1.35	4.53	336.27
17	Aspect ratio AR	0.39	5.83	1.83	0.92	49.92
18	Shear span ratio M/Vl_w	0.39	4.10	1.80	0.73	40.80
19	Drift ratio θ (%)	0.028	4.85	0.87	0.75	86.82

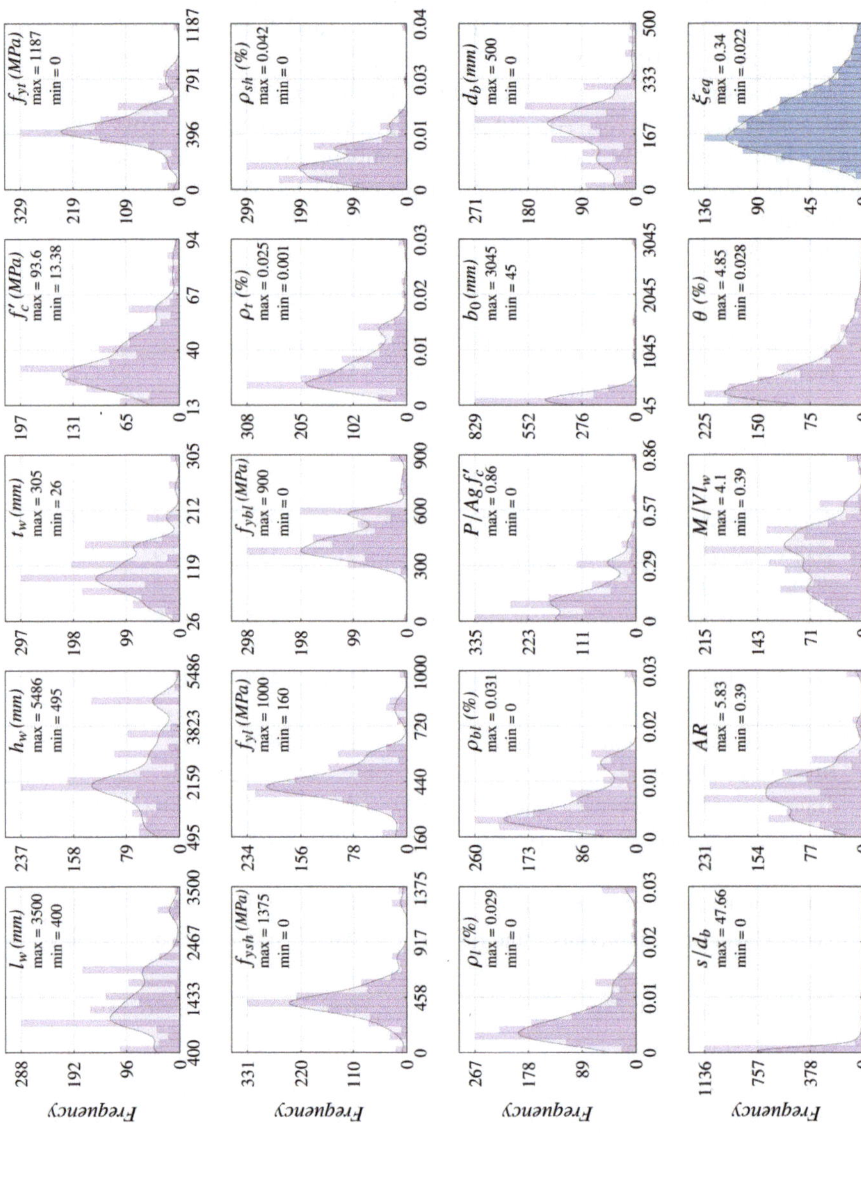

Fig. 3 Frequency distribution of input and output parameters

4 Sensitivity Analysis

To obtain sensitivity analysis results, the variability of the input variables must be characterized and a representative model output is selected for evaluation. In this study, a single ML technique (NCA) was used to perform the sensitivity analysis. ML, a subfield of artificial intelligence, infers patterns from input features to predict the response variable. Among the available methods, NCA is adopted here as a supervised feature-weighting algorithm that learns non-negative importance weights for the inputs by maximizing predictive performance, thereby quantifying each variable's influence without imposing a fixed functional form. The objective is to rank the input features according to their influence on the estimation of ξ_{eq}. Accordingly, NCA is used to rank features via the learned weights [42]. NCA was originally designed to improve k-nearest neighbors (KNN) performance through stochastic neighborhood assignment and optimizing a leave-one-out (LOO) regression objective on the training set [43]. Figure 4 shows weights of 19 features in a heat map across trials for the NCA method.

The *x*-axis indicates the feature IDs for the 19 input features, consistent with Table 1, while the *y*-axis presents the results of each trial. Note that the heat map colors turn yellow as the feature weight increases and blue as the weight decreases. That is, the yellow-colored feature is ranked first, whereas the blue-colored feature is ranked last. To understand each feature's effect on predicting ξ_{eq}, the ranked features are introduced into the nonlinear models one at a time in descending order (starting with the highest feature weight). Figure 5 presents the RF model trend, showing how the number of ranked features included at each run of ξ_{eq} estimation affects prediction results in terms of R^2 and RMSE. Evaluating all features in terms of R^2 and RMSE was performed by following the NCA-derived ordering. The model was first fit using only the top-ranked feature, drift ratio and its performance was recorded, after which additional features were introduced sequentially in descending order of NCA weights. This cumulative inclusion shows that adding ρ_{l}, ρ_{bl}, $P/(A_{\mathrm{g}}f_{\mathrm{c}}')$, f_{c}', AR, l_{w}, s/d_b, b_0,

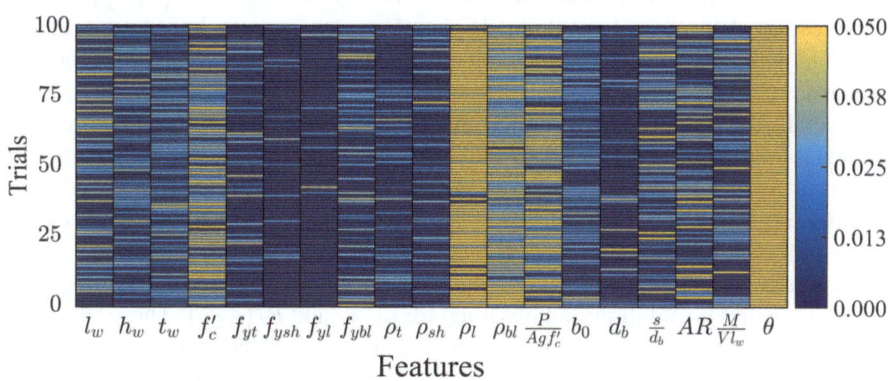

Fig. 4 NCA feature weights over 100 trials

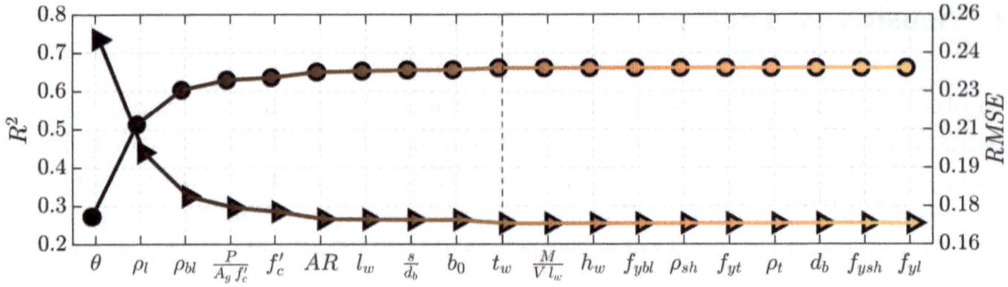

Fig. 5 Mean R^2 and RMSE obtained from RF models, using the NCA-derived feature ranking averaged over 100 trials

and t_w, yields substantial gains, evidenced by increases in R^2 and corresponding decreases in RMSE, indicating strong influence on the estimation of ξ_{eq}.

By contrast, subsequent inclusion of M/Vl_w, h_w, f_{ybl}, ρ_{sh}, f_{yt}, ρ_t, d_b, f_{ysh}, and f_{yl} produces no meaningful improvement (negligible changes in R^2 and RMSE), implying limited contribution relative to the more influential variables.

Imposed drift is the dominant driver of ξ_{eq}: larger cyclic drifts produce fuller hysteresis loops and higher ξ_{eq}, whereas smaller drift and pinched cycles reduce it, consistent with displacement-based design and equivalent linearization formulations [3, 44]. Geometric Properties: aspect ratio governs behavior: squat walls are shear-dominated, and show limited ductility and pinched loops (lower ξ_{eq}), while slender walls are flexure-dominated, and permit larger cyclic deformations and fuller loops (higher ξ_{eq}). Wall thickness also modulates ξ_{eq} via its effects on confinement efficacy, shear transfer, and stiffness degradation [45–47]. Reinforcement: a higher ρ_{bl} enhances deformation capacity and stabilizes flexural plasticity, yielding less pinching and higher ξ_{eq}. Prior work often finds limited sensitivity to ρ_l [48, 49]. Confinement (s/d_b), smaller spacing-bar-diameter ratios improve boundary confinement, increase compressive strain capacity, restrain spalling, and yield larger loop areas relative to elastic energy [50]. Axial load and strength: $P/(A_g f_c')$ and f_c' shape cyclic response, moderate axial load can stabilize cycles, higher $P/(A_g f_c')$ promotes compression damage and pinching. Higher f_c' raises strength but may accelerate crushing without adequate confinement, effects on ξ_{eq} depend on detailing and drift demand.

Overall, the wall design variables identified as influential by the ranking ρ_l, s/d_b, axial load ratio, and to a lesser extent ρ_l and f_c' are physically consistent with mechanisms that govern loop fullness and therefore with expected trends in ξ_{eq}.

Figure 6 shows a polar heatmap of pairwise Pearson correlations between RC shear wall parameters and ξ_{eq}. Most relationships are modest in magnitude, with a few variables standing out. In particular drift is directly correlated with damping. Smaller positive correlations are observed for d_b, t_w and $P/(A_g f_c')$. These trends are consistent with evidence that thicker webs and larger bars delay damage localization and pinching, while moderate axial compression stabilizes the compression zone, effects that enlarge the hysteresis loop area and increase ξ_{eq}.

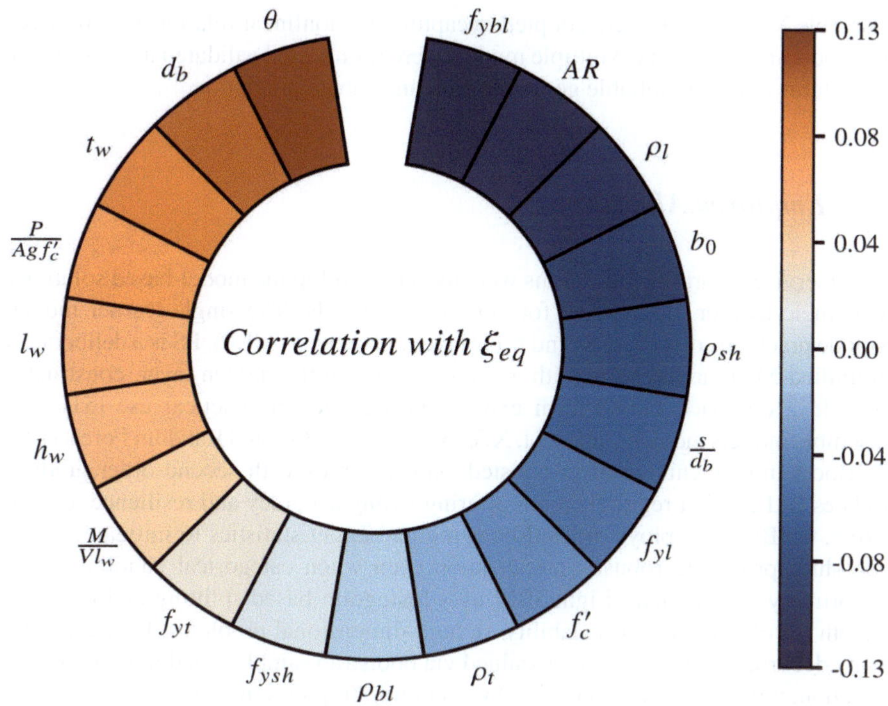

Fig. 6 Polar heatmap of Pearson correlations between RC shear wall parameters and ξ_{eq}

On the negative side, several parameters show weak-to-moderate inverse relationships with ξ_{eq}. The most negative correlation is with boundary bar f_{ybl} followed closely by AR, ρ_l, b_o, and ρ_{sh}.

s/d_b and f'_c trend slightly negative, as do steel yield strengths f_{yl}, f_{yt}, and f_{ysh}, higher strengths and detailing can stiffen and pinch hysteresis at a given drift, reducing ξ_{eq}. Slender walls likewise concentrate curvature near the base, shrinking the stabilized loop area and ξ_{eq}. Other geometry terms, l_w, h_w, and M/Vl_w show weak pairwise correlations, with effects likely conditional on reinforcement, axial load and drift that emerge in multivariate or nonlinear models. Coefficients are pairwise, so read signs as tendencies, the strong role of θ *(drift ratio)* should be verified and retained via feature importance and local effect diagnostics.

4.1 Solutions Development

A data-driven framework was developed to predict the ξ_{eq} of RC shear walls. The dataset comprises geometric characteristics, material strengths, reinforcement detailing, and loading protocols as input features, with ξ_{eq} as the output. Single and ensemble ML methods were adopted to capture the nonlinear relationships between the inputs and the output. Multiple models were trained and validated using standard procedures to ensure reliable generalization and robust prediction.

4.2 Employed Algorithms

Six supervised learning algorithms were used to develop the model-based solutions: two single-learner models and four ensemble models. The single learner models include practical solution (PS) and Multilayer Perceptron (MLP). PS is a deliberately simplified MLP architecture with a single neuron in the hidden layer, constructed to yield a compact, closed-form expression suitable for practical use [51]. The ensemble models include CatBoost, XGBoost, LightGBM, and Random Forest (RF). XGBoost implements gradient-boosted decision trees with second-order gradient updates and explicit regularization, offering strong accuracy and resilience to overfitting. CatBoost employs ordered boosting and target statistics to mitigate prediction shift, providing robust generalization even when categorical handling is not the primary requirement. LightGBM uses histogram-based splitting and leaf-wise growth for efficiency and scalability on high-dimensional problems. RF aggregates many decorrelated decision trees trained via bootstrap sampling and random feature selection, with predictions obtained by averaging (regression) across trees.

Hyperparameters were tuned with Optuna using a Bayesian strategy to efficiently explore the search space based on prior evaluations. After identifying the best configurations, models were retrained with the selected hyperparameters to confirm performance. The dataset was randomly partitioned into training, validation, and testing subsets in 70%, 15% and 15% proportions, and each algorithm was trained for 500 trials to obtain stable, and accurate solutions.

4.3 Practical Solution

A feedforward MLP with a single neuron in the hidden layer is adopted to provide a practical, closed-form equation [52, 53]. The network comprises three components (i) an input layer that accepts the features, (ii) one hidden neuron that introduces nonlinearity, and (iii) a linear output that returns the prediction.

For an input vector $x = [x_1, x_2, x_3, \ldots, x_d]$, the hidden neuron's weighted sum is calculated by Eq. (2).

$$z = \sum_{n=1}^{d} w_{i,xi} + b_a = w^T x + b_b, \tag{2}$$

where w and b_a are the trained weights and bias of the hidden layer, and b_b is the bias of the output layer. To capture the nonlinear interaction between the inputs and the output, the neuron output is obtained via a log-sigmoid activation.

$$y = \sigma(z) = \frac{1}{1 + \mathrm{e}^{-z}}. \tag{3}$$

Here y represents the neuron's activation, and σ is the activation function. Choosing a smooth, differentiable f let's gradient propagation occur reliably, enabling stable and effective training.

To enhance transparency and enable closed-form use, the MLP was kept to one hidden neuron. All variables were scaled to [− 1, 1] prior to training. This improves numerical conditioning and allows the final expression to be applied directly once the scaling constants are known.

Let $\tilde{x} \in [-1, 1]$ denote the scaled inputs and $\overline{y} \in [-1, 1]$ scaled target. The forward mapping is:

$$\hat{y}_{\mathrm{PS}} = w_{\mathrm{OL}} y + b_{\mathrm{OL}}. \tag{4}$$

Here, y is the hidden neuron's activation, and w_{OL} and b_{OL} are the output-layer weights and bias, respectively.

The dataset was randomly split into training (70%), validation (15%), and testing (15%) subsets. The training and testing sets were used to fit and evaluate PS, yielding the hidden and output-layer weights and biases. With these learned parameters, a closed-form expression for the normalized ξ_{eq} is obtained and reported below:

$$\begin{aligned} z = 0.055(l_{\mathrm{w}}^{*}) + 0.018(h_{\mathrm{w}}^{*}) - 0.271(t_{\mathrm{w}}^{*}) - 0.145(f_{\mathrm{c}}^{\prime *}) + 0.826\left(f_{\mathrm{yt}}^{*}\right) \\ - 0.114\left(f_{\mathrm{ysh}}^{*}\right) + 0.129\left(f_{\mathrm{yl}}^{*}\right) - 1.610\left(f_{\mathrm{ybl}}^{*}\right) + 0.013(\rho_{\mathrm{t}}^{*}) \\ - 0.849(\rho_{\mathrm{sh}}^{*}) - 0.639(\rho_{\mathrm{l}}^{*}) - 0.949(\rho_{\mathrm{bl}}^{*}) + 0.094(P/(A_{\mathrm{g}} f_{\mathrm{c}}^{\prime *})), \\ - 1.035(\mathrm{b}_{\mathrm{o}}^{*}) + 0.114(d_{\mathrm{b}}^{*}) - 1.185(s/d_{\mathrm{b}}^{*}) \\ - 1.701(AR^{*}) + 0.363(M/Vl_{\mathrm{w}}^{*}) + 3.177(\theta^{*}) \end{aligned} \tag{5}$$

$$y = \mathrm{Log} - \mathrm{sigmoid} = \frac{1}{1 + \mathrm{e}^{-z}}, \tag{6}$$

$$\xi_{\mathrm{eq,PS}} = (0.861)\,(y) + (-0.686) \quad \in [-1, 1]. \tag{7}$$

Each term in Eq. (5) is a scaled input multiplied by its hidden-layer weight, plus the hidden-layer bias. In Eq. (6), this weighted sum is passed through the logistic (log-sigmoid) activation function. Several activation functions were used, but the logistic (log-sigmoid) provided the most stable and accurate fit [54]. The output layer then applies its learned weight and bias to the activated response, yielding the scaled prediction in Eq. (7), bounded within [− 1, 1]. To report the result in physical units, the normalized output is inverse-transformed using the observed minimum and maximum values of the original ξ_{eq}. Eq. (8), giving the final estimate $\xi_{eq,\,PS}$. For clarity, a PS flow chart is shown in Fig. 7.

$$\xi_{eq} = \frac{\xi_{eq,PS} + 1}{2}(\xi_{max} - \xi_{min}) + \xi_{min} \tag{8}$$

The above equation rescales the PS model's normalized output ($\xi_{eq,\,PS}$) from [−1, 1] back to the physical ξ_{eq} using the maximum ξ_{max} and minimum ξ_{min} given in Table 1.

4.4 Performance Metrics

To evaluate the performance of single and ensemble models, predictive power was assessed using standard metrics: the coefficient of determination (R^2) which indicates explained variance (higher is better), root mean squared error (RMSE), which indicates residual dispersion (smaller is better); and mean absolute error (MAE), the average error magnitude in target units (smaller is better). The mean error (ME), define as the signed average of residuals, reflects systematic bias (values near zero are ideal). Metrics were reported separately for the training, validation, and testing subsets to distinguish in-sample fit from generalization. A split-weighted Overall Balanced Grade (OBG) was also computed, combining MAE, RMSE, and (R^2) with weights proportional to subset size, lower OBG indicates better overall fit. Together, R^2, RMSE, MAE, ME, and OBG provide a transparent basis for comparing and ranking models.

$$R^2 = 1 - \sum_{i=1}^{n}\left(y_i^{'} - y_i\right)^2 / \sum_{i=1}^{n}(y_i - \overline{y_i})^2, \tag{9}$$

$$\text{RMSE} = \sqrt{\frac{\sum_{i=1}^{n}\left(y_i^{'} - y_i\right)^2}{n}}, \tag{10}$$

$$\text{MAE} = \frac{1}{m}\sum_{i=1}^{m}|\hat{y}_i - y_i|, \tag{11}$$

$$\text{ME} = \frac{1}{n}\sum_{i=1}^{n}(y_i^{\prime} - y_i), \tag{12}$$

$$\text{OBG} = \sum_{s\in(\text{test})}\left(\frac{w_{s.(\text{MAE}_s + \text{RMSE}_s)}}{w_s.R^2 + 1}\right). \tag{13}$$

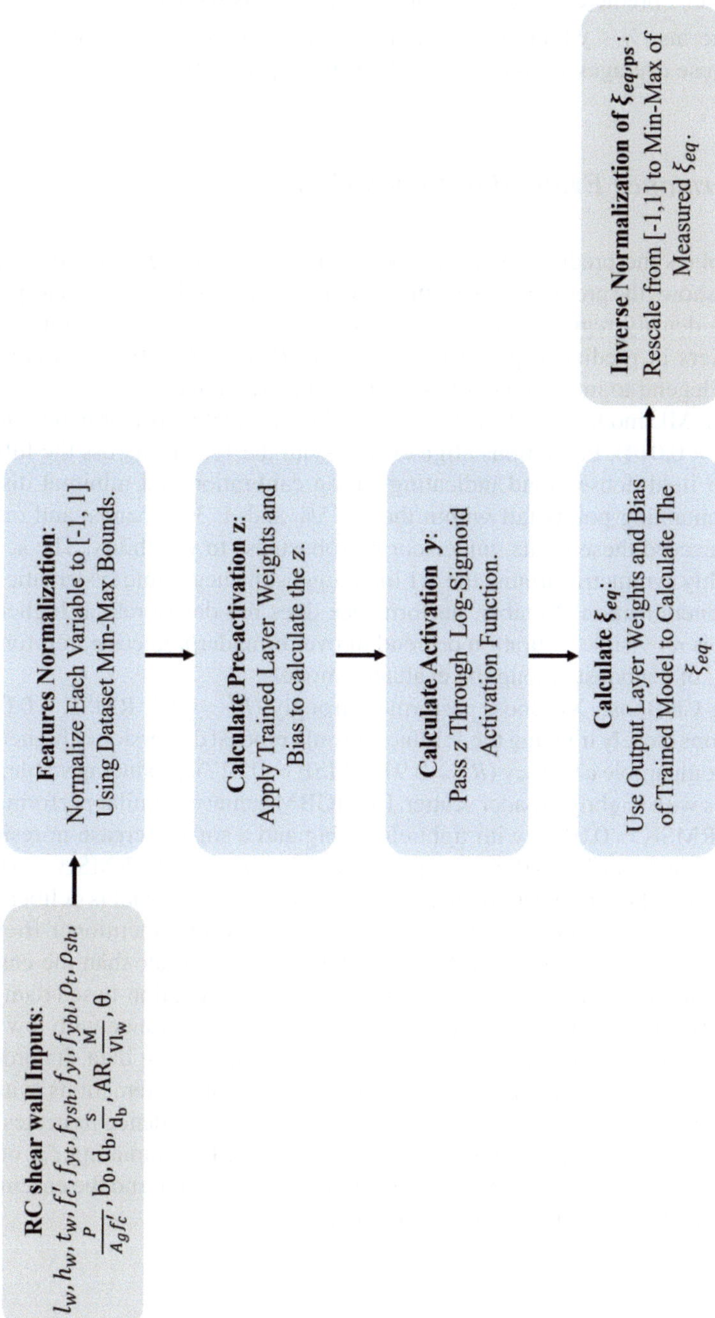

Fig. 7 Flowchart of the developed PS solution for estimating ξ_{eq} of RC shear walls

Here, y_i is the observed (true) target for sample i, $\hat{y}_i$ (also written as the prediction y_i') is the model's predicted value for that sample and $\overline{y}$ is the mean of all observed targets. The terms $|\hat{y}_i - y_i|$ and $\left(y_i' - y_i\right)^2$ are the per-sample absolute and squared residuals, whose averages yield MAE and RMSE, respectively.

4.5 Performance Evaluation Comparison

Figure 8 displays the predicted versus experimental ξ_{eq} for the developed models; each subplot shows the predictive performance of the solutions. Predictive data points are shown as density regions, and $\pm 15\%$ and $\pm 30\%$ error bands are included to identify outliers in predictive performance. Model R^2 and RMSE values are also shown in the legend to indicate goodness of fit and prediction error.

Among the ML models, CatBoost shows the best predictive performance ($R^2 = 0.95$, RMSE $= 0.014$). Predictions align closely with the 1:1 line across the full ξ_{eq} range, with a tight dense cloud indicating strong calibration and minimal dispersion. Most remaining points fall within the $\pm$ 15% and $\pm$ 30% bands, and only a few outliers exceed these limits, underscoring robustness to variability. The scatter appears roughly symmetric around the 1:1 line, suggesting negligible systematic bias and stable generalization. Notably, performance does not deteriorate at higher ξ_{eq} values; there is no visible saturation or trend of over or underprediction, reinforcing the reliability of Catboost among the evaluated models.

Following CatBoost, XGBoost performes strongly (R^2 =0.92, RMSE =0.017), with predictions closely tracking the 1:1 line and only modest dispersion at higher ξ_{eq}. RF showed comparable accuracy ($R^2 = 0.91$, RMSE $= 0.017$), producing stable, low bias estimates with slightly broader scatter. LightGBM achieved similar performance ($R^2 = 0.91$, RMSE $= 0.018$), with tight clustering and a small increase in residual spread. MLP matched LightGBM in aggregate metrics ($R^2 = 0.91$, RMSE $= 0.018$) but showed mildly larger dispersion near the extremes. The PS model is deliberately constrained to a single hidden neuron to yield a closed-form; practitioner-friendly expression; it attains $R^2 = 0.63$ and RMSE $= 0.035$, less accurate than the ensembles but transparent, lightweight, and acceptable for quick, equation-based damping estimates. CatBoost maintains high accuracy across the full ξ_{eq} range with low bias and variance, showing stable generalization on validation and test data. Its ordered boosting and robust regularization capture nonlinear feature interactions without overfitting, yielding well-calibrated predictions. Overall, the evidence indicates that ML approaches offer an accurate and practical alternative for estimating ξ_{eq} of RC shear walls, with the PS model serving as an interpretable baseline and the ensembles providing state of-the-art predictive performance.

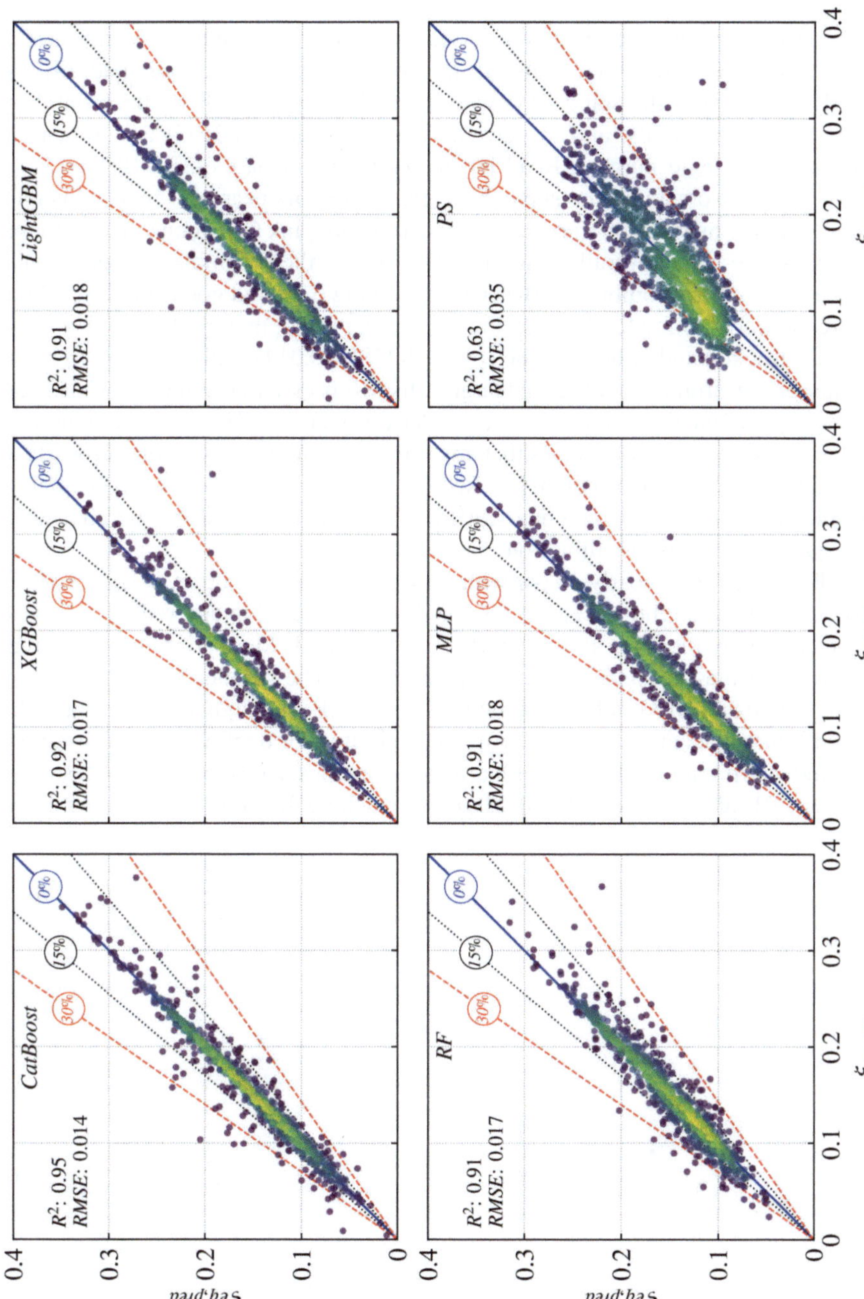

Fig. 8 Predicted versus experimental ξ_{eq} density coded scatter plot

4.6 Accuracy Evaluation of the Developed Model

In this section, model accuracy is assessed using four complementary indicators. The results are plotted in Fig. 9, RMSE to capture overall error magnitude, ME to quantify signed bias, the standard deviation of ME (σ_{ME}) to reflect the variability of that bias across splits, and the Overall Balanced Grade (OBG) to summarize fit quality in a single number. Read left to right, shorter bars on RMSE and σ_{ME} signify tighter errors and more stable bias, green ME bars near zero indicate negligible systematic over or underprediction, and a smaller pink OBG consolidates these advantages into better overall suitability. As a group, the ensemble learners dominate the Pareto front, combining low error, vanishing bias, and good stability, while the intentionally minimalist PS baseline lags, as expected, though it still keeps bias small. CatBoost stands out with the lowest RMSE (0.014) indicating the tightest error dispersion around the observed targets. Its ME is essentially zero, so CatBoost does not systematically overestimate or under estimate ξ_{eq}. The σ_{ME} is among the smallest in the set, which indicates that any tiny bias remains very stable across data splits and folds. This combination, low RMSE, negligible ME, and low σ_{ME} explains its leading OBG (0.003). In practice, CatBoost would be expected to generalize well and require minimal post-calibration.

XGBoost performs nearly on par with CatBoost in absolute error, posting RMSE (0.017). Its ME is slightly negative (− 0.001), a very small underprediction that is practically negligible, σ_{ME} is low, mirroring the overall low variability, implying that the small bias is consistent rather than erratic. The balanced nature of these statistics yields an OBG of 0.003, joint best with CatBoost, signaling a dependable trade-off among accuracy, bias control, and stability. For deployments prioritizing speed and interpretability tooling built around ensembles, XGBoost remains a strong candidate. RF also achieves RMSE (0.017), matching XGBoost on average error

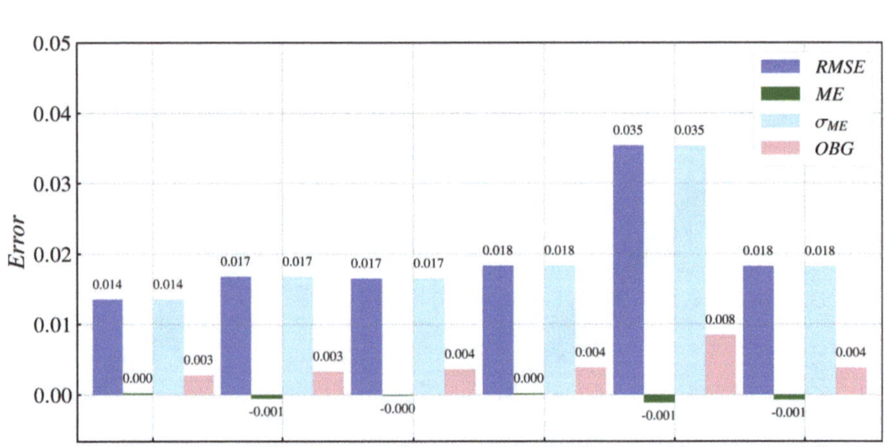

Fig. 9 Performance assessment across models: RMSE, ME, σ_{ME}, and OBG comparison

magnitude. Its ME is essentially zero, so RF predictions center well on the truth. σ_{ME} is low, indicating stable behavior across splits, with no sign that bias swings from fold to fold. The OBG (0.004) compared with the top two, reflects a marginally less balanced profile, not a qualitative deficiency. LightGBM follows closely with RMSE (0.018) only a small step behind the previous ensembles. ME remains near zero, so systematic error is limited and σ_{ME} is low enough to suggest reliable repeatability across resamples. Its OBG of 0.004 is identical to RF, reinforcing that LightGBM offers a solid balance of accuracy and stability. In scenarios with large datasets where training efficiency matters, LightGBM computational profile can make it an attractive alternative with only a minor trade-off in error.

The PS baseline shows the expected constraints of its deliberately minimal capacity, RMSE rises to 0.035 and σ_{ME} to 0.035, the largest in the test set. Those larger bars indicate both wider scatter in errors and greater variability in any residual bias across splits. Even so, ME itself remains close to zero (− 0.001), so the model is not systematically skewed, it is simply underpowered for the complexity of ξ_{eq}. Its higher OBG (0.008) is consistent with this diagnostic and makes PS a transparent but limited benchmark rather than a production model. The MLP lands between LightGBM and PS with RMSE (0.018) and ME near zero (− 0.001), demonstrating competitive average accuracy and negligible systematic bias. σ_{ME} is modest, pointing to reasonably stable behavior across folds. Its OBG of 0.004 matches RF and LightGBM, suggesting that with careful regularization and tuning the MLP can emulate the ensembles balance, though here it does not surpass them. In contexts where neural networks are preferred for integration or downstream differentiability, this profile is reassuring. Taken together, OBG ranks the models almost exactly as RMSE does, with CatBoost and XGBoost leading (0.003), RF, LightGBM, and MLP forming a tight second tier (0.004), and PS trailing (0.008). Among these, CatBoost is the best performer, it combines the smallest error (RMSE = 0.014), negligible bias (ME = 0), and low bias variability σ_{ME} into the best overall balance (OBG = 0.003), making it the most accurate, most stable, and most deployment-ready option for estimating ξ_{eq}.

Radar chart aggregating of five metrics (R^2, RMSE, MAE, ME, σ_{ME}) are shown in Fig. 10 with directional scaling chosen so that a larger radial extent always denotes better performance. Specifically, the R^2 axes increase from 0 at the center to 1 at the perimeter, whereas the error/bias axes (RMSE, MAE, ME, σ_{ME}) are inverted, the outer rings correspond to smaller values and inner rings to larger errors. With this scheme, the most desirable model traces a polygon that stays closest to the outer boundary on all axes.

Under this unified scale, the ensemble methods delineate the outer envelope of the chart. CatBoost exhibits the widest profile, combining the highest R^2 with the smallest RMSE, and MAE and near-zero ME and σ_{ME}, indicating high accuracy, minimal bias, and stable residuals. XGBoost and RF form a close second tier; their R^2 values are only marginally lower and their RMSE/MAE radii are slightly pulled inward relative to CatBoost, while their ME and σ_{ME} remain near the outer rings. LightGBM and MLP follow with high R^2 and slightly larger error radii; their ME and σ_{ME} stay close to the outer boundary, suggesting limited bias and good consistency.

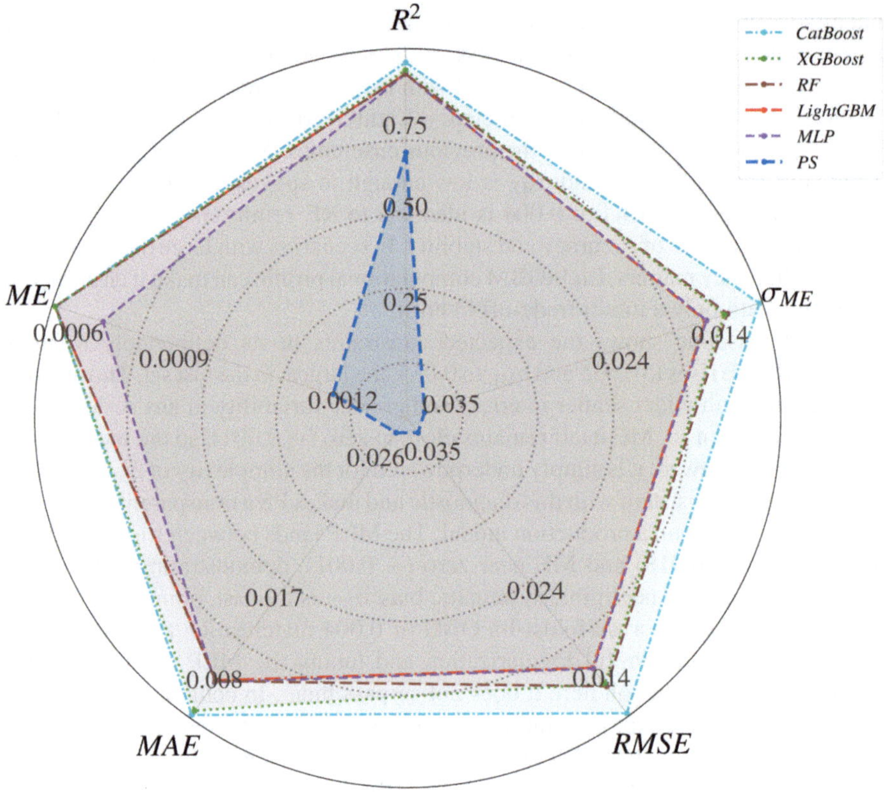

Fig. 10 Models comparison radar chart

In contrast, PS contracts markedly toward the center on the error axes and shows a shorter R^2 radius, reflecting comparatively lower predictive accuracy, nevertheless, its ME radius remains near the outer ring, indicating little systematic bias.

Overall, the radar chart makes the trade-offs clear shows that CatBoost is the most balanced and accurate model, and XGBoost and RF are close behind with slightly higher errors, LightGBM and MLP deliver solid accuracy with negligible bias, and PS is a simple baseline with modest variance control but lower R^2. The outward-inward pattern shows that gains come mainly from lower RMSE/MAE and steadier residuals, while bias is small for all ML models.

5 Conclusions and Outlook

This research developed single and ensemble learning predictive models to estimate the equivalent viscous damping ratio ξ_{eq} of RC shear walls using a comprehensive dataset of 200 experimental results. Nineteen design parameters, including drift ratio, were selected, and sensitivity analysis was conducted to identify which parameters are less influential in predicting damping. According to the predictive outcomes in this research, the conclusions are as follows:

- Based on the selected parameters, the sensitivity of each parameter was analyzed by obtaining parameter weights using the widely used NCA method. The parameters were then ranked, and subsets were created from the highest to lowest weights. Subset performance was evaluated using R^2 and RMSE, and the results indicate that the parameters (ρ_l, ρ_{bl}, $P/(A_g f_c')$, f_c', AR, l_w, s/d_b, b_0, and t_w) have the highest influence on the prediction, as adding these parameters to the subsets showed a gradual increase in R^2 and a decrease in RMSE.
- In the comparison of single and ensemble learning models, the results show that CatBoost achieved superior performance compared with the others models. CatBoost has the lowest errors (RMSE = 0.014), ME = 0.0002 and $\sigma_{ME} = 0.014$, indicating that the model is unbiased and highly reliable.
- CatBoost achieved the highest R^2 value of 0.95 with an OBG score of 0.003, making the model accurate and robust in predicting the ξ_{eq}.
- Practical solution (PS) achieved $R^2 = 0.63$ and RMSE $= 0.035$ with the error between experimental and predicted values ME $= -0.001$. It should be noted that PS was intentionally developed to provide a practical mathematical equation for practical use using a simplified MLP architecture. Users can estimate the ξ_{eq} using the trained weights and biases of the network.

Since the predictive models showed stable and accurate performance in this research, future work could expand the database to include additional wall types, higher drift levels, and different loading protocols. Researchers are encouraged to add uncertainty estimates (prediction interval) so engineers can gauge confidence in each result. Using hybrid ML algorithms can also be considered.

AI Use Disclosure Statement The authors used ChatGPT to enhance the clarity and readability of the manuscript. All content was thoroughly reviewed and revised by the authors, who assume full responsibility for its accuracy, integrity, and final form.

Acknowledgments This research was supported by the National Natural Science Foundation of China (NSFC) under grant numbers U2139209, and 52378528. The author, Atta Ullah Noor, also gratefully acknowledges the China Scholarship Council (CSC) for providing financial support for his postgraduate studies.

References

1. Miranda E, Ruiz-García J (2002) Evaluation of approximate methods to estimate maximum inelastic displacement demands. Earthq Eng Struct Dyn 31:539–560

2. Dwairi H, Kowalsky M (2004) Investigation of Jacobsen's equivalent viscous damping approach as applied to displacement-based seismic design. In: Proceeding of 13th world conference on earthquake engineering, pp 1–6
3. Priestley MN, Calvi GM, Kowalsky MJ, Powell G (2007) Displacement-based seismic design of structures. IUSS Press Pavia, Italy
4. Priestley MN (2003) Myths and fallacies in earthquake engineering, revisited: the ninth mallet Milne lecture. IUSS press Pavia, Italy
5. Zaharia R, Taucer F (2008) Equivalent period and damping for EC8 spectral response of SDOF ring-spring hysteretic models
6. Sullivan TJ (2013) Highlighting differences between force-based and displacement-based design solutions for reinforced concrete frame structures. Struct Eng Int 23:122–131
7. Housner GW (1956) Limit design of structures to resist earthquakes. Proc World Conf Earthq Eng 61956
8. Trifunac MD, Brady AG (1975) A study on the duration of strong earthquake ground motion. Bull Seismol Soc Am 65:581–626
9. Akiyama H (1985) Earthquake-resistant limit-state design for buildings. University of Tokyo Press
10. Jafari A, Sadeghi M, Alaee FJ, Bengar HA (2024) Determining the optimal layout of energy-dissipating mechanisms for hybrid self-centering walls. In: Structures. Elsevier, p 105854
11. Benavent-Climent A, López-Almansa F, Bravo-González DA (2010) Design energy input spectra for moderate-to-high seismicity regions based on Colombian earthquakes. Soil Dyn Earthq Eng 30:1129–1148
12. Erberik MA, Sucuoğlu H, Acun B (2012) Inelastic displacement response of RC systems with cyclic deterioration—an energy approach. J Earthq Eng 16:937–962
13. Dindar AA, Yalçın C, Yüksel E, Özkaynak H, Büyüköztürk O (2015) Development of earthquake energy demand spectra. Earthquake Spectra 31:1667–1689
14. Jafari A, Naserpour A, Zhou Y, Dhakal RP (2025) Trade-off between post-earthquake serviceability and energy dissipation capacity in RC rocking walls. Eng Struct 345:121425
15. Abdollahi A, Jafari A, Dehcheshmeh EM (2024) Optimal energy dissipation ratio of base-rocking walls. J Build Eng 87:109034
16. Jacobsen LS (1930) Steady forced vibration as influenced by damping: an approximate solution of the steady forced vibration of a system of one degree of freedom under the influence of various types of damping. Trans Am Soc Mech Eng 52:169–178
17. Jacobsen LS (1960) Damping in composite structures. In: Proceedings of 2nd WCEE, Tokyo and Kyoto
18. Rosenblueth E, Herrera I (1964) On a kind of hysteretic damping. J Eng Mech Div 90:37–48
19. Hudson D (1965) Equivalent viscous friction for hysteretic system with earthquake-like excitations. 3 WCEE. New Zealand
20. Iwan W (1980) Estimating inelastic response spectra from elastic spectra. Earthquake Engineering & Structural Dynamics 8:375–388
21. Fardis M, Panagiotakos T (1996) Hysteretic damping of reinforced concrete elements. In: 11th World conference on earthquake engineering, Acapulco, Mexico
22. Blandon CA, Priestley M (2005) Equivalent viscous damping equations for direct displacement based design. J Earthq Eng 9:257–278
23. Buttmann P (1983) Experimental determination of damping factors for masonry walls during dynamic load with special regard to nuclear power plants. In: Annual meeting on nuclear technology'83-Reactor conference, pp 816–9
24. Jeon JS, Shafieezadeh A, Des RR (2014) Statistical models for shear strength of RC beam-column joints using machine-learning techniques. Earthq Eng Struct Dyn 43:2075–2095
25. Chou J-S, Pham T-P-T, Nguyen T-K, Pham A-D, Ngo N-T (2020) Shear strength prediction of reinforced concrete beams by baseline, ensemble, and hybrid machine learning models. Soft Comput-A Fusion Found, Methodol Appl:24
26. Song I, Cho IH, Wong RK (2020) An advanced statistical approach to data-driven earthquake engineering. J Earthq Eng 24:1245–1269

27. Vu D-T, Hoang N-D (2016) Punching shear capacity estimation of FRP-reinforced concrete slabs using a hybrid machine learning approach. Struct Infrastruct Eng 12:1153–1161
28. Siam A, Ezzeldin M, El-Dakhakhni W (2019) Machine learning algorithms for structural performance classifications and predictions: application to reinforced masonry shear walls. In: Structures. Elsevier, pp 252–265
29. Mangalathu S, Jang H, Hwang S-H, Jeon J-S (2020) Data-driven machine-learning-based seismic failure mode identification of reinforced concrete shear walls. Eng Struct 208:110331
30. Jafari A, Akbarzadeh Bengar H, Hassanli R, Nazari M, Dugnani R (2021) The response of self-centering concrete walls under quasi-static loading. Bull Earthq Eng 19:2893–2917
31. Shahmansouri AA, Jafari A, Bengar HA, Zhou Y, Taciroglu E (2025) A scaling-based generalizable integrated ML-mechanics model for lateral response of self-centering walls. Eng Struct 336:120326
32. Xiong Q, Xiong H, Kong Q, Ni X, Li Y, Yuan C (2022) Machine learning-driven seismic failure mode identification of reinforced concrete shear walls based on PCA feature extraction. Structure 44:1429–1442
33. Jafari A, Preti M, Beheshti M, Dugnani R (2021) Self-centering walls strengthening by high-performance concrete: a feasibility study. Mater Struct 54:117
34. Yaghoubi ST, Deger ZT, Taskin G, Sutcu F (2023) Machine learning-based predictive models for equivalent damping ratio of RC shear walls. Bull Earthq Eng 21:293–318
35. Dazio A, Beyer K, Bachmann H (2009) Quasi-static cyclic tests and plastic hinge analysis of RC structural walls. Eng Struct 31:1556–1571
36. Sengupta P, Li B (2014) Hysteresis behavior of reinforced concrete walls. J Struct Eng 140:04014030
37. Deger ZT, Basdogan C (2019) Empirical expressions for deformation capacity of reinforced concrete structural walls. ACI Struct J 116
38. Orakcal K, Massone LM, Wallace JW (2009) Shear strength of lightly reinforced wall piers and spandrels. ACI Struct J 106:455
39. Shegay A, Motter C, Henry R, Elwood K (2015) A database for investigating NZS3101 structural wall provisions. In: Proceedings on 10th Pacific conference on earthquake engineering. Australian Earthquake Engineering Society, Melbourne
40. Nie X, Wang J-J, Tao M-X, Fan J-S, Mo Y, Zhang Z-Y (2020) Experimental study of shear-critical reinforced-concrete shear walls under tension-bending shear-combined cyclic load. J Struct Eng 146:04020047
41. Wallace JW (2012) Performance of structural walls in recent earthquakes and tests and implications for US building codes. In: Proceedings of the 15th world conference on earthquake engineering
42. Yang W, Wang K, Zuo W (2012) Neighborhood component feature selection for high-dimensional data. J Comput 7:161–168
43. Goldberger J, Hinton GE, Roweis S, Salakhutdinov RR (2004) Neighbourhood components analysis. Adv Neural Inf Proces Syst 17
44. Blandon CA, MJN P (2005) Equivalent viscous damping equations for direct displacement based design. J Earthq Eng 09:257–278
45. Engineers ASoC (2017) Seismic evaluation and retrofit of existing buildings. American Society of Civil Engineers
46. Oesterle R, Fiorato A, Corley W (1980) Reinforcement details for earthquake-resistant structural walls. Portland Cement Association, Skokie
47. Hube MA, Marihuén A, de la Llera JC, Stojadinovic B (2014) Seismic behavior of slender reinforced concrete walls. Eng Struct 80:377–388
48. Sittipunt C, Wood SL (1995) Influence of web reinforcement on the cyclic response of structural walls. SJ 92:745–756
49. Hidalgo PA, Ledezma CA, Jordan RM (2002) Seismic behavior of squat reinforced concrete shear walls. Earthquake Spectra 18:287–308
50. Tasnimi AA (2000) Strength and deformation of mid-rise shear walls under load reversal. Eng Struct 22:311–322

51. Jafari A, Shahmansouri AA, Taslimi A, Naser M, Zhou Y (2025) Shear capacity of FRP-RC beams: XML modeling and reliability evaluation. Elsevier, Structures, p 110150
52. Jafari A, Mollaali M, Ma L, Shahmansouri AA, Zhou Y, Dugnani R (2025) Brittle fracture strength prediction via XML with reliability considerations. Eng Fract Mech 111555
53. Jafari A, Mirrashid M, Hoult RD, Zhou Y (2025) Plastic hinge length of RC shear walls: practical approximation via machine learning and probabilistic assessment. Eng Fail Anal 169:109179
54. Shahmansouri AA, Jafari A, Bengar HA, Zhou Y (2025) Steel rebar bond-slip behavior in CFRP-strengthened low-strength concrete: ML-driven solution and reliability analysis. Constr Build Mater 490:142302

Smart Sensing with AI-Based Control Strategies for Responsive Façade Monitoring

Amir Faraji, Elnaz Ghasemi, Shima Homayoon Arya, Tahereh Meydani Haji Agha, and Bahar Mehdizadeh

Abstract The increasing complexity of modern buildings and the growing emphasis on energy efficiency, occupant comfort, and environmental sustainability have driven the evolution of façades from static barriers to intelligent, responsive systems. This paper explores the integration of smart sensing technologies and AI-based control strategies in the monitoring and operation of responsive façades. It provides a comprehensive review of sensing modalities, including environmental, optical, occupancy, and structural sensors, and examines their deployment within IoT-enabled networks for real-time data acquisition and performance optimization. Advanced AI techniques, such as machine learning, reinforcement learning, and predictive modeling, are discussed in the context of adaptive control for shading, glazing, ventilation, energy generation, and façade maintenance. Case studies demonstrate the benefits of AI-enhanced façades, including improved thermal and visual comfort, energy savings, and operational resilience. The paper also identifies current limitations, such as sensor reliability, computational demands, and economic barriers, and highlights future directions involving digital twins, edge AI, and self-learning façades. By unifying sensing, computation, and actuation, AI-driven responsive façades represent a transformative approach to high-performance, sustainable building envelopes.

A. Faraji · E. Ghasemi · S. H. Arya
Construction Project Management Department, Faculty of Architecture, Khatam University, Tehran, Iran
e-mail: a.faraji@khatam.ac.ir; a.faraji@westernsydney.edu.au

E. Ghasemi
e-mail: el.ghasemi@khatam.ac.ir

S. H. Arya
e-mail: sh.homayoon@khatam.ac.ir

A. Faraji · B. Mehdizadeh (✉)
School of Engineering, Design and Built Environment, Western Sydney University, Penrith, NSW, Australia
e-mail: B.miyandehi@westernsydney.edu.au

T. M. H. Agha
Architecture Department, Faculty of Architecture, Khatam University, Tehran, Iran
e-mail: t.meydani@khatam.ac.ir

A. Azizi et al. (eds.), *Intelligence at the Interface*, Emerging Trends in Mechatronics,
https://doi.org/10.1007/978-981-95-7584-8_7

Keywords Smart sensing systems · AI-based control strategies · Responsive building façades · Façade performance monitoring · Intelligent building envelopes · Real-time data analytics

1 Introduction

A building façade serves as a vital interface between indoor and outdoor environments, playing a central role in shaping thermal comfort, visual quality, structural performance, and architectural expression. Since façades are in direct dialogue with their climatic, cultural, and functional context, they have historically exhibited a wide range of styles, materials, and technological configurations. Over the years, methods and techniques used in façade design have changed significantly, reflecting the evolving relationship between architecture and technology [1]. Early façade design relied heavily on manual calculations, craftsmanship, and empirical material knowledge. During the twentieth century, the introduction of computer-aided design (CAD) revolutionized architectural practice by enabling greater geometric complexity and precision. The subsequent development of building information modeling (BIM) further advanced façade design through collaborative, data-rich, and multidisciplinary workflows [2]. More recently, the integration of artificial intelligence (AI), machine learning, and big data has marked a new shift toward highly intelligent and performance-driven design strategies [3].

Traditional static façades, however, often exhibit limited energy and daylight performance under varying environmental conditions [4]. Their inherent limitations, such as high rates of heat transfer through conduction, radiation, and convection; excessive daylight admission; moisture migration; and restricted aesthetic modulation, pose significant challenges in achieving optimal comfort and energy efficiency [5]. These constraints underscore the need for façades that can adapt dynamically to changing outdoor and indoor conditions [6].

Over the past 70 years, advancements in computer science, electrical control systems, AI, cybernetics, and material science have initiated a paradigm shift in façade design [7]. These technological developments enabled the emergence of dynamic façade systems capable of modifying their functions, behaviors, and physical properties over time. Such systems support richer interactions between occupants, façades, and their surrounding environment, enhancing both environmental performance and user experience [8]. The introduction of dynamic components and sensors has demonstrated substantial performance benefits; studies indicate that advanced technological interventions can improve active façade performance by 40–65% compared to traditional static envelopes [9].

Control technologies, in particular, provide essential attributes that define responsive façade systems, including flexibility [10], adaptability [11], multifunctionality [12], scalability [13], controllability, and predictability [14]. These attributes allow façades to intelligently regulate daylight levels, glare, lighting energy use, thermal resistance, solar heat gain, passive ventilation, and occupant comfort [15]. By overcoming the limitations of static systems, responsive façades can achieve up to a 20% reduction in carbon emissions, 50% energy savings, and a 30% improvement in visual comfort [16]. Additional research shows that implemented responsive technologies can reduce cooling loads by 25%, decrease heat gain by up to 80%, and lower cooling costs by 15–20% [8]. Moreover, intelligent daylight modulation has been shown to reduce artificial lighting demand by approximately 11%. These improvements are facilitated by the integration of sensors, actuators, and transducers into façade materials and structural components, enabling continuous monitoring and adaptive control [17].

Within this evolutionary trajectory, AI technologies, including machine learning, computer vision, predictive analytics, and optimization algorithms, have become powerful tools for enhancing façade intelligence [18]. It should be highlighted that machine learning models have been successfully employed in engineering and construction-related applications to predict and optimize complex material and performance characteristics [19–24]. By enabling pattern recognition, learning from experience, and autonomous decision-making, AI supports the operation, inspection, and optimization of façade systems across domains such as building physics, structural engineering, and architectural design [25]. Modern buildings increasingly rely on such technologies to achieve higher levels of energy efficiency, occupant well-being, operational flexibility, and environmental performance [26].

As a result, façades have evolved from static building skins into responsive, intelligent systems capable of adjusting their optical, thermal, and geometric characteristics in real time. These systems depend on integrated sensing networks, data processing pipelines, and AI-driven control strategies [27]. This paper explores the interplay between smart sensing technologies and AI-based control approaches for monitoring and managing responsive façades, highlighting their transformative potential for future building envelopes.

2 Responsive Façades: Concepts and Evolution

2.1 From Static to Responsive Systems

Early façade designs functioned primarily as passive environmental modifiers, fixed barriers separating indoor and outdoor conditions [28]. Their performance relied almost entirely on material properties and geometric configuration, offering limited capacity to respond to fluctuating climatic stimuli. Such static façades could provide basic protection and enclosure, but they lacked the ability to adapt to changing solar radiation, temperature variations, or evolving occupant needs. As a result, their effectiveness often varied significantly across seasons, times of day, and geographical contexts [29].

The limitations of static envelopes gradually motivated the exploration of more adaptive strategies. Beginning in the mid-twentieth century, architects and engineers introduced operable components such as adjustable louvers, shading devices, vents, and movable panels. Although these elements increased flexibility, they were still largely dependent on manual operation or simple preprogrammed mechanisms. Nevertheless, this shift marked an important conceptual transition: façades were no longer seen as fixed surfaces but as systems with the potential to change form or behavior in response to external conditions [30].

With advancements in materials science, computer-aided engineering, and cybernetic thinking, kinetic façade systems emerged as a new category of dynamic interfaces. These systems employed mechanical movement, variable geometry, or transformable materials to modulate solar exposure, daylight penetration, and air flow. Kinetic façades demonstrated that performance-oriented transformations could be embedded directly into the building envelope, enabling more precise control over environmental quality. However, most early kinetic systems still lacked real-time intelligence; their responses were typically predefined rather than contextually optimized [31].

Responsive façades represent the next major evolutionary step in this progression. Unlike static, manually adjustable, or purely kinetic systems, responsive façades integrate sensing, computation, and actuation into a unified, autonomous framework. Embedded sensors continuously monitor environmental parameters such as solar irradiance, temperature, wind velocity, humidity, and occupancy patterns [32]. Computational engines, often incorporating AI, predictive models, and multi-objective optimization, interpret these data streams, generating control strategies that balance competing performance goals. Actuators then execute these decisions by adjusting shading elements, modifying optical properties, opening ventilation pathways, or altering geometric configurations [33].

Through this closed-loop feedback mechanism, responsive façades achieve a level of environmental intelligence unattainable in earlier façade types. They can regulate daylight to reduce glare, optimize thermal transfer to minimize heating, ventilation, and air conditioning (HVAC) loads, enhance natural ventilation, and maintain visual comfort based on real-time conditions. Moreover, their capacity to learn from historical data and anticipate future conditions positions them as key components in high-performance, low-energy, and user-centric buildings [34].

As buildings strive for greater sustainability, adaptability, and resilience, the transition from static to responsive façades marks a critical paradigm shift, one in which the envelope becomes an active contributor to energy efficiency, environmental quality, and occupant well-being [35].

2.2 Functional Objectives of Responsive Façades

Before evaluating the specific functions that responsive façades are designed to achieve, it is essential to understand the broad performance agenda that drives their development. Unlike traditional static envelopes, responsive systems operate as intelligent environmental mediators—continuously sensing external conditions, interpreting data, and adjusting their configuration to maintain optimal indoor quality. These façades balance multiple, sometimes competing, objectives such as thermal regulation, glare control, natural ventilation, and energy efficiency. By leveraging embedded sensors and AI-driven control logic, they can adapt in real time to changing weather patterns, occupancy behavior, and building energy demands [36]. Table 1 summarizes the core functional objectives that guide the design and operation of responsive façade systems.

Table 1 Functional objectives of responsive façades

Objective	Description	Key responsive strategies	References
Thermal comfort optimization	Regulates heat transfer by controlling solar gains and airflow to maintain comfortable indoor temperatures	Dynamic shading, adjustable louvers, smart glazing, automated ventilation panels	[37, 38]
Visual comfort enhancement	Maintains optimal daylight levels while minimizing glare and visual disturbances	Electrochromic glass, responsive shading devices, light-redirecting systems, AI-based glare detection	[39, 40]
Energy demand reduction	Lowers cooling/heating loads and decreases peak electrical consumption through adaptive façade behavior	Predictive control algorithms, occupancy-based actuation, integration with HVAC systems	[41]
Active ventilation and air quality management	Enhances indoor environmental quality by regulating natural or hybrid ventilation based on real-time air parameters	Motorized openings, CO_2 and VOC sensors, adaptive airflow optimization	[3, 38]
On-facade energy generation	Supports sustainable energy production by incorporating active systems such as photovoltaics	Building-integrated photovoltaics (BIPV), solar tracking mechanisms	[37, 42]
User control and personalization	Allows occupants to interact with or override façade behavior for customized comfort settings	Mobile app interfaces, local control panels, adaptive user-preference algorithms	[42, 43]
Data-driven performance management	Ensures continuous performance improvement through sensing and data interpretation	AI analytics, sensor fusion, machine learning-based system tuning	[3, 44]

2.3 Gap Analysis

The comparative review of previous studies on smart sensing and AI-based control strategies for responsive façade monitoring reveals both notable advancements and persistent limitations in the current literature. While significant progress has been made in environmental sensing, AI-assisted analysis, and façade performance assessment, these components are often investigated in isolation rather than within integrated, real-time control frameworks. Many existing approaches rely on rule-based or predefined optimization methods, with limited implementation of adaptive or self-learning AI models capable of evolving under dynamic environmental and occupancy conditions. Moreover, long-term operational performance, interoperability with building management systems, and occupant-centered feedback mechanisms remain underexplored. Collectively, these gaps emphasize the need for holistic, scalable, and intelligent façade monitoring systems that unify sensing, AI-driven control, and lifecycle performance management. Table 2 presents a synthesized gap analysis of the reviewed studies, highlighting key research focuses and unresolved challenges.

2.4 Technologies Employed in Responsive Façade Systems

Responsive façades operate as integrated, intelligent systems that rely on the seamless interaction of control, sensing, actuation, structural, and material technologies. Each system plays a distinct role: sensing technologies detect environmental and occupant-related conditions, control technologies interpret these signals and generate appropriate responses, actuators execute the physical transformations, and materials and structural elements enable and support these adaptive behaviors. The combination of these systems allows responsive façades to continuously adjust their performance in real time, optimizing thermal comfort, visual quality, energy efficiency, and occupant satisfaction [1]. Table 3 summarizes the key systems, their classifications, and the primary functions they serve in responsive façades.

3 Smart Sensing Technologies for Façade Monitoring

3.1 Types of Sensors Used in Responsive Façades

Responsive façades rely on a diverse range of sensors to perceive environmental conditions, building performance, and occupant-related factors in real time. Environmental sensors are commonly used to measure external and internal parameters such as solar radiation, illuminance, temperature, humidity, wind speed, and air quality, providing essential inputs for shading, ventilation, and thermal control strategies [36]. Optical sensors, including light and glare sensors as well as cameras,

Table 2 Summary of previous studies on smart sensing and AI-based control for responsive façade

No.	References	Year	Article type	Façade focus	Sensing technologies	AI/Digital techniques	Key contributions
1	[45]	2018	Review	Adaptive façade monitoring	Multi-sensor systems	Control logic, early AI approaches	Identifies technical, AI, and operational challenges
2	[1]	2019	Comparative review	Responsive and adaptive façades	Environmental sensors (solar, temperature, daylight)	Limited automation, rule-based control	Classifies responsive façade technologies and sensing mechanisms
3	[46]	2023	Perspective chapter	Data-driven façades	Façade performance sensors	Data analytics, early AI reasoning	Emphasizes data-informed façade design and feedback loops
4	[47]	2024	Methodological study	Smart-material façades	Material performance data (simulated/ experimental)	AI-driven decision-making, ML-based selection	Introduces AI for intelligent material selection in façades
5	[48]	2024	Conceptual framework	Active, multifunctional façades	IoT sensor networks (environmental and occupancy)	IoT platforms, AI analytics	Highlights IoT–AI synergy for dynamic façade performance
6	[49]	2024	Applied AI study	Façade inspection and maintenance	Vision sensors, image datasets	Computer vision, deep learning (CNNs)	Demonstrates AI-based automated defect detection
7	[50]	2025	Conceptual framework	AI-ready adaptive façades	Anticipated sensor infrastructures	Conceptual AI integration models	Establishes design principles for AI-compatible façades
8	[3]	2025	Systematic review	Smart and intelligent façades	Sensor-based datasets	ML, ANN, GA, optimization algorithms	Comprehensive overview of AI methods applied to façade design and control

Table 3 Technological systems in responsive façades

Category	Summary description	Sub-types/classifications	References
Control technologies	Core decision-making systems that interpret environmental and user data and translate them into actuation commands Performance depends on algorithmic complexity, learning ability, predictability, and user integration	• Extrinsic versus intrinsic control (external controller versus material-based self-regulation) • Open-loop (passive) versus closed-loop (active) • Centralized versus distributed control • Control mechanisms: Homeostasis, reflexive, emergent • Complexity levels: Direct control, reactive control, system-based/ AI control • Advanced types: Multi-input, ubiquitous multi-input, intelligent, heuristic	[10, 12, 51–53]
Sensing technologies	Systems that detect climatic, indoor, or occupant-related conditions and convert physical stimuli into measurable electrical signals for the control unit	• Types of sensors: Temperature, humidity, solar radiation, wind, pressure, light, air quality, occupancy • Sensor integration: Embedded in hard body, applied to soft body, or material-as-sensor • Sensing levels: Environmental boundary conditions, indoor conditions, occupant preference • Evolution: No-sensor systems → central sensors → distributed sensors → material sensors • Usage of smartphones for decentralized user control	[10–12, 53]

(continued)

Table 3 (continued)

Category	Summary description	Sub-types/classifications	References
Actuating technologies	Components that convert signals into mechanical or material movements, enabling deformation or reconfiguration of façade elements	• Types of actuators: Mechanical, electrical, pneumatic, hydraulic, material-based, self-actuating • Motion types: Translational, rotational, folding, sliding, expanding (inflating/deflating), morphing • Smart material actuation: Shape memory alloys/ polymers, electroactive polymers, phase-change materials • Active versus passive actuation	[8, 53–56]
Structural technologies	Structural systems that support or integrate responsive elements, evolving from secondary skins to lightweight self-supporting envelopes	• Cable networks & cable structures • Structural glass façades, curtain walls • Space frames • Lightweight, single-layer, self-loading skins	[57, 58]
Material technologies	Materials that determine the façade's environmental responsiveness and physical transformation capabilities. Include rigid conventional materials and soft smart materials	• Rigid materials: Wood, metals, glass, polymers, composites soft smart materials: Chromic materials (photo-, thermo-, electrochromic), conductive polymers, shape memory alloys/ polymers, electroactive polymers, phase-change materials • Classification by behavior: Property-changing versus energy-changing • Shape-change versus shape-memory materials	[11, 35, 59–63]

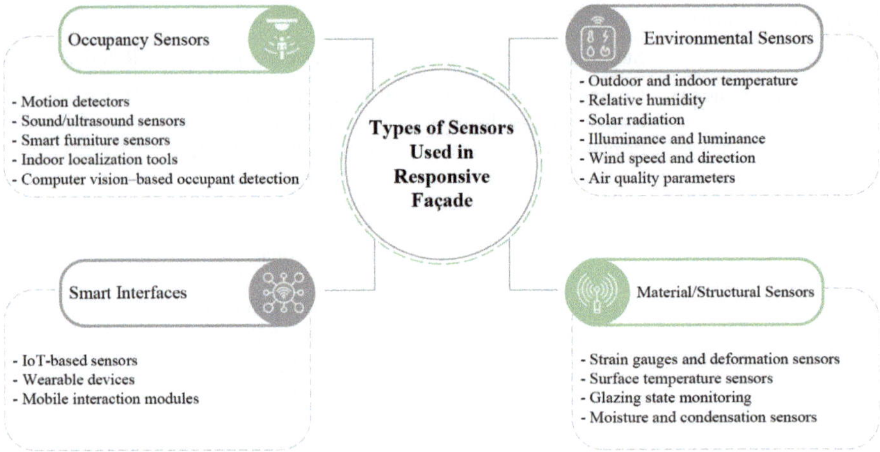

Fig. 1 Types of sensors used in responsive façades

support daylight regulation and visual comfort assessment [64]. Occupancy and user-related sensors, such as motion detectors, CO_2 sensors, and presence sensors, enable façades to respond dynamically to actual space usage and indoor environmental quality requirements [65]. In addition, structural and operational sensors, including strain gauges, vibration sensors, and actuator position sensors, are employed to monitor façade integrity, movement accuracy, and system health [66]. The integration of these heterogeneous sensor types forms the foundation for intelligent, adaptive façade behavior by supplying high-resolution, real-time data to control and AI-based decision-making systems [67]. Figure 1 highlights the categories of various sensors used in responsive façades.

3.2 *Sensor Networks and IoT Integration*

The integration of sensor networks with the Internet of Things (IoT) has emerged as a critical enabler for responsive façades, providing a robust infrastructure for continuous environmental monitoring and adaptive control [68]. In IoT-enabled façades, distributed sensing nodes are deployed across the building envelope, capable of measuring a wide range of parameters including temperature, humidity, solar radiation, daylight levels, air quality, and occupancy patterns. These nodes communicate with each other and with central or distributed control units via wireless communication protocols such as Zigbee, Bluetooth Low-Energy (BLE), LoRaWAN, Wi-Fi, or Narrowband IoT (NB-IoT). The choice of protocol depends on factors such as coverage area, data transmission rate, power consumption, and integration requirements with existing building systems [69].

The deployment of IoT-enabled sensor networks provides several operational advantages. Firstly, it enables real-time monitoring of both environmental conditions and façade performance, allowing the system to respond immediately to changing climatic or user-driven inputs. Secondly, many IoT nodes are designed for low-power operation, facilitating long-term deployment without frequent maintenance or battery replacement. Thirdly, the collected data can be transmitted to cloud-based analytics platforms, where advanced algorithms, including machine learning and predictive modeling, analyze patterns, detect anomalies, and optimize façade performance over time. Moreover, these networks can be seamlessly integrated with existing building management systems (BMS), creating a unified platform that connects lighting, HVAC, energy, and façade subsystems to optimize overall building efficiency. Finally, IoT-enabled monitoring supports predictive maintenance, allowing building managers to anticipate component wear, detect failures early, and schedule interventions before critical issues arise, reducing downtime and maintenance costs [70].

The incorporation of IoT and sensor networks significantly enhances a façade's adaptive capacity. Beyond immediate adjustments to lighting, shading, and ventilation, the system can also optimize long-term operational strategies based on historical performance data, seasonal patterns, and predictive analytics. This dual capability, reactive and anticipatory, enables responsive façades to maintain high levels of thermal comfort, visual quality, energy efficiency, and occupant satisfaction, while also contributing to the building's sustainability and operational intelligence. As a result, IoT integration transforms the façade from a passive envelope into an active, intelligent interface that continuously interacts with its environment and users [71].

IoT-enabled sensing plays a central role in transforming building façades into responsive and intelligent systems by enabling continuous interaction with environmental and operational conditions. Real-time monitoring allows façade elements to immediately adjust to variations in climate, daylight, and occupancy, while low-power operation ensures long-term functionality with minimal energy consumption [72]. Sensor data are processed through cloud-based analytics, supporting advanced performance assessment, optimization, and data-driven decision-making. Integration with Building Management Systems (BMS) enables coordinated control between the façade, HVAC, and lighting systems, improving overall building efficiency. Furthermore, predictive maintenance based on sensed data trends helps identify potential failures in advance, extending the façade's service life and reducing operational costs [73]. Figure 2 illustrates the multifaceted role of IoT-enabled sensing in responsive façade systems, highlighting how continuous data acquisition supports both immediate and long-term adaptability.

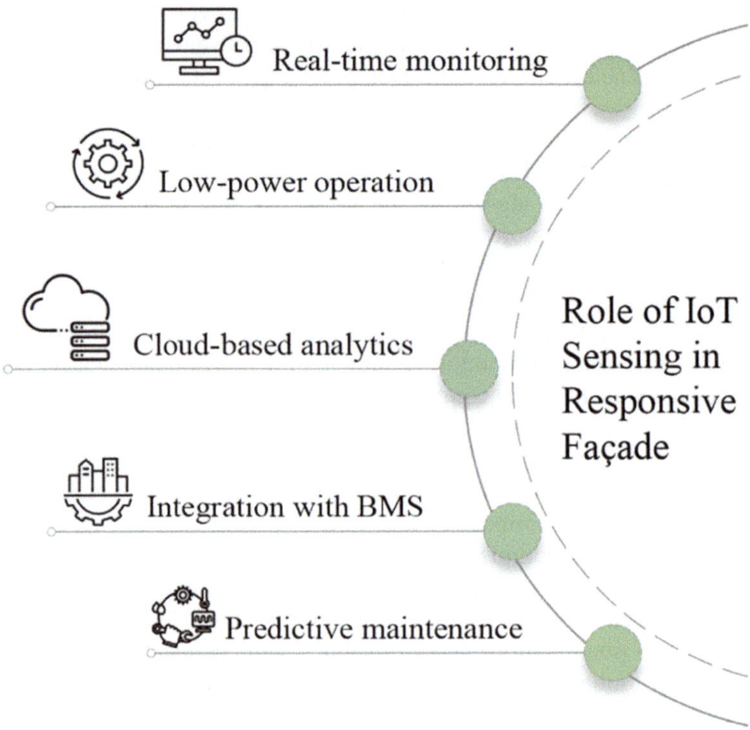

Fig. 2 Role of IoT sensing in responsive façades

4 Data Acquisition, Processing, and Fusion

Data serves as the cornerstone for AI-enabled façade systems, with the success of AI applications heavily dependent on both the accessibility and quality of the information collected. Within responsive façades, data can generally be divided into structured and unstructured categories. Structured data is systematically organized, often in tabular formats, making it straightforward to store, manage, and process using conventional databases. This includes numerical information such as material characteristics, dimensions of façade elements, or environmental parameters like temperature, humidity, and solar radiation. Categorical data involves classifications, for instance, energy performance ratings, material types, or building usage categories [74]. Time series data captures sequential measurements over time, encompassing environmental factors such as air quality and temperature, as well as structural health indicators like vibration or tilt. Spatial data, meanwhile, incorporates geolocated information including latitude, longitude, and elevation, which is essential for mapping façade performance across different geographic contexts [75].

In contrast, unstructured data does not conform to standard database schemas and requires specialized analytical methods. This type of data encompasses images and videos, such as high-resolution photographs, drone imagery, or street-level views of façades used for monitoring and inspection purposes. Audio data, including recordings of environmental or mechanical sounds, can provide supplementary insights into structural or operational conditions. Textual data, such as architectural documentation, historical records, maintenance logs, and sensor metadata, also falls into this category. By systematically acquiring, processing, and integrating both structured and unstructured datasets, AI algorithms are able to identify meaningful patterns, develop predictive models, and enhance the real-time operational performance of responsive façade systems [3].

4.1 Data Preprocessing

Raw sensor data collected from responsive façades require careful preprocessing to ensure reliability, consistency, and usability for AI-driven control systems. Sensor measurements are often noisy due to environmental variability, hardware limitations, or transmission errors, making it essential to apply filtering and smoothing techniques. Methods such as Kalman filters or moving averages are commonly used to reduce noise while preserving the underlying signal trends. Outlier detection is another critical step, as extreme or erroneous measurements can skew analyses and compromise control decisions; these outliers can be identified using statistical methods or machine learning-based approaches [76]. Once cleaned, data often undergo feature extraction, which involves deriving meaningful parameters from raw signals, such as daylight illuminance levels, solar heat gain coefficients, or occupancy patterns, that directly influence façade performance. Additionally, time-series normalization is applied to standardize the data range and temporal resolution, enabling accurate comparisons across multiple sensors and over different periods. Through these preprocessing steps, raw sensor readings are transformed into high-quality datasets suitable for advanced AI algorithms, ensuring robust prediction, decision-making, and façade responsiveness [45].

4.2 Sensor Data Fusion

Responsive façades often rely on multiple sensors capturing diverse environmental and occupant-related data streams. To generate a coherent and actionable understanding of the building context, these data streams must be fused into a unified representation [10]. Sensor data fusion combines information from different sources, enhancing the overall situational awareness and reliability of the façade system. Techniques for data fusion range from rule-based approaches, where predefined logic integrates signals, to probabilistic methods such as Bayesian inference, which accounts for uncertainty and varying sensor reliability [77, 78]. Machine learning-based fusion methods, including neural networks and gradient boosting, allow the system to learn complex patterns and interdependencies between sensor inputs, improving decision-making in dynamic environments [79]. Additionally, combining computer vision with environmental sensing enables façades to interpret visual information, such as glare or shading conditions, alongside quantitative measurements like temperature or irradiance. Effective data fusion significantly increases the accuracy and robustness of AI-driven control, particularly under complex or rapidly changing conditions, allowing responsive façades to maintain thermal comfort, optimize daylighting, and efficiently manage energy use [80].

5 AI-Based Control Strategies for Responsive Façades

AI-based control strategies enable responsive façades to move beyond static or rule-based operation and toward intelligent, adaptive performance. By leveraging machine learning, deep learning (DL), computer vision, and reinforcement learning (RL), these systems can interpret complex environmental inputs, anticipate future conditions, and optimize façade behavior in real time [81]. AI algorithms learn from historical and live data, such as illuminance levels, occupant presence, thermal patterns, and weather forecasts, to make informed decisions that balance competing objectives, including energy efficiency, visual comfort, thermal comfort, and daylight quality. Predictive models forecast solar radiation and demand patterns, while optimization techniques like genetic algorithms explore the most effective façade configurations across multiple criteria [82]. RL agents further enhance adaptability by refining control actions through continuous interaction with the environment. Together, these AI-driven approaches transform façades into intelligent, self-regulating systems capable of achieving performance levels unattainable through conventional control methods [83].

5.1 Rule-Based Versus AI-Driven Control

Traditional façade control systems are typically based on fixed, predefined rules such as "close the blinds when solar radiation exceeds a specified threshold." While these rule-based strategies are simple to implement and require minimal computational resources, they lack the ability to manage dynamic, nonlinear, or conflicting performance requirements. Such systems cannot adapt to variations in occupant behavior, changing weather patterns, or multi-objective demands, such as balancing daylight availability with thermal comfort and energy efficiency. As a result, rule-based control often leads to suboptimal or rigid façade behavior, limiting the overall performance of responsive building systems [41].

In contrast, AI-driven control strategies introduce adaptive, data-driven intelligence that allows façades to respond more effectively to complex and evolving conditions. By learning from historical and real-time data, AI models can recognize patterns, predict future environmental states, and make informed decisions that optimize façade operation across multiple performance metrics [84]. These systems integrate a range of computational methods, including supervised learning for predicting illuminance or thermal loads, unsupervised learning for identifying occupant behavior clusters or environmental patterns, and RL for autonomously discovering optimal control actions through continuous interaction with the environment. Genetic algorithms (GA) further enhance decision-making by searching for optimal façade configurations in multi-objective scenarios, while DL enables advanced pattern recognition from sensor data, images, or complex environmental datasets [85].

Together, these AI techniques allow responsive façades to move beyond rigid rule-based actions and evolve into intelligent, predictive, and self-optimizing systems capable of delivering superior comfort, energy performance, and user adaptability.

5.2 Reinforcement Learning for Façade Control

RL offers a powerful framework for enabling façades to autonomously determine optimal control strategies through continuous interaction with their environment. Unlike rule-based or supervised learning approaches, RL does not rely on predefined input–output pairs. Instead, an RL agent learns by exploring different actions, receiving feedback in the form of rewards or penalties, and gradually improving its policy to maximize long-term performance. This makes RL particularly well suited for building façade systems, where the relationship between variables, such as solar irradiance, occupant comfort, energy demand, and outdoor climate, is highly dynamic and often nonlinear [86].

In practical façade applications, RL has been used to govern a variety of adaptive components. For instance, shading devices can learn when and how to adjust their position to balance daylight availability with solar heat gain reduction [87]. Electrochromic glazing systems can dynamically modify their tint levels to regulate glare and thermal loads without relying on static thresholds. RL-based controllers can also manage natural ventilation openings or hybrid ventilation systems, determining when to allow outdoor air to enter based on indoor air quality, temperature, and weather forecasts. Because these decisions evolve through experience rather than fixed programming, RL systems continually refine their behavior as environmental patterns or occupant preferences change [82].

One of the most significant advantages of RL in façade control is its ability to handle conflicting performance objectives. Buildings frequently encounter trade-offs, for example, maximizing daylight penetration while simultaneously minimizing cooling loads, or enhancing thermal comfort without increasing energy consumption. Traditional control mechanisms struggle with these competing priorities, but

RL naturally accommodates them by optimizing cumulative reward functions that balance multiple criteria. As a result, RL enables façade systems to operate with higher efficiency, adaptability, and resilience, making it one of the most promising approaches for next-generation intelligent façades [88].

5.3 Predictive Control Using Machine Learning

Predictive control strategies leverage machine learning models to forecast future environmental and operational conditions, allowing façade systems to make proactive rather than reactive adjustments [89]. By analyzing historical data, ML algorithms can anticipate variations in solar angle and radiation, shifts in weather patterns, fluctuations in occupant presence, and upcoming changes in building energy demand. These predictions enable the façade to prepare in advance for periods of high solar gain, increased glare risk, or cooling and heating loads, ultimately improving comfort and reducing energy consumption. When integrated into a model predictive control (MPC) framework, ML-based forecasts guide the optimization process by providing a forward-looking view of system behavior [90]. The MPC continuously evaluates multiple future scenarios and selects the façade actions that achieve the best overall performance across the prediction horizon. This combination of ML forecasting and MPC decision-making allows the façade to operate with greater intelligence, stability, and efficiency, ensuring that shading devices, glazing systems, and ventilation elements respond not only to current conditions but also to those expected in the near future [91].

5.4 Genetic Optimization for Multi-objective Control

Genetic algorithms (GAs) are powerful evolutionary optimization techniques well suited for managing the inherently multi-objective nature of responsive façade systems. These algorithms operate by simulating natural selection, iteratively evolving a population of candidate façade control strategies, such as shading louver angles, glazing tint levels, ventilation openings, or actuator speeds, toward increasingly optimal configurations [45]. In responsive façade applications, GAs are particularly effective because they can simultaneously optimize multiple, often competing objectives, including reducing energy consumption, enhancing visual and thermal comfort, maximizing occupant satisfaction, and prolonging material lifespan by minimizing excessive actuator movement or stress [10]. Unlike conventional optimization methods that may become trapped in local optima or struggle with nonlinear relationships, genetic algorithms efficiently explore a broad and complex design space [92], identifying high-performing solutions even under dynamic environmental conditions. Their robustness and flexibility make them especially suitable for façade systems with numerous interdependent parameters, enabling designers and control systems to generate balanced, high-performance responses that account for the full spectrum of environmental, operational, and human-centric criteria [93].

6 AI-Enhanced and Sensor-Integrated Façade Applications

Responsive façade technologies have increasingly moved from experimental prototypes to real-world implementations, where AI, sensor networks, and advanced materials collectively enhance building performance. These applications demonstrate how AI-driven strategies can optimize daylighting, thermal behavior, ventilation, and user comfort under dynamic environmental and occupancy conditions. The following case examples highlight key domains where intelligent façade systems have shown measurable benefits and continue to shape the future of high-performance building envelopes.

6.1 Electrochromic Glazing with AI Prediction

Electrochromic glazing systems, which modulate their tint in response to electrical stimuli, have been significantly enhanced through the integration of predictive machine learning algorithms. Instead of reacting solely to instantaneous sensor readings, AI models forecast upcoming glare levels, solar radiation intensity, cooling loads, and sky conditions using historical data, weather forecasts, and occupant behavior patterns [94]. These predictive controls allow the glazing to pre-emptively adjust its tint state to maintain visual comfort while reducing HVAC demand and preventing overheating. Research has shown notable improvements in daylight autonomy, substantial reductions in peak cooling loads, and smoother transitions between tint states, minimizing visual disturbances for occupants [95]. As a result, AI-augmented electrochromic façades demonstrate superior performance compared to conventional threshold-based or manual controls [96].

6.2 Responsive Shading Panels with Reinforcement Learning

Kinetic shading façades controlled by RL represent one of the most advanced applications of AI in building envelopes. RL agents learn optimal shading strategies by continuously interacting with real or simulated environments, evaluating the effects of each action on metrics such as daylight distribution, glare probability, cooling loads, and occupant comfort. Over time, the system autonomously converges toward control policies that outperform static or rule-based shading strategies, especially under highly variable conditions [97]. Case studies have shown that RL-driven shading panels can optimize façade performance on a minute-by-minute basis, adapting to cloud movement, changing solar angles, or occupancy fluctuations. This approach not only enhances energy efficiency but also supports personalized comfort by responding to user feedback and preference patterns [98].

6.3 Sensor-Enhanced Natural Ventilation

AI-enabled natural ventilation systems leverage comprehensive sensor networks to regulate façade openings, including operable windows, louvers, or vents, based on real-time and predicted indoor and outdoor conditions. These systems process data such as indoor CO_2 concentration, temperature, humidity, occupancy density, and anticipated weather patterns to determine optimal ventilation actions [43]. By predicting wind speed, external air quality, and upcoming thermal loads, AI can modulate opening sizes to ensure effective cross-ventilation while preventing discomfort or heat loss. Buildings equipped with such intelligent systems have demonstrated significant reductions in mechanical HVAC usage, improved indoor air quality, and enhanced occupant well-being. Moreover, sensor-enhanced ventilation strategies contribute to resilience by enabling passive cooling during peak energy periods or system failures, further reinforcing the value of AI-driven adaptive façades [99].

6.4 AI-Optimized Photovoltaic Façades

BIPV can benefit from AI algorithms that dynamically optimize panel orientation, tilt angle, or façade segment activation to maximize solar energy yield [100]. Machine learning models forecast irradiance patterns, shading duration, and seasonal variations, enabling façade-integrated PV modules to adjust their configuration or prioritize energy flow. This leads to higher annual energy production and improved grid interaction through predictive load management [101].

6.5 Dynamic Thermal Mass and Phase-Change Materials (PCM) Control

AI systems can regulate how thermal storage materials integrated into façades absorb or release heat [102]. By predicting temperature swings and occupancy trends, machine learning models determine the ideal timing for charging or discharging thermal mass or activating PCM-based ventilation strategies. This enhances thermal comfort, delays peak HVAC loads, and reduces energy consumption [38].

6.6 AI-Assisted Façade Maintenance and Fault Detection

Computer vision and DL enable automated façade inspection using drones, cameras, or fixed imaging sensors. These systems identify cracks, material degradation, moisture damage, or mechanical failures in shading devices long before they become critical. Predictive maintenance models estimate component lifespan, reducing operational costs and improving long-term façade reliability [103].

6.7 Smart Daylighting Redirection Systems

Adaptive daylighting elements, such as heliostats, light-redirecting louvers, or prismatic panels, use AI-driven control to maintain optimal daylight distribution throughout interior spaces. Real-time data on cloud movement, sun position, and indoor illuminance conditions allow the system to reposition elements to reduce glare while maximizing natural light penetration [97].

6.8 AI-Controlled Acoustic Façades

In noise-sensitive urban environments, responsive façade elements equipped with smart materials or movable panels modulate acoustic transmission. AI models analyze real-time noise profiles, identify patterns (e.g., traffic peaks, construction noise), and adjust façade components to maintain comfortable sound levels indoors [104].

6.9 Adaptive Green Façades and Bio-responsive Systems

AI algorithms can manage irrigation, shading, and plant growth patterns in living façades. Sensors track soil moisture, solar radiation, temperature, and plant health indicators, adjusting water supply and shading to optimize vegetation performance [105]. This enhances air quality, thermal buffering, and aesthetic value [106].

6.10 AI-Supported Fire and Safety Management in Façades

Intelligent façades can respond dynamically to emergencies. Sensor fusion systems detect smoke, temperature anomalies, or hazardous gases, enabling façade units to open for smoke extraction, close for compartmentalization, or guide occupants through visual signaling. AI enhances response speed and reduces human error in life-safety situations [3].

6.11 Morphing Façades with Shape-Memory Alloys and Smart Materials

Façades integrating shape-memory alloys (SMA), electroactive polymers (EAP), or thermochromic/photovoltaic nanomaterials use AI to regulate their behavior based on environmental cues [107]. Machine learning models optimize transitions in material states to adjust shading, insulation, and ventilation [42].

7 Case Studies

The following table presents selected case studies where AI-enhanced and sensor-integrated façade systems have been implemented in real-world buildings and experimental testbeds. These examples illustrate a range of façade technologies, from kinetic shading and adaptive louvers to smart glazing and modular photovoltaic panels, all equipped with sensors to monitor environmental and occupancy conditions. By integrating AI-based control strategies, including predictive algorithms, machine learning, RL, and optimization techniques, these systems dynamically adjust to changing conditions, improving daylighting, thermal comfort, energy efficiency, and overall occupant satisfaction. Table 4 highlights the diversity of approaches and measurable outcomes achieved across different geographic and institutional contexts.

8 Challenges and Limitations

Despite the significant potential of AI-enhanced and sensor-integrated façades, the challenges below highlight the gap between experimental research and widespread adoption. Technical issues, such as sensor reliability and real-time data processing, can limit the consistent performance of intelligent façades, while AI-related concerns, including model transparency and computational demands, may hinder user trust and practical implementation. Economic and operational barriers, from high upfront costs to the need for specialized expertise, further constrain scalability. Addressing these limitations requires a holistic approach that combines robust sensor networks, efficient AI algorithms, cost-effective materials, and standardized operational frameworks to ensure that responsive façades can deliver reliable, sustainable, and occupant-centered performance in real-world buildings. Figure 3 summarizes the

Table 4 Case studies of responsive façades

Project		Location	Façade technology	Sensors/automation	AI/Control strategy	Purpose
1	Al Bahr towers	Abu Dhabi, UAE	Kinetic mashrabiya shading façade	Solar radiation sensors, daylight sensors	Automated tracking algorithms (sun-path based)	Reduce solar gain, improve shading performance, lower cooling load
2	Media-TIC building	Barcelona, Spain	ETFE cushion façade with automated pneumatic control	Irradiance, temperature, indoor light sensors	Rule-based + predictive façade management system	Dynamic envelope responding to climate, reducing cooling demand
3	Tower at PNC plaza	Pittsburgh, USA	Automated double-skin façade with operable floppers/poppers	Environmental sensors controlling façade elements	Automated façade ventilation control	Enhances natural ventilation, thermal comfort, and energy performance
4	The edge	Amsterdam, Netherlands	Smart building envelope with sensor-integrated systems (blinds, lighting)	~28,000 IoT sensors (light, temp, occupancy, etc.)	Data-driven optimization integrated with BMS	Enhances energy efficiency, occupant comfort, daylighting & 5HVAC optimization
5	Council house 2 (CH2)	Melbourne, Australia	Kinetic shading façade with automated louvers	Environmental sensors (sun position, temp)	Automated control based on environmental input	Reduces solar heat gain, enhances daylight and natural ventilation
6	Institut du monde Arabe	Paris, France	Motor-controlled aperture façade	Light-sensitive diaphragms/motor actuators	Pre-programmed automatic response to sun	Controls light intrusion, cultural expression
7	Bund finance centre	Shanghai, China	Kinetic "veil" façade with layered metal panels	Motorized layers adjust opacity	Automated adjustment based on sunlight & interior conditions	Dynamic façade improves occupant comfort, shading, and aesthetic impact

(continued)

Table 4 (continued)

Project		Location	Façade technology	Sensors/automation	AI/Control strategy	Purpose
8	Milwaukee art museum (Quadracci pavilion)	Milwaukee, USA	Movable sun-shield/ brise-soleil	Actuators respond to sunlight and wind	Automatic mechanical movement control	Reduces solar gain, enhances visitor comfort
9	SDU Kolding	Kolding, Denmark	Climate-responsive triangular panels	Temperature and light sensors	Automated control responding to heat/light	Optimizes daylight penetration and solar shading
10	Kiefer technic showroom	Austria	Motorized adjustable façade panels	Sun position sensors, interior temp	Automated panel adjustment for shading and comfort	Reduces heat gain, improves daylighting
11	Lisbon wood	Lisbon, Portugal	Kinetic louvres	Sun and privacy sensors	Automated louvre control	Adaptive solar shading, visual privacy
12	Hedge building	Rostock, Germany	Responsive vegetative screens	Environmental sensors (temperature, sunlight)	Automated adjustment based on environment	Microclimate control, shading, occupant comfort

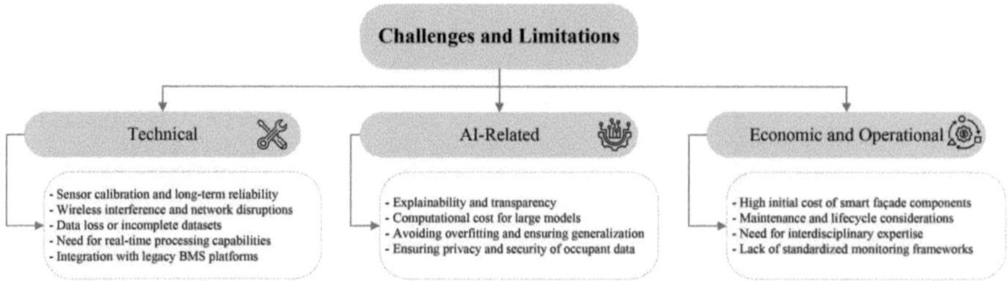

Fig. 3 Challenges and imitations of smart sensing with AI-based control strategies in responsive façades

challenges and limitations of smart sensing with AI-based control strategies in responsive façades.

9 Future Directions

The future of responsive façades lies in the convergence of advanced sensing, AI, and energy systems, enabling buildings to become truly adaptive and self-optimizing. Digital twins will provide continuous virtual feedback, allowing designers and operators to anticipate performance issues and test optimization strategies before implementation [108]. Edge AI will enable real-time decision-making on-site, minimizing latency and enhancing responsiveness to changing environmental and occupancy conditions. Self-learning façades will leverage historical data, climate patterns, and user interactions to continuously refine their behavior, while integration with renewable energy sources and storage systems will allow façades to actively contribute to a building's energy balance. Together, these innovations point toward highly resilient, energy-efficient, and occupant-centric building envelopes. Figure 4 illustrates the future directions of responsive façades integrating AI, advanced sensing, digital twins, and renewable energy systems for adaptive, self-optimizing building envelopes.

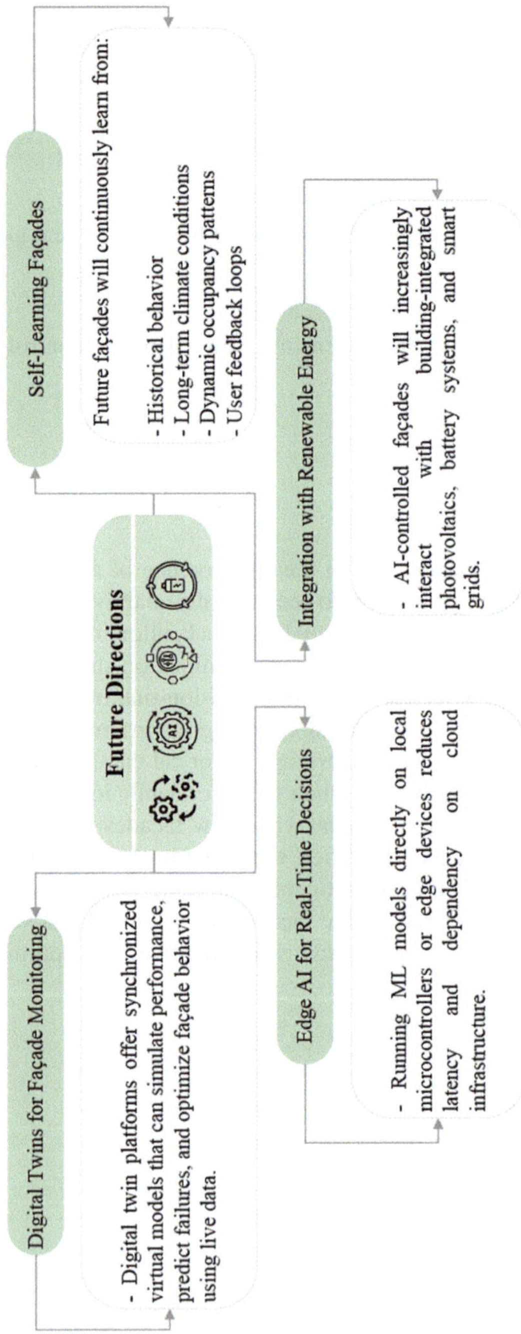

Fig. 4 Future directions of responsive façades

10 Conclusion

Responsive façades equipped with smart sensors and AI-based control strategies represent a significant advancement in architectural and building technology, enabling façades to actively respond to dynamic environmental and occupancy conditions. By integrating heterogeneous sensing networks with machine learning, reinforcement learning, predictive control, and optimization algorithms, these systems achieve superior thermal comfort, visual quality, energy efficiency, and occupant satisfaction compared to traditional static or rule-based façades.

Real-world case studies highlight the practical benefits of AI-driven façades in energy management, daylight optimization, natural ventilation, and maintenance efficiency. However, challenges remain in ensuring sensor reliability, computational efficiency, model transparency, and economic feasibility. Future developments, including edge AI, digital twins, self-learning systems, and integration with renewable energy sources, are poised to overcome these limitations, paving the way for fully adaptive, self-optimizing building envelopes. Ultimately, AI-enhanced responsive façades exemplify the convergence of architecture, technology, and sustainability, marking a paradigm shift in how buildings interact with their environment and occupants.

AI Use Disclosure Statement The authors did not use any generative AI or AI-assisted technologies in the creation of this manuscript.

References

1. Heidari Matin N, Eydgahi A (2022) Technologies used in responsive facade systems: a comparative study. Intell Build Int 14(1):54–73
2. Faraji A, Homayoon Arya S, Ghasemi E, Rahnamayiezekavat P, Perera S (2024) Building information modeling (BIM), blockchain, and LiDAR applications in construction lifecycle: bibliometric, and network analysis. Buildings 14(4):919
3. Duran A, Waibel C, Piccioni V, Bickel B, Schlueter A (2025) A review on artificial intelligence applications for facades. Build Environ 269:112310
4. Selkowitz S, Aschehoug O, Lee ES (2003) Advanced interactive facades-critical elements for future green buildings?
5. Kim K, Jerratt C (2011) Energy performance of an adaptive façade system. J Archit Res 179–186
6. Sørensen LS (2013) Heat transmission coefficient measurements in buildings utilizing a heat loss measuring device. Sustainability 5(8):3601–3614
7. Pask G (2018) The architectural relevance of cybernetics
8. Kolarevic B, Parlac V (2015) Building dynamics: exploring architecture of change. Routledge
9. Dewidar K, Mahmoud AH, Magdy N, Ahmed S (2010) The role of intelligent façades in energy conservation. In: International conference on sustainability and the future: future intermediate sustainable cities (FISC 2010)
10. Loonen RCGM, Favoino F, Hensen JLM, Overend M (2017) Review of current status, requirements and opportunities for building performance simulation of adaptive facades. J Build Perform Simul 10(2):205–223

11. Aelenei D, Aelenei L, Vieira CP (2016) Adaptive façade: concept, applications, research questions. Energy Procedia 91:269–275
12. Moloney J, Globa A, Wang R (2016) Hybrid environmental-media facades. KnE Eng 190–196
13. Decker M (2013) Emergent futures: nanotechology and emergent materials in architecture. In: Conference of tectonics of teaching: building technology educators society (BTES). Rhode Island, Newport
14. Ogwezi B, Bonser R, Cook G, Sakula J, Happold B (2011) Multifunctional, adaptable facades. In: TSBE EngD conference. University of Reading
15. Faraji A, Homayoon Arya S, Rashidi M, Noroozinejad Farsangi E, Meydani Haji Agha T, Ghasemi E (2024) Critical considerations for maximizing decarbonization through prefabrication in the construction industry. In: International conference on construction applications of virtual reality. Springer, pp 107–122
16. Grobman YJ, Capeluto IG, Austern G (2017) External shading in buildings: comparative analysis of daylighting performance in static and kinetic operation scenarios. Archit Sci Rev 60(2):126–136
17. Adriaenssens S et al (2014) Dialectic form finding of passive and adaptive shading enclosures. Energies (Basel) 7(8):5201–5220
18. Karanouh A, Kerber E (2015) Innovations in dynamic architecture. J Facade Des Eng 3(2):185–221
19. Mehdizadeh B, Fakharian P, Nouri Y, Afrazi M, Samali B (2026) Machine learning-assisted analysis of fracture energy in externally bonded reinforcement on groove bond strength prediction. Buildings 16(5):1070
20. Ziaie A, Mehdizadeh B, Safi Jahanshahi F, Ahmadi N, Ghanizadeh AR (2026) Prediction of Liquefaction-Induced lateral displacements using hybrid GBRT and EOA. J Soft Comput Civil Eng 10(1):e225827
21. Shan F, He X, Armaghani DJ, Zhang P, Sheng D (2022) Success and challenges in predicting TBM penetration rate using recurrent neural networks. Tunn Undergr Space Technol 130:104728
22. Yong W, Zhou J, Jahed Armaghani D, Tahir MM, Tarinejad R, Pham BT, Van Huynh V (2021) A new hybrid simulated annealing-based genetic programming technique to predict the ultimate bearing capacity of piles. Eng Comput 37(3):2111–2127
23. Paji MK, Gordan B, Biklaryan M, Armaghani DJ, Zhou J, Jamshidi M (2021) Neuro-swarm and neuro-imperialism techniques to investigate the compressive strength of concrete constructed by freshwater and magnetic salty water. Measurement 182:109720
24. Hosseini S, Armaghani DJ, He X, Pradhan B, Zhou J, Sheng D (2024) Fuzzy cognitive map for evaluating critical factors causing rockbursts in underground construction: a fundamental study. Rock Mech Rock Eng 57(11):9713–9738
25. Faraji A, Homayoon Arya S, Ghasemi E, Abagheri Mahabadi H, Perera S (2025) Exploring the metaverse: integrating emerging technologies for Industry 5.0 advancement. Smart Sustain Built Environ 1–36
26. Gh SH, Miyandehi BM, Khotbehsara MM, Tarbiat M (2015) Study quality steel mill slag for use in concrete containing metakaolin. In: International conference on advances in engineering, Tehran, Iran
27. Faraji A, Ghasemi E, Arya SH, Mahabadi HA (2024) Prediction of effects of industry 5.0 concepts on construction 5.0. In: 3rd International conference on recent advances in engineering, innovation & technology [online]. https://www.researchgate.net/publication/385848718_Prediction_of_Effects_of_Industry_50_Concepts_on_Construction_50. 08 Dec 2024
28. Faraji A, Arya SH, Ghasemi E (2023) Investigation of the prospective developments in the architectural design process in the fourth industrial revolution
29. Velikov K, Thün G (2013) Responsive building envelopes: characteristics and evolving paradigms. In: Design and construction of high-performance homes. Routledge, pp 75–92
30. Cocho-Bermejo A (2025) Adaptive architectural facades: review 1985–2024. A comparative analysis of media-TIC, the Arab institute, and Al Bahar towers dynamic facades approaches. Nexus Netw J 27(4):925–945

31. Andrade T, Beirão JN, de Arruda AJV, Eysen C (2021) Toward adaptable and responsive facades: using strategies for transforming of the material and bio-based materials in favor of sustainability. Cuadernos del Centro de Estudios en Diseño y Comunicación. Ensayos 149:37–59
32. Brzezicki M (2021) A typology of adaptive façades. An empirical study based on the morphology of glazed facades. Cogent Arts Humanit 8(1):1960699
33. Loonen RCGM, Trčka M, Cóstola D, Hensen JLM (2013) Climate adaptive building shells: state-of-the-art and future challenges. Renew Sustain Energy Rev 25:483–493
34. Iqbal W et al (2025) Optimizing energy efficiency: louver systems for sustainable building design. Buildings 15(7):1183
35. Perino M, Serra V (2015) Switching from static to adaptable and dynamic building envelopes: a paradigm shift for the energy efficiency in buildings. J Facade Des Eng 3(2):143–163
36. Zhang X, Zhang H, Wang Y, Shi X (2022) Adaptive façades: review of designs, performance evaluation, and control systems. Buildings 12(12):2112
37. Ünlütürk MS (2024) Integration thermal and daylight performance of responsive facades: a comprehensive literature review. A+ Arch Des Int J Arch Des 10(2):114–130
38. Gonçalves M, Figueiredo A, Almeida R, Vicente R (2024) Dynamic façades in buildings: a systematic review across thermal comfort, energy efficiency and daylight performance. Renew Sustain Energy Rev 199:114474
39. Mehrvarz F et al (2024) Designerly approach to design responsive façade for occupant visual comfort in different latitudes. J Daylighting 11(1):149–164
40. Konstantoglou M, Tsangrassoulis A (2016) Dynamic operation of daylighting and shading systems: a literature review. Renew Sustain Energy Rev 60:268–283
41. Attia S, Bertrand S, Cuchet M, Yang S, Tabadkani A (2022) Comparison of thermal energy saving potential and overheating risk of four adaptive façade technologies in office buildings. Sustainability 14(10):6106
42. Baraboi D-R et al (2025) Smart 3D-printed facades: a review of innovations, materials, and sustainable performance. Front Sustain Cities 7:1610729
43. Luna-Navarro A, Loonen R, Juaristi M, Monge-Barrio A, Attia S, Overend M (2020) Occupant-facade interaction: a review and classification scheme. Build Environ 177:106880
44. Abu-Shaikha M (2025) AI-driven parametric facade design for adaptive architectural performance. In: 24th International scientific conference engineering for rural development
45. Attia S (2018) Challenges and future directions of smart sensing and control technology for adaptive facades monitoring. In: COST action TU1403–adaptive facades network final conference, Lucerne University of Applied Sciences and Arts. Lucerne, Switzerland
46. Ibrahim MR (2023) Perspective chapter: from data to design–leveraging façade sensors for intelligent architecture. In: Façade design-challenges and future perspective. IntechOpen
47. Amer N, Nagy G (2024) Smart materials for modern facades: an AI-powered selection process. In: IOP conference series: earth and environmental science, vol 1396. IOP Publishing, p 012014
48. Giovanardi M, Baietta A, Belletti F, Magnani S, Casadei O, Pracucci A (2024) Exploiting the value of active and multifunctional façade technology through the IoT and AI. Appl Sci 14(3):1145
49. Silva AS, Melo RRS, Costa DB (2024) Automated facade inspection: application and challenge in using artificial intelligence for construction defect recognition. In: XX international conference on building pathology and constructions repair, pp 705–718
50. Pantilimonescu F, Tudora G, Tudora AC (2025) Designing buildings with potential for artificial intelligence integration: a conceptual study on adaptive façades. In: Journal of physics: conference series, vol 3071. IOP Publishing, p 012016
51. Barozzi M, Lienhard J, Zanelli A, Monticelli C (2016) The sustainability of adaptive envelopes: developments of kinetic architecture. Procedia Eng 155:275–284
52. Velasco R, Brakke AP, Chavarro D (2015) Dynamic façades and computation: towards an inclusive categorization of high performance kinetic façade systems. In: International conference on computer-aided architectural design futures. Springer, pp 172–191

53. Gu N, Watanabe S, Erhan H, Haeusler MH, Huang W, Sosa R (2014) Controlling kinetic cladding components in building facades: a case for autonomous movement. In: Proceedings of the 19th international conference of the association of computer-aided architectural design research in Asia, pp 129–138
54. Matin NH, Eydgahi A, Shyu S, Matin P (2018) Evaluating visual comfort metrics of responsive facade systems as educational activities. In: 2018 ASEE annual conference & exposition
55. Addington M, Schodek D (2012) Smart materials and technologies in architecture. Routledge
56. Neugebauer J, Wallner-Novak M (2018) Let thin glass in the façade move: Thin glassnew possibilities for glass in the facade. In: Facade tectonics 2018 world congress
57. Sharaidin K, Burry J, Salim F (2012) Integration of digital simulation tools with parametric designs to evaluate kinetic façades for daylight performance. In: Digital physicality-proceedings of the 30th eCAADe conference, Czech Technical University in Prague, Faculty of Architecture Prague, Czech, pp 701–709
58. Khoo CK, Burry J, Burry M (2011) Soft responsive kinetic system: an elastic transformable architectural skin for climatic and visual control
59. Tashakori M (2014) Design of a computer controlled sun-tracking façade model
60. Kuru A, Oldfield P, Bonser S, Fiorito F (2020) A framework to achieve multifunctionality in biomimetic adaptive building skins. Buildings 10(7):114
61. Eftekhar Afzali S, Ghasemi M, Rahimiratki A, Mehdizadeh B, Yousefieh N, Asgharnia M (2025) Compaction and compression behavior of waste materials and fiber-reinforced cement-treated sand. J Struct Des Constr Pract 30(2):04025007
62. Han S et al (2023) Sustainable design of self-consolidating green concrete with partial replacements for cement through neural-network and fuzzy technique. Sustainability 15(6):4752
63. Movagharnejad K, Zareei F, Mehdizadeh B, Salahi S, Lashkenari MS (2015) Prediction of the surface tension of hydrocarbons. Pet Sci Technol 33(9):1008–1016
64. Ji W, Yang L, Liu Z, Feng S (2023) A systematic review of sensing technology in human-building interaction research. Buildings 13(3):691
65. Li Z, Tian J, Ji G, Cheng T, Loftness V, Han X (2025) Reinforcement learning-enabled adaptive control for climate-responsive kinetic building facades. Buildings 15(16):2977
66. Matin NH, Eydgahi A, Shyu S (2017) Comparative analysis of technologies used in responsive building facades. In: 2017 ASEE annual conference & exposition
67. Faraji A, Arya SH, Ghasemi E, Shiri H (2024) The role of artificial intelligence (AI) in advancing construction 4.0 practices. Appl Res Sci Eng
68. Faraji A, Arya SH, Ghasemi E, Soleimani H (2024) IoT core applications in AEC: investigation of the most compatible technologies of industry 4.0. Berlin, Germany, ResearchGate
69. Alsafery W, Rana O, Perera C (2023) Sensing within smart buildings: a survey. ACM Comput Surv 55(13s):1–35
70. Villegas W, Gutiérrez R, Govea J (2025) Modular middleware for IoT: scalability, interoperability and energy efficiency in smart campus. Front Commun Netw 6:1672617
71. Inthasuth T, Kaewjumras Y, Sahapong Somwong WB (2025) Comparative analysis of ZigBee, LoRa, and NB-IoT in a smart building: advantages, limitations, and integration possibilities. Int J Reconfigurable Embed Syst 14(1):165–175
72. Sabit H, Tun T (2024) IoT integration of failsafe smart building management system
73. Poyyamozhi M, Murugesan B, Rajamanickam N, Shorfuzzaman M, Aboelmagd Y (2024) IoT—a promising solution to energy management in smart buildings: a systematic review, applications, barriers, and future scope. Buildings 14(11):3446
74. Ma Z, Jørgensen BN, Ma ZG (2024) A systematic data characteristic understanding framework towards physical-sensor big data challenges. J Big Data 11(1):84
75. Liang X, Xie J, Zhao T, Stouffs R, Biljecki F (2025) OpenFACADES: an open framework for architectural caption and attribute data enrichment via street view imagery 230. arXiv preprint arXiv: 2504.02866
76. Aggarwal CC (2016) An introduction to outlier analysis. In: Outlier analysis. Springer, pp 1–34
77. Chen T (2016) XGBoost: a scalable tree boosting system. Cornell University

78. Mehdizadeh B, Vessalas K, Castel A, Asadi H, Ben B, Deilami S (2024) Correlation behaviours for accelerated and natural carbonation testing of Fly ash and ground granulated blast-furnace slag-based concrete mixes. Concr Aust 50:30–38
79. Goodfellow I, Bengio Y, Courville A, Bengio Y (2016) Deep learning, 1, 2. MIT Press Cambridge
80. Szeliski R (2022) Computer vision: algorithms and applications. Springer Nature
81. Wei T, Wang Y, Zhu Q (2017) Deep reinforcement learning for building HVAC control. In: Proceedings of the 54th annual design automation conference 2017, pp 1–6
82. Vázquez-Canteli JR, Nagy Z (2019) Reinforcement learning for demand response: a review of algorithms and modeling techniques. Appl Energy 235:1072–1089
83. Drgoňa J et al (2020) All you need to know about model predictive control for buildings. Annu Rev Control 50:190–232
84. Kong W, Dong ZY, Jia Y, Hill DJ, Xu Y, Zhang Y (2017) Short-term residential load forecasting based on LSTM recurrent neural network. IEEE Trans Smart Grid 10(1):841–851
85. Deb C, Zhang F, Yang J, Lee SE, Shah KW (2017) A review on time series forecasting techniques for building energy consumption. Renew Sustain Energy Rev 74:902–924
86. Ji Y, Wang J, Xu J, Fang X, Zhang H (2019) Real-time energy management of a microgrid using deep reinforcement learning. Energies (Basel) 12(12):2291
87. Sutton RS, Barto AG (1998) Reinforcement learning: an introduction, 1, 1. MIT Press Cambridge
88. Park JY, Dougherty T, Fritz H, Nagy Z 2018 LightLearn: occupant centered lighting controller using reinforcement learning to adapt systems to humans
89. Amasyali K, El-Gohary NM (2018) A review of data-driven building energy consumption prediction studies. Renew Sustain Energy Rev 81:1192–1205
90. Voyant C et al (2017) Machine learning methods for solar radiation forecasting: a review. Renew Energy 105:569–582
91. Mehdizadeh Miyandehi B, Vessalas K, Castel A, Mortazavi M (2023) Investigation of carbonation behaviour in high-volume GGBFS concrete for rigid road pavements. ASCP (Australian Society for Concrete Pavements)
92. A. Faraji, S. H. Arya, E. Ghasemi, and P. Rahnamayiezekavat (2024) Historical analysis of architectural design paradigms across three generations: byproducts of industrial revolution
93. Nguyen A-T, Reiter S, Rigo P (2014) A review on simulation-based optimization methods applied to building performance analysis. Appl Energy 113:1043–1058
94. Isaia F, Fiorentini M, Serra V, Capozzoli A (2021) Enhancing energy efficiency and comfort in buildings through model predictive control for dynamic façades with electrochromic glazing. J Build Eng 43:102535
95. Teixeira H, Moret Rodrigues A, Aelenei D, Gomes MG (2024) Simulation-based evaluation of the impact of an electrochromic glazing on the energy use and indoor comfort of an office room. Energies (Basel) 17(9):2110
96. Faraji A, Rezaei F, Rahnamayiezekavat P, Rashidi M, Soleimani H (2023) Subjective and simulation-based analysis of discomfort glare metrics in office buildings with light shelf systems. Sustainability (Switzerland) 15(15). https://doi.org/10.3390/su151511885
97. Takhmasib M, Lee HJ, Yi H (2023) Machine-learned kinetic façade: construction and artificial intelligence enabled predictive control for visual comfort. Autom Constr 156:105093
98. Dai S, Kleiss M, Alani M, Pebryani N (2022) Reinforcement learning-based generative design methodology for kinetic facade. In: Proceedings of the 27th conference on computer aided architectural design research in Asia (CAADRIA), pp 151–160
99. Andreeva D, Nemova D, Kotov E (2022) Multi-skin adaptive ventilated facade: a review. Energies (Basel) 15(9):3447
100. Qiu H, Ma Z, Hu Y, Wu D, Zhang L, He G (2025) Multi-performance prediction and optimization for building-integrated photovoltaics facades with passive design via explainable machine learning. Sol Energy 301:113871
101. Dong K et al (2025) Advancing building facade solar potential assessment through AIoT, GIS, and meteorology synergy. Adv Appl Energy 17:100212

102. Tariq R, Recio-Garcia JA, Cetina-Quiñones AJ, Orozco-del-Castillo MG, Bassam A (2025) Explainable artificial intelligence twin for metaheuristic optimization: double-skin facade with energy storage in buildings. J Comput Des Eng 12(3):16–35
103. Zheng Y, Sun B, Guan Y, Yang Y (2025) Artificial intelligence empowering the transformation of building maintenance: current state of research and knowledge. Buildings (2075–5309) 15(22)
104. Cuce PM, Cuce E (2025) Ventilated facades for low-carbon buildings: a review. Processes 13(7):2275
105. Mohseni E, Yazdi MA, Miyandehi BM, Zadshir M, Ranjbar MM (2017) Combined effects of metakaolin, rice husk ash, and polypropylene fiber on the engineering properties and microstructure of mortar. J Mater Civ Eng 29(7):04017025
106. Ziaie A, Mehdizadeh B, Safi Jahanshahi F, Ahmadi N, Ghanizadeh AR (2026) Prediction of liquefaction-induced lateral displacements using hybrid GBRT and EOA. J Soft Comput Civ Eng 10(1):e225827
107. Mohammadifar L et al (2022) Properties of lime-cement concrete containing various amounts of waste tire powder under different ground moisture conditions. Polymers (Basel) 14(3):482
108. Faraji A, Arya SH, Ghasemi E, Soleimani H (2023) Considering the digital twin as the evolved level of BIM in the context of construction 4.0. In: Proceedings of the 1st national & international conference on architecture, advanced technologies and construction management (IAAC), Tehran, Iran

Supplementary Cementitious Materials (SCMs) in Concrete Performance Reviewing Material Behavior to Advance Uncertainty Analysis in Design

Bahareh Mehdizadeh, Amir Faraji, Tahereh Meydani Haji Agha, Shima Homayoon Arya, and Hannan Abagheri Mahabadi

Abstract Supplementary cementitious materials (SCMs) stand at the forefront of sustainable concrete development by delivering both environmental benefits and improved performance. The purpose of this review is to evaluate the mechanical properties, durability, and environmental impact of the principal supplementary cementitious materials, including agricultural ashes, natural pozzolans, industrial by-products, and the latest pioneer waste-based recycled materials. From the integration of data from multiple past studies, it is evident that SCM has significantly improved resistance to chlorides, shrinkage, thaw cycles, and increased compressive and flexural strength over a very long term, as well as reduced permeability, among other durability aspects. These beneficial performances are linked to natural pozzolanic reactions, microfillers, and enhanced pore structure refinement brought about by increased levels of calcium silicate hydrate C-S-H gel gelation. Reduced early age strength, workability, and variability remain concerns, especially with the more recent SCMs from various generated wastes. This review emphasizes also the role of performance metrics, research data integration, and meta-analysis in enhancing decisiveness under challenging circumstances. Glass powder materials, ceramic wastes, and agricultural ashes are examples of innovative SCMs that hold tremendous promise for

B. Mehdizadeh (✉) · A. Faraji
School of Engineering, Design and Built Environment, Western Sydney University, Penrith, NSW, Australia
e-mail: B.miyandehi@westernsydney.edu.au

A. Faraji
e-mail: a.faraji@khatam.ac.ir; a.faraji@westernsydney.edu.au

A. Faraji · T. M. H. Agha · S. H. Arya · H. A. Mahabadi
Construction Project Management Department, Faculty of Architecture, Khatam University, Tehran, Iran
e-mail: t.meydani@khatam.ac.ir

S. H. Arya
e-mail: sh.homayoon@khatam.ac.ir

H. A. Mahabadi
e-mail: H.Abagheri.8754@khatam.ac.ir

A. Azizi et al. (eds.), *Intelligence at the Interface*, Emerging Trends in Mechatronics,
https://doi.org/10.1007/978-981-95-7584-8_8

near-future applications, provided there is further refinement of processing methods. The conclusions of this study confirm that SCM significantly contributes to material recyclability, durability performance enhancement, and carbon decarbonization. In this way, this concept introduces SCMs as a key innovation route to ensuring high-performance, greener, and sustainable concreting technology.

Keywords Supplementary cementitious materials (SCMs) · Concrete performance · Material behavior · Sustainable concrete design · Mechanical and durability properties · Performance-based design

1 Introduction

With the increasing need for sustainable construction materials, on one hand, and the limitations that come along with conventional Portland cement, on the other, SCMs are being recognized as a primal need from a scientific, as well as an industrial, standpoint. The increasing incorporation of SCMs represents a paradigm shift in concrete technology, a need that goes beyond improvement for performance, development, and advances, but is also driven by forces of environment and carbon reduction. As more research emerges, more evolution is expected in the concept of SCM, from a supplement to a primal parameter, influencing the design, assessment, and application of concrete mixtures for various structural purposes.

1.1 Background of SCM

Concrete is one of the most widely used building materials because of its strength, durability, and adaptability. However, a significant amount of CO_2 emissions and nearly 8% of the world's carbon output come from the production of Portland cement, the primary binder in concrete [1]. Numerous studies into alternative materials that can reduce the carbon footprint of concrete have been spurred by this environmental issue. One of the most promising solutions is supplementary cementitious materials (SCMs).

SCMs are substances that react chemically with the calcium hydroxide ($Ca(OH)_2$) produced during cement hydration when they are added to concrete. The main substance that gives concrete its binding strength, calcium silicate hydrate (C-S-H) gel, is produced as a result of this reaction. By reducing the quantity of Portland cement needed, the use of SCMs lowers CO_2 emissions associated with the production of cement [2]. SCMs are crucial parts of contemporary sustainable concrete technology because, in addition to their positive effects on the environment, they can greatly improve mechanical performance and long-term durability.

SCMs come from different origins and have different chemical compositions. Some of the most commonly applied SCMs are FA, ground granulated blast-furnace slag (GGBFS), and silica fume, which are all industrial by-products. These waste materials from industry often have a high pozzolanic or hydraulic potential. FA has been used for many years to enhance durability and workability while reducing cement consumption [3]. GGBFS, a product of the production of steel, enhances the long-term strength of concrete and its resistance to sulfate attack [4]. Natural pozzolans like volcanic ash, pumice, and diatomaceous earth have pozzolanic activity intrinsically. These materials have traditionally been used in construction and often contribute to enhanced durability and chemical resistance as well as improved performance in aggressive environments [5].

In recent years, agricultural waste-based SCMs like rice husk ash (RHA), sugarcane bagasse ash (SCBA), or palm oil fuel ash have gained considerable attention. These are readily found in agricultural areas and provide a feasible alternative for traditional SCMs. According to research, agricultural waste can increase workability as well as mechanical properties, mostly at a larger ratio of cement substitution [6]. Some potential and newly developed SCMs include waste glass powder, ceramic waste powder, and red mud. These are mostly generated through industrial processes and possess substantial potential as a sustainable resource, though further investigation is required for a complete perception of their performance capability within the construction sector [7].

Incorporation of SCMs in concrete mixtures has several advantages. First, there will be a reduction in carbon emissions. This reduction will attain about 20% if 30% Portland cement replacement with fly ash is achieved, among other emissions [8]. In addition, there will be improved concrete durability due to reduced permeability of concrete, which will withstand chemical attacks such as those caused by sulfur and chlorine. In this regard, concrete made from SCMs will perform in a superior manner when subjected to freeze/thaw cycling, and there will be less potential to suffer damage due to cracking, among other effects [9]. Silica fumes will greatly reduce susceptibility to ASR, which can cause concrete to undergo damage due to extensive cracking caused by ASR reactions in the concrete mixture [4].

The SCMs may also improve the mechanical characteristics. Silica fume and metakaolin have been found to improve early-age strength properties owing to their high pozzolanic activity, producing a denser C-S-H matrix [2]. The GGBFS is added to improve the long-term strength characteristics, making it suitable for large structures requiring high durability [9, 10]. The SCMs are also added to improve the workability property, making processes such as mixing, placing, and finishing easier. The effect on the hydration reactions reduces shrinkage and thermal cracking, making the structure more durable [11].

In spite of these benefits, there are still issues in the general application of SCMs. These are typically found in the chemical composition, especially concerning industrial by-products and waste. For instance, the chemical composition of fly ash may vary widely, possibly influencing the pozzolanic property and strength [12]. There are also potential issues concerning the reduction of early strength, which may not be suitable in fast-track projects that demand early strength development [3]. In addition, it has not yet been determined whether the properties concerning newly introduced SCMs, such as glass powder or ceramic waste, are amendable or not [13].

1.2 Performance Indicators of SCM-Modified Concrete

The use of SCMs in concrete mixes has also received considerable attention for the beneficial dual role it plays in improving the performance characteristics of concrete and lowering its negative impact on the environment. Discussion in the previous Sect. 1.1, highlighted that the common advantages of the use of SCMs include a reduction in the emission of CO_2 gases along with improvement in the durability of concrete. This section, however, focuses on the use of performance indicators to measure the effect of SCMs on the concrete properties of strength and durability.

The key performance parameter, from the perspective of SCM-modified concrete, is the value of its compressive strength, since it plays a crucial role in determining the structural integrity and ultimate load-carrying capacity of the structural member. Although the effect of certain SCMs, such as fly ash and slag, on the value of the compressive strength may result because of the decrease in the hydration rate at the lower age, the respective characteristics of the aforementioned materials make way for an accelerated increment in the value of the compressive strength. It has been observed that the use of fly ash makes a substantial contribution to the development of strength because of the formation of C-S-H gel [14]. On the contrary, the major contribution to the preparation for the value of the early age of strength is attributed to the silica fume because of its micro-fine particles, primarily because it fills the pores [15]. Another tangible parameter is the value of the flexural strength, especially in the case of structural members such as slabs and beams.

The use of SCMs such as GGBFS, silica fume, and metakaolin has proved to enhance the flexural strength by optimizing the micro-homogeneity and refining the structure of the pores. GGBFS is effective in optimizing the micro-homogeneity of the concrete, hence its effects on the flexural strength of the concrete [12]. Additionally, metakaolin with its improved properties has proved effective in enhancing the tensile strength of the concrete when used in large dosages [16]. Besides the above-stated properties, the tensile strength of the concrete is paramount, especially with regard to its tendency to crack due to shrinkage. SCMs such as silica fume and fly ash can minimize the porosity in concrete; this enhances the tensile strength and resistance to deformation [17, 18]. The above properties can be considered especially advantageous for minimizing the creation of micro-cracks caused by shrinkage in concrete subjected to tensile loading. Additionally, the inclusion of metakaolin and GGBFS is intended to minimize the brittleness commonly encountered in cement mixes. It is well recognized that common Portland cement mixes exhibit a brittle performance [19].

The performance of concrete containing SCMs over time remains a topic of great research interest. Permeability is a very important property, as reduced permeability indicates increased resistance to the penetration of chloride ions, sulfur attacks, and carbonation. Silica fume and fly ash were found to reduce permeability significantly by improving pore connect ability, thus making SCM concrete more resistant to severe environmental conditions, especially in marine environments [20]. In addition, GGBFS and metakaolin were found to increase resistance to sulfur attacks, thus extending the useful life of structures exposed to chemical attacks [4]. The other important parameter that affects SCM concrete performance is its shrinkage mechanism; a higher shrinkage rate may possibly cause early-age cracking. Fly ash and metakaolin were found to help counteract drying shrinkage, thus significantly minimizing the possibility of crack formation due to increased shrinkage rates associated with mass concrete structures [2]. In addition, improved resistance to freeze-thaw cycles, which are very important, especially in colder climates, was attained with the use of silica fume and GGBFS as SCMs. These materials assist in generating a compact pore structure, which prevents water passage or, consequently, crack formation due to freezing [21].

Finally, the resistance to cracking is the result of a synergistic interaction between the increased mechanical properties, shrinkage, hydration processes, and microstructure compaction [22]. The shrinkage-reducing/crack-reducing admixtures reduce the growth and delay the onset of the initiation of cracks by lessening the temperature gradient caused by hydration heat and the size of the capillary pores. Fly ash reduces the hydration heat, hence the temperature differences and the shrinkage cracking caused by the latter [3]. The combination of metakaolin and silica fume improves resistance to work fractures by filling micro-voids, which hinders the growth of cracks [23]. Based on the performance, the research question has to be framed to study the impact caused by the various shrinkage-reducing/crack-reducing admixtures on concrete.

- How do various SCMs influence the mechanical and flexural performance of concrete?
- How do different SCMs compare in terms of overall performance across mechanical and durability criteria?
- How do key behavioral characteristics of SCMs (such as pozzolanic activity, reactivity, microstructure, and performance consistency) influence uncertainty in concrete design?

To address these issues, this study aims to explore the combined effects of various SCMs on concrete performance. By answering these questions, it will provide a clearer understanding of how SCMs can be optimized for different types of concrete applications and environmental conditions.

2 Supplementary Cementitious Materials (SCMs)

Figure 1 shows a more organized categorization of SCMs according to source and processing method. Materials are further divided into four groups: by-products from industry, natural pozzolans, calcined or processed pozzolans, and waste-based or emerging SCMs. The materials in each category possess unique chemical and physical properties that can be used, to some degree, as substitutes for Portland cement. This categorization aids in highlighting the range of SCM choices available and

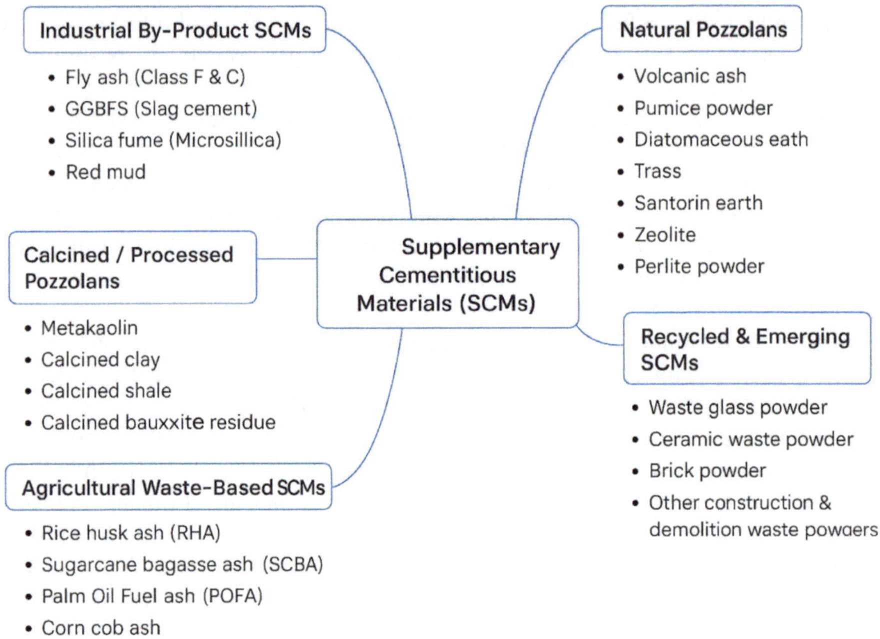

Fig. 1 Major groups of supplementary cementitious materials used in concrete

enables a more cogent understanding of what effects could be incurred upon concrete strength, durability, and sustainability.

2.1 Industrial By-Products

- **Fly Ash**: Fly ash is one of the most important replacement materials for a portion of Portland cement obtained from the combustion of coal in thermal power plants and is divided into two classes: F (low calcium, mainly pozzolanic) and C (high calcium, self-setting) [24]. Class F is produced from bituminous coal and class C from lignite; both particles are spherical and glassy, but class C has a higher initial reactivity due to 15–30% CaO and hardens even without additional calcium hydroxide. Its global production is about 1 billion tons per year, of which 70% is recycled. Its use reduces costs by up to 30% and CO_2 emissions by up to 50%, improves concrete fluidity, reduces permeability by up to 50%, increases resistance to sulfate, chloride, and alkali-silica reactions, and increases long-term strength by up to 20%. In ultra-high-strength concrete (UHPC), it also increases durability by up to 20%. However, the initial strength (7 days) is 10–20% lower, setting time is delayed by several hours, especially in class F, and at low temperatures or above 40%, problems of segregation or excessive slowness may occur.

Class F is more suitable for high-durability structures, such as bridges, and class C for applications requiring rapid initial strength (such as flooring). Typical replacement percentages are 15–30% for class F and 20–40% for class C, and it is best added at the dry mix stage. With the gradual reduction of coal-fired power plants, the recycling of old fly ash will become more important in the future [25–31].

- **GGBFS (Slag Cement)**: Ground granulated blast-furnace slag (GGBFS) is a by-product of the blast furnace of iron production, which, after rapid granulation and grinding, is converted into a very fine powder (specific surface area 400–600 m^2/kg) and can replace up to 60% of cement. Due to its 35–45% CaO, it has both pozzolanic and cementitious properties. Its use reduces CO_2 emissions by up to 50%, greatly increases durability (excellent resistance to sulfate, chloride, and shrinkage), reduces the heat of hydration, has excellent performance at high temperatures (up to 800°), and increases the long-term compressive strength by up to 40%. However, the early strength (7 days) is 10–20% lower, the setting is delayed by 1–3 h, and requires an accelerator at low temperatures or high percentages. The best replacement percentage is 30–60% and is added at the dry mix stage. Its most suitable application is in marine structures, massive concrete, and heat-resistant concrete [32–36].
- **Silica Fume (Microsilica)**: Silica fume is a by-product of silicon and ferrosilicon production, obtained as an ultrafine gray powder (less than 1 μm) with more than 90% amorphous silica and a specific surface area of 15,000–30,000 m^2/kg. It is one of the strongest pozzolans available and, by replacing only 3–10% of cement, significantly densifies concrete, reduces chloride infiltration by up to 66.7%, increases compressive and flexural strengths, increases peak tensile stress in UHDCC by more than 330%, and provides excellent mechanical and environmental performance in permeable and geopolymer concretes. It is produced from electric arc furnace exhaust gases and has a high recycling rate, although global production is limited. Its biggest advantage is the creation of a very dense structure, reduced porosity, excellent resistance to thermal damage, and prevention of the formation of delayed ettringite, but its biggest disadvantage is a severe reduction in fluidity (necessary need for superplasticizer) and the possibility of a drop in strength at percentages above 9%. It accelerates setting slightly and is best added to the dry mix before water. Overall, silica fume is the king of admixtures for high-strength concretes, marine structures, and any project where durability and mechanical strength are an absolute priority [37–40].
- **Red Mud**: Red mud has been found to be an alkaline alumina industry waste with low Cementous potential and the presence of toxic heavy metals. Its composition primarily consists of iron oxide, alumina, silica, and sodium [41]. The worldwide production of this waste substance is immense, and storage has resulted in severe environmental concerns like contamination of soil and groundwater. Literature has reported that when it is mixed with other cementitious materials like Portland cement, steel slag, fly ash, and desulfurized gypsum, it enhances the mechanical strengths of concrete. Analyzing the positive and negative properties of this waste substance in concrete technology, it has been found that it reduces Portland cement demand and CO_2 emissions, enhances microstructure densification, and prolongs

concrete durability. However, it has disadvantages like low activity, high alkalinity, the presence of toxic heavy metals, and variations in properties with source. Based on positive properties, it promotes hydration products like C-S-H gel and enhances flexural and compressive strengths. Based on negative properties, it decreases concrete workability and prolongs setting time. The setting time in case of red mud alone is slow; however, when mixed with cement and gypsum, ettringite phases hasten setting. Its optimum utilization has to be done at a replacement level of 20–40% when added simultaneously with other SCMs during the initial mix processing of concrete. Initially, it has to be pulverized and dried to increase its specific surface area. Utilization of this waste substance, besides minimizing negative environmental impacts, has helped in waste management in industry and environmental risks generated by toxic heavy metals, which are fixed in hydration products [42–45] (Table 1).

2.2 *Natural Pozzolans*

- **Volcanic Ash**: The complex chemical composition and amorphous nature, reducing substantial CO_2 emissions with no effect on the long-term strength of concrete, have urged volcanic ash as a natural pozzolana, and thus a great alternative to be employed as a partial substitute for cement clinker. Basically, it was found to contain a combination of silica, alumina, and iron oxides. The reactivity is said to be a function of the volume of glass content, particle size, and activation technique. Mechanical grinding and sieving can be used to prepare volcanic ash, and other activation techniques will mostly be thermal and chemical in order to raise the surface area and reactivity. The employment of volcanic ash also highlights a lot of advantages, including the improvement of long-term concrete durability, decrease in permeability, and augmentation of resistance to chemical attacks, apart from difficulties such as early strength reduction, loss of fresh workability, and variability and unreliability of reactivity, which can be controlled using appropriate substitution levels (10–30%) and activation techniques. It also exhibits a pozzolanic reaction with calcium hydroxide from cement to form secondary C–S–H, which can result in a denser matrix and may thus play an important role in geopolymer reactions. They need to be added into concrete during initial mix preparation, and in this respect, a proper source analysis, including chemical and mineral analysis, should be quite essential. Finally, since there are immense environmental prospect, formulation, and adoption of uniform testing and activation procedures using pilot projects should be done to ensure effective utilization of this material in the concrete and cement industry [46–50].
- **Pumice Powder**: Pumice powder, being a natural material with pozzolanic properties and an additive for cement, is more useful when it is used for sustainable concretes. The primary material composing the chemical structure of pumice powder is silica and alumina. In addition, because it is an amorphous material, it

Table 1 Comparison between 4 concrete SCMs (industrial by-product SCMs)

Comparison criteria	Fly ash	GGBFS	Silica fume	Red mud
Original source	Coal power plant waste	Steel mill waste	Silicon/ ferrosilicon production waste	Bayer alumina process waste
Optimal replacement percentage (%)	Class F: 15–30 Class C: 20–40	30–60	3–10	20–40 (with other SCM)
Maximum CO_2 reduction	Up to 50%	Up to 50%	Low (due to low volume)	Up to 30%
Pozzolanic activity	Class F: High Class C: High + cementitious	Very high + cement	Extremely high	Very low to moderate (with activation)
Effect on early strength (7 days)	10–20% less	10–20% less	Almost unchanged or slightly higher	15–30% reduction
Effect on final strength	+ 15–20%	+ 30–40%	+ 50–100%	85–100% (with activation)
Effect on concrete fluidity	Improvement (especially class F)	Slight reduction	Severe reduction (requires superplasticizer)	High reduction + high SP requirement
Effect on durability and permeability	Excellent (50% reduction in permeability)	Excellent (best in sulfate and chloride)	King of durability (66% chloride reduction)	Improvement with activation; reduced permeability
Processing requirement	Not usually	No (mill only)	No	Milling + drying + activation with gypsum/slag
Approximate cost	Very cheap	Cheap	Expensive	Almost free to very cheap
Best application	Massive structures, dams, bridges	Marine structures, bulky, heat-resistant	UHPC, marine structures, ultra-strong	Bulk concretes, blocks, geopolymers + heavy metal immobilization

also participates in the formation of the secondary C-S-H phases in the alkaline environment of the concrete. On the basis of the findings cited from different studies conducted by researchers, the amount of 15–25% by the weighing of the cement replaced by the use of the pumice powder improves the microstructure and density without any considerable decrease in the values of the compressive strength. Although the values of the compressive strength of the concrete made with the use of the pumice powder are lower in the early stages compared to the conventional concretes, it is capable of providing higher values of the mechanical strength with improved durability performance in the later stages. The permeability values related to the concretes are lowered by the use of the pumice powder with improved levels of endurance against the chemical attacks. In addition, it performs with improved results under the conditions of corrosive environment. The use of the pumice powder reduces the permeability values related to the concretes with improved levels of endurance against the chemical attacks. In addition, it performs with improved results under the conditions of the corrosive environment. The use of the pumice powder decreases the permeability values related to the concretes with improved levels of endurance against the chemical attacks. In addition, it performs with improved results under the conditions of the corrosive environment. Recently, studies provide equal importance toward revealing that the use of admixtures for the pumice powder with different materials such as fibers or different materials having complementary properties reduce the drawbacks found with the use of the pumice powder. On the other hand, it enhances the properties of the mechanical strength along with the values of the durability of the concretes [51–53].

- **Diatomaceous Earth**: Diatomaceous earth (DE) is a natural amorphous silica-rich pozzolan; it is produced by the fossilized remains of diatom blooms and is widely used as a supplementary cementitious material (SCM). The replacement level of Portland cement by DE is around 5–20% and succeeds in greatly reducing CO_2 emissions, making its production based on the recycling of industrial wastes, like that of beer, wine, and oil industries, due to its high amounts of active SiO_2, porosity, and high surface area, which, due to grinding, drying, and especially calcinations between 500 and 900 °C, exhibits high pozzolanic activity, thus getting rid of excess organic impurities naturally present. Its production is based mainly on natural extraction/recycling of its wastes like DE-LD (low-density variety) and DE-HD (High-density variety). The advantages of this admixture mainly depend upon the removal of excess organic impurities and mainly involve: An increase in strength by a high amount of 53% within 15% replacement, an improvement in long-term strength by a high of 50% at a period of 180 days, compaction of the microstructure, decreases the values of water absorption and ensures appropriate values of permeability in road permeable concretes, whereas the disadvantages involve: An increase in values of water required by a high amount of 1.7 times, decreases the values of fresh workability like slump, increases the values of final setting time and decreases early strength, especially no calcined ones, which can be coped with by the addition of water-reducing agents. The positive attributes, like enhanced resistance, high durability, and high mechanical strength, especially in the long periods, seem more important, whereas the negative attributes can be prominently observed within the early periods of hydration and workability. The mechanism is based upon very slow pozzolanic reactions involving the production of hydrated silica compounds and $Ca(OH_2)$. Setting time and addition-related procedures seem to directly replace around 10–15% of calcined powder within the stage of batching, which succeeds greatly within the periods of 28–180 days and greatly increases strength values. Finally, diatomaceous earth also has enormous potential in green, permeable, and circular concretes and, through the use of waste materials, also makes a significant contribution to the resolution of the lack and the carbon reduction demand in the construction sector regarding traditional SCMs, although an optimization concerning pretreatment and economic studies is required [51, 54–56].

- **Trass**: Some recent genuine scientific publications have focused on the natural pozzolanic activity of Trass in concrete, where it has been considered alongside other-available SCMs like fly ash and slag. Trass, which comes from the extraction of natural volcanic tuffs, has a medium reactivity and produces secondary hydration compounds like C-S-H, which render the concrete dense with lower permeability. Research has indicated the positive effect of Trass on the durability of concrete exposed to aggressive environments like sulfates and acidic conditions, and also its viability for permeable concrete construction projects. However, the major disadvantage or drawback of Trass may include the lower early strength development of concrete, which comes with its lower reactivity compared to other reactive SCMs like silica fume, such that the subsequent strengthening process takes longer periods during the initial days. From the environmental aspect, the supplementary or partial use of Trass instead of ordinary cement reduces the production of carbon dioxide, making it environmentally friendly for utilization where there are rich natural resources found in the form of volcanoes. According to articles reviewed on the research work, the supplementation or mixing or combining altogether other SCMs like fly ash or slag renders the best results on the utilization and supplementation of Trass since the two collectively can neutralize the lower effect on the early strength developments of the concrete. Also, the mixing or combining process with Trass can be compared to the mixing or combining procedure for all supplementary types of materials where it is added partly instead in a mixture or blend substituting up to a certain percentage starting either from 10 or 30% during the mixing or combining procedure. Apart from improving or enhancing the mechanical properties, durability, and strength, supplementing or substituting Trass further helps in the cost and resource reduction through the utilization or extraction displayed by natural resources found either through volcanoes or any other natural resource source found on natural resources [43, 57, 58].
- **Santorin Earth**: Santorin earth, a natural volcanic pozzolan, originates from the Greek island of Santorini and is one of the oldest recorded SCM, used as a historic concrete material like the Roman "opus cementitious" and more recently for the construction of the Suez Canal. Its high amorphous content of silicon, aluminum, and iron oxides makes it a good pozzolan material that will react with the calcium hydroxide produced during the hydration process to form calcium silicates. Its production also exhibits low environmental impact since the material is extracted from volcanic deposits and ground into a fine powder, and calcination can also be involved depending on the material's properties, thus not contributing significantly to the carbon dioxide emissions during the production process like the Portland cement. The most significant advantage and benefit derived from the use of the Santorin Earth SCM include enhanced durability and resistance to chemical attacks such as acid and sulfur compounds, improved resistance to

alkali-silica reaction (ASR), and improved malleability and flexibility for structures. Its major disadvantage and limitations include a significant reduction in the initial workability and the initial compressive strength, which also requires the use of superplasticizers. In its application as a SCM, the most significant advantages include enhanced flexural and tensile strengths and lower water absorption capacity, while the most significant disadvantages include increased shrinkage and porosity, which may require further studies and investigations. Its setting process is based on the slow reaction between the material and the water, creating a slow pozzolanic reaction process that completes the hydration process and makes the material form a dense mass, even in wet or underwater settings. Its addition as a SCM also takes a process that entails the use of a 10–30% replacement-level method as a cement material at the batching process level, which increases the material's strengths in a span of 28–90 days but significantly delays the initial setting process. Its method and process for possible future use and application as a SCM also entails a mill process together with the clinker material or a separate process based on its performance evaluations and the isothermal calorimeter method based on its reactivity rates. In summary and conclusion, the Santorin Earth SCM exhibits a high potential and future applications as a material in the development of future SCM materials, especially for the reduction of carbon emissions by the year 2050, and its past studies and investigations continue to offer current inspiring trends and motivations for the analysis and investigation for the development of future SCM materials. It can be compared to volcanic pumice powder and can be applied as a material for green concrete in road construction and marine structures [59, 60].

- **Zeolite**: Natural zeolite, acting as a pozzolanic additive in addition to cement, is effectively contributing to enhancing the durability properties of sustainable concrete. The aluminosilicate skeleton with high porosity and high surface area of zeolite results in its chemical interaction with the calcium hydroxide in the high pH environment of concrete to form secondary C-S-H hydrated phases. This results in the reduction of the concrete pore size, making the concrete denser with reduced permeability, resistant to aggressive ions such as chloride and sulfate. Research has revealed that the addition of 5–20% of zeolite to the cement has been effective in enhancing the durability properties of the concrete without compromising the ultimate compressive strength. However, the high-water absorption characteristics of zeolite result in the reduced workability of concrete; hence, its compressive strength is low compared to the normal concrete. This problem can be corrected by proper mix design and the use of superplasticizers. In the context to the environment, zeolite is more preferable to cement since zeolite does not need the energy-intensive curing stage. Natural zeolite, acting as a natural SCM, is effective in conserving cement, enhancing the durability properties, and achieving the goal for low-carbon concrete. This can be achievable with proper mix design performance [61, 62].
- **Perlite Powder**: Perlite powder, essentially in the form of waste non-expanded perlite (WPP) or ground expanded perlite (GWEP), represents an excellent pozzolanic cementitious material obtained from the thermal expansion process

of raw perlite, thereby giving it a potential to replace 5–30% Portland cement, that decreases the release of CO_2 emissions, adding to greater stability. Its specific surface area was found to be relatively small, possess low to medium activity index values between 13 and 83%, and consist of an amorphous silica-alumina component that, in reaction with calcium hydroxide, forms an supplementary C-(A)-S-H gel, resulting in enhanced values of compressive and bending strengths and smaller size of large pores. It can be produced in a simple and less energy-abundant manner from the industrial waste product after merely grinding industrial wastes without the necessity of high-temperature incineration, thereby making perlite powder an optimal choice to pursue a circular economy approach. Because of its regained usage, the initial set of advantages encompasses large beneficial aspects like high natural ability to reduce the adverse effects of environment, reduce burial in landfills, sustain the properties of the construction material in the reclaimed concrete, and enhance the downtrodden conditions of concrete environment in terms of improved mechanical properties in the extended period. Disadvantages encompass an increased mixture hydration, reduction of initial mechanical strength, accelerated setting time, and expansion of porosity, which, after careful analysis, was found to be counterbalanced by its ability to add additives like polymer additives or superplasticizers, where in concrete, major advantages will include enhanced density and confined good mechanical properties, but in case of reduced strengths or enhanced waters, it expects to adhere to construction related to circular concretes. During set departure, reliance lies in the slowness of pozzolanic reaction that finally attains hydration in cements to form strengthened construction, especially when mixed in combination form or in mixture form with varied other industrial wastes. In terms of addition in relation to time passing, addition occurs, relating to replacement in concrete mixture, in direct form replacing 10–15%, repeating the enhanced values in about 28–180 days, accordingly. Ultimately, perlite powder represents an extremely promising choice to pursue next generation concretes in a more sustainable manner, pending future modifications to their mixing ratios in exhibiting potential setbacks of ASRs, perlite microstructures in case of recycling perlite powder to act as second materials, assuming an extremely vital role in minimizing adverse effects of construction activities on our environment [63–65] (Table 2).

2.3 Engineered/Processed Pozzolans

- **Metakaolin**: It is one of the most reactive and strongest available pozzolans, produced by controlled calcination of kaolin clay at 600–900 °C, which appears in form of a white-gray substance with above 90% content of ($SiO_2 + Al_2O_3$) and a high surface area of 10,000–20,000 m^2/kg. At 5–15% to 30% replacements of cement, it greatly improves the compressive, tensile, and flexural strengths, in both high-performance concrete (HPC) and ultra-high-performance concrete (UHPC),

Table 2 Comparison between 7 concrete SCMs (natural pozzolans SCMs)

Comparison criteria	Volcanic ash	Pumice powder	Diatomaceous earth	Trass	Santorin earth	Zeolite	Perlite powder
Original source	Natural volcanic ash	Natural volcanic pumice powder	Diatomaceous earth (fossil diatoms, often industrial waste)	Terrace (volcanic tuff)	Santorini volcanic ash (Greece)	Natural zeolite	Perlite powder (unexpanded or expanded milled waste)
Optimal replacement percentage (%)	10–30	15–25	10–15	10–30	10–30	5–20	10–15 (up to 30% possible)
Maximum CO_2 reduction	High (up to 30% with maximum replacement)	High (low energy, up to 25%)	Very high (waste recycling, up to 20%)	Medium to high (up to 30%)	High (low energy, up to 30%)	High (no energy-intensive process)	Very high (waste recycling, circular economy)
Pozzolanic activity	High (dependent on glass phase and processing)	High (amorphous silica-alumina)	High (after calcination)	Medium	High (volcanic amorphous)	High (porous aluminosilicate)	Medium to high (13–83% activity index)
Effect on early strength (7 days)	Significant reduction	Reduced	Reduction (especially without calcination)	Decrease	Reduced	Reduced	Decrease
Effect on final strength	Good long-term enhancement	Increased or near normal	High increase (up to 50–53%)	Medium long-term increase	Good long-term increase	Good long-term increase	Increases at later ages
Effect on concrete fluidity	Reduction (reduced slump, increased water demand)	Reduced (reduced workability, increased setting time)	Significant reduction (water requirement up to 1.7 times)	Decrease	Reduced (requires superplasticizer)	Reduced (high water absorption)	Decreases (increases water demand)

(continued)

Table 2 (continued)

Comparison criteria	Volcanic ash	Pumice powder	Diatomaceous earth	Trass	Santorin earth	Zeolite	Perlite powder
Effect on durability and permeability	Excellent (reduced permeability, chemical resistance/ASR)	Good (reduced permeability, chemical resistance)	Excellent (reduction of water absorption, microstructure compaction)	Good (sulfate/ acid resistance)	Excellent (reduced permeability, chemical resistance)	Excellent (chloride/sulfate resistance, reduced permeability)	Good (reduces large pores, long-term durability)
Processing requirement	Mill, sometimes thermal/chemical activation	Mechanical grinding	Calcination necessary (500–900 °C)	Medium grinding	Mill, sometimes calcined	Low grinding	Waste mill (low)
Approximate cost	Low (abundant natural resources)	Low to medium (low energy)	Low (waste recycling)	Low (local volcanic sources)	Low (simple extraction)	Medium (extraction and grinding)	Very low (waste by-product)
Best application	Sustainable concrete, geopolymer, and chemically resistant structures	Green concrete, microstructure improvement	Permeable green concrete, road, circular	Sulfate-resistant concrete, permeable projects	Marine concrete, road, long-term carbon reduction	Chloride/ sulfate-resistant concrete, low-carbon	Circular concrete, non-structural, recycled

and significantly reduces permeability and porosity, shows high resistance to chloride and chemical attacks, and also has a highly dense matrix. Being highly reactive, it speeds up hardening and secondary C-S-H gel formation, however, it also decreases the paste fluidity, and above 20% of cement content, it also decreases overall strength. It is relatively energy-efficient, readily available in plenty of resources, though with high temperature requirements. The major disadvantages are that it is costly in comparison with fly ash and slag, and less efficient, though this shortcoming is overcome with high-performance superplasticizers. The most effective mixing ratio is in the dry mixture phase, before adding water. For those construction projects that are most focused on high durability, high mechanical strength, and lower CO_2 emissions, like offshore construction, bridges, and high-performance UHPC, metakaolin is one of the most advanced and best available materials, which, in combination with other multi-taskers, always demonstrates outstandingly high-performance exceeding expectations [66–68].

- **Calcined Clay**: Calcined clay is one of the most promising and promising cement substitutes in the coming decades. It is obtained by simply heating ordinary clay at 700–900° (optimum 800–850°) and then finely grinding, creating a powder with more than 70% ($SiO_2 + Al_2O_3$) and a specific surface area of 5000–15,000 m^2/kg. By replacing 15–30% of cement, it reduces CO_2 emissions by up to 40%, increases long-term compressive strength by 15–20%, significantly reduces permeability and water absorption, and offers performance very close to pure Portland cement by forming C-A-S-H and carboalumination phases (especially in the well-known LC3 system with limestone and gypsum). Its production energy consumption is much lower than clinker and is provided by abundant and cheap clay available in most countries. The main disadvantage is reduced fluidity (need for lubricant) and a slight loss of early strength, which is almost eliminated in the LC3 system; at percentages above 30%, strength also decreases. It slightly accelerates setting and is best added in the dry mix stage, before water. In developing countries with both abundant clay availability and an urgent need to reduce carbon emissions, calcined clay (especially in LC3 form) could soon reduce conventional cement consumption by 30–50% and be one of the most important tools for decarbonizing the concrete industry globally [69–73].
- **Calcined Shale**: Shale is a very abundant and important sedimentary rock that is used both in geology and in the construction and energy industries. Due to its high silica and alumina content, when calcined (calcined shale), it becomes an active pozzolanic material and can be used as an SCM in concrete or for soil stabilization. Calcined shale is a pozzolanic material obtained by heating natural shales at temperatures of about 750–850 °C and has the ability to react with calcium hydroxide cement due to its high silica and alumina content. It is used as an SCM in concrete and also for stabilizing sandy soils. Its chemical composition includes SiO_2 about 50–66% and Al_2O_3 about 14–18%, while the CaO content is low and Fe_2O_3 is relatively high. Its production requires careful control of the calcination temperature because excessive temperature increases reduce the specific surface area and pozzolanic activity. The advantages of calcined shale include reduced Portland cement consumption, reduced CO_2 emissions by about 16%, improved

strength and durability of concrete, and an abundance of natural resources. In contrast, its disadvantages include low activity at early ages, increased setting time, and variability of properties depending on the mineral composition. In concrete, its positive characteristics include the formation of C-S-H and C-A-H gels, increased compressive strength, and reduced permeability, but negative characteristics include reduced initial strength and the possibility of increased porosity at high percentages. The setting process of concrete containing calcined shale is initially slow, but at higher ages, pozzolanic reactions increase strength. The best time to add it is during the initial mixing stage of concrete and as a partial replacement of cement (10–30%). Finally, the use of calcined shale, in addition to its technical advantages, reduces the cost of cement production and increases the environmental compatibility of green and porous concretes [74–79].

- **Calcined Bauxite Residue**: Calcined bauxite residue (or calcined red mud) is one of the world's largest industrial wastes, with over 60 million tons produced annually, of which less than 5% has ever been recycled. Due to its high iron, alumina, silica, and titanium content, it is converted into a reactive supplementary cementitious material with an amorphous content of up to 45% after calcination at 750–850 °C (especially when co-calcined with 20–40% kaolinite or cheap clay). By replacing 30% of Portland cement, 28-day compressive strength is achieved between 88 and 101% of the control sample and even surpasses metakaolin at early ages (2–7 days), while CO_2 emissions are reduced by up to 35%. Its most important advantage is the drastic reduction of heavy metal leaching (up to 90%) and excellent durability against sulfate and freeze-thaw cycles, but the presence of dissolved sodium causes a decrease in fluidity and the need for more superplasticizers, and the possibility of efflorescence. The method of use is very simple: direct replacement of 20–30% by weight of cement in the dry mix phase. In geopolymer applications, a compressive strength of 50–70 MPa and a fire resistance of REI 120 min are also achieved. Overall, CBR is one of the most stable, least expensive, and most available SCMs of the new generation, which simultaneously solves the huge environmental problem of red mud and offers very good technical performance in concrete and cement products [80–83] (Table 3).

2.4 Agricultural Waste-Based SCMs

- **Rice Husk Ash (RHA)**: Rice husk ash (RHA) is one of the world's most sustainable and cost-effective cementitious additives, derived entirely from agricultural waste (rice husk). When rice husk is burned at a controlled temperature of 500–700 °C (optimally 650 °C) and then finely ground, it produces a grey powder with over 90% amorphous silica and a very high specific surface area (200,000–500,000 m^2/kg) whose pozzolanic activity is even stronger than that of many industrial pozzolans. Replacing 10–20% of cement with RHA can reduce CO_2 emissions by up to 82% (at a water-to-binder ratio of 0.5), increase long-term

Table 3 Comparison between 4 concrete SCMs (calcined/processed pozzolans)

Comparison criteria	Metakaolin	Calcined clay	Calcined shale	Calcined bauxite residue
Original source	Kaolin clay	Regular clay	Natural shale sedimentary rock	Same as red mud but calcined at 750–850 °C
Optimal replacement percentage (%)	5–15 (max 30%)	15–30	10–30	20–30
Maximum CO_2 reduction	Up to 40%	Up to 40%	Up to 16–20%	Up to 35%
Pozzolanic activity	Strongest synthetic pozzolan	High	Good to excellent	Excellent (Chapelle > 1200)
Effect on early strength (7 days)	Slightly less	A little less	10–20% reduction	Increase or equal (even better than metakaolin)
Effect on final strength	+ 40–80%	+ 15–20%	95–110%	88–101% (often > 100% in 90 days)
Effect on concrete fluidity	Strong reduction	Reduction	Moderate reduction	Moderate reduction + SP requirement
Effect on durability and permeability	Excellent	Excellent	Permeability reduction + excellent durability	Strong reduction of metal leaching + excellent sulfate/ice durability
Processing requirement	Yes (careful calcination)	Yes (calcination)	Precision calcination 750–850 °C + grinding	Calcinated 750–850 °C (preferably with 20–40% kaolinite)
Approximate cost	Relatively expensive	Cheap	Cheap	Almost free (waste)
Best application	HPC, UHPC, special structures	The future of cement (LC3), everywhere	Conventional concrete, green concrete, soil stabilization	Conventional concrete, UHSC, fire-resistant geopolymer, solving the red mud environmental dilemma

compressive strength by up to 31.7%, significantly reduce permeability and water absorption, provide excellent resistance to chloride intrusion and chemical attack, and greatly densify the concrete microstructure by forming additional C-S-H. This material has also shown very high predictability in machine learning models ($R^2 > 0.94$). Its main disadvantage is the decrease in flowability (slump loss) and the need for superplasticizers, the loss of initial strength, and especially the sharp loss of strength at percentages above 20%; its quality is also highly dependent on the precise control of the burning temperature (temperatures above 700° make silica crystalline and inactive). It slightly accelerates setting, and the best method of use is to add it to the dry mix before water. When combined with silica fume, an increase in strength of up to 250% has been reported. In countries such as Iran, India, China, Vietnam, and Thailand, where millions of tons of rice husks are discarded or burned annually, RHA is not only an excellent technical solution, but also a huge economic and environmental opportunity to significantly reduce cement consumption and decarbonize the concrete industry [84–88].

- **Sugarcane Bagasse Ash**: Sugarcane bagasse ash (SCBA) is one of the most abundant agricultural wastes in major sugarcane producing countries (Brazil, India, Pakistan, Thailand, Mexico, etc.) which is classified as a class F pozzolan with 55–82% amorphous silica, but only after proper processing (washing, re-firing at 600–650 °C and fine grinding) gives excellent performance. Optimal replacement of 5–15% of cement results in a significant increase in compressive, tensile, and flexural strength at high ages, greater C-S-H formation, a sharp reduction in porosity, excellent acid resistance, and even a reduction in thermal conductivity. In its raw state, it is almost useless and severely reduces fluidity, reduces early strength, and may increase final water absorption; therefore, success depends entirely on the quality of processing. The initial setting is slower, but in the long term, it provides much higher density and durability. The best method is to add it dry with cement, and by incorporating silica fume, up to 40% of cement can be replaced without loss of strength. In countries where millions of tons of bagasse are burned or buried annually, SCBA is a cheap, green, and highly effective solution to significantly reduce cement consumption, manage waste, and produce resistant concrete in corrosive environments [89–96].
- **POFA**: Palm oil fuel ash (POFA) is one of the most abundant agricultural wastes in palm oil producing countries (Malaysia, Indonesia, Thailand, Nigeria, Colombia, etc.) which, with 43–71% silica, can be a very effective cementitious additive, but only after careful processing (ultrafine grinding to UPOFA + heating to 500 °C to remove carbon). In this case, a 10–20% (and up to 30% for UPOFA) replacement significantly increases the ultimate compressive strength, greatly reduces permeability and dry shrinkage, increases the ultrasonic pulse velocity (UPV), and significantly improves the durability of concrete against chemical attacks. Its biggest advantage is the transformation of a huge and environmentally problematic waste into a cheap and green source to reduce cement consumption and CO_2 emissions, but in its raw state, it is almost useless and greatly reduces early strength and slows down setting. Therefore, success is entirely dependent on high-quality processing. The best method of addition is dry, simultaneously with the cement. In areas where millions of tons of palm shells and bunches are burned or buried annually, POFA is not only a technical solution, but also a great economic and environmental opportunity to produce green concrete and move toward a true circular economy [97–100].
- **Corn Cob Ash**: Corn cob ash (CCA) is a green and very cheap SCM obtained by burning corn cob waste at 600–800 °C and contains 40–60% amorphous silica.

By replacing 10–20% of Portland cement, the 28-day compressive strength is maintained by 85–100% of the control sample and is usually higher at 90 days; in ultra-high strength concrete (UHSC), it even reaches 140 MPa. CO_2 emissions are reduced by 25–35% and concrete costs are reduced by 15–20%; chloride and water infiltration are reduced by 50%; sulfate and frost resistance are also excellent. In roller compacted concrete (RCC) with jute fiber, cracking is improved by 40% and abrasion by 30%. Its main disadvantages are the loss of fresh flowability (requires more superplasticizer), accelerated setting due to high potassium, and the possibility of efflorescence; therefore, no more than 20–25% replacement is recommended. The method of use is very simple: direct replacement of 10–20% by weight of cement in the dry mix phase, together with the superplasticizer PCE. Overall, CCA is considered one of the most economical and sustainable agricultural SCMs of the new generation, especially in large corn-producing countries (USA, China, India, Africa), and can eliminate millions of tons of waste from the environment annually. Corn Cob Ash (CCA) is a green and very cheap SCM obtained by burning corn cob waste at 600–800 °C and contains 40–60% amorphous silica. By replacing 10–20% of Portland cement, the 28-day compressive strength is maintained at 85–100% of the control sample and is usually higher at 90 days; in ultra-high strength concrete (UHSC), it even reaches 140 MPa. CO_2 emissions are reduced by 25–35% and concrete costs by 15–20%, and chloride and water penetration are reduced by 50%; sulfate and frost resistance are also excellent. In roller compacted concrete (RCC) with jute fiber, cracking is improved by 40% and abrasion by 30%. Its main disadvantages are the loss of fresh flowability (requires more superplasticizer), accelerated setting due to high potassium, and the possibility of efflorescence; therefore, no more than 20–25% replacement is recommended. The method of application is very simple: direct replacement of 10–20% by weight of cement in the dry mix phase, combined with PCE superplasticizer. Overall, CCA is considered one of the most economical and sustainable new-generation agricultural SCMs, especially in major corn-producing countries (USA, China, India, Africa), and can remove millions of tons of waste from the environment annually [101–104].

- **Wheat Straw Ash**: Wheat straw ash (WSA) is a very stable and inexpensive SCM obtained from the controlled combustion of wheat straw at 550–700 °C and contains 65–94% amorphous silica. By replacing 10–20% of Portland cement, 85–95% of the control compressive strength is maintained at 28 days and typically exceeds 100% at 90 days; it also performs excellently in ultra-high strength concrete (UHSC) and recycled aggregate concrete (RAC). CO_2 emissions are reduced by 25–38% and concrete costs by 18–22%; chloride and water penetration are reduced by 40% and durability against sulfate, frost, and ASR is significantly improved. Combination with silica fume or coconut fiber creates a synergistic effect and reduces cracking by up to 35%. Its main drawbacks are the loss of fresh flow (need for superplasticizer), accelerated setting due to high potassium, and the possibility of efflorescence; therefore, a replacement of more than 20% is not recommended. The method of use is very simple: direct replacement of 10–20% by weight of cement in the dry mix phase, together with PCE or naphthalene superplasticizer. Overall, WSA is one of the best new generation agricultural SCMs, especially in large wheat-producing countries (India, China, Pakistan, Europe), which solves the problem of straw burning and offers very high technical performance and stability in concrete [105–108] (Table 4).

2.5 Emerging SCMs

- **Waste Glass Powder**: The waste glass powder (WGP) is a supplementary cementitious material in concrete technology; it can be produced via the recycling of post-consumer glass materials such as bottles, which have high amounts of amorphous silica (approximately 70%) and possess high pozzolanic activity; as a consequence, it can significantly contribute to the reduction of carbon dioxide emissions associated with Portland cement manufacturing. By partially substituting cement (usually in amounts of 10–20%), it can enhance the mechanical properties of concrete regarding compressive and tensile strength as well as improve their durability performance in relation to environmental attacks (frost resistance, sulfates, and chlorides). The cement substitution process in the WGP technology entails glass collection, washing, primary crushing followed by precise grinding down to less than 75 μm in particle size; this process significantly reduces greenhouse gas emissions by 35% compared to cement manufacturing. The main advantages include the recycling of glass materials in landfills; it also improves the fresh mix workability because of the spherical particles; in addition, it enhances the concrete microstructure properties as a consequence of increased amounts of calcium silicate hydrate. However, it also has several constraints regarding increased initial setting times; in addition, it may cause alkali-silica reactions in the coarse form; finally, it reduces concrete strength in amounts in excess of 30% of the mix design; these problems can be addressed by grinding the initial glass materials into finer powders either separately or by mix blending with fumed silica. During the concrete setting process, it reduces the concrete structure's porosity by reacting with the cement's calcium hydroxides and it reduces the concreting heat of hydration; it can be significantly beneficial in high-rise concrete structures. Usually, it can be mixed at the dry mixture processing phase with cement as well as aggregate; the ideal time should be before combining with water. Also, supplementing it with other supplementary cementitious materials can improve its performance; in addition, studies have identified an increased thermal resistance capability up to 400 °C in stable concrete. Generally, not only will it be useful in environmental conservation but also improve the construction-facilitated economy recycling process and the supplement substitution rates can be determined by conducting performance tests after 28–90 days. In conclusion, additional comments stress that WGP can be used in geopolymer concrete too, and future studies will be on examining long-term durability in WGP in actual conditions [109–111].

Table 4 Comparison between 5 concrete SCMs (agricultural waste-based SCMs)

Comparison criteria	RHA	POFA	SCBA	Corn cob ash	Wheat straw ash
Original source	Rice agricultural waste	Palm oil factory waste	Sugar factory waste	Burnt corn cob	Controlled burnt wheat straw
Optimal replacement percentage (%)	10–20	10–20 (up to 30% with UPOFA)	5–15	10–20	10–20
Maximum CO_2 reduction	Up to 82%	Up to 70–80% (with processing)	Up to 70–80% (with processing)	Up to 35%	Up to 38%
Pozzolanic activity	Extremely high	High (only after processing)	High (only after processing)	Good to excellent	Excellent (65–94% amorphous silica)
Effect on early strength (7 days)	Slightly lower	Significant reduction	Significant reduction	5–15% reduction	5–15% reduction
Effect on final strength	+ 20–35%	+ 20–40% (with processing)	+ 25–50% (with processing)	85–100% → often > 100% in 90 days	85–95% → often > 100% in 90 days
Effect on concrete fluidity	Strong reduction	Strong reduction	Strong reduction	Moderate to high reduction + SP required	Moderate to high reduction + SP required
Effect on durability and permeability	Excellent	Excellent (after processing)	Excellent (especially acid resistance)	50% reduction in chloride and water penetration	40% reduction in chloride and water penetration
Processing requirement	Yes (temperature control + grinding)	Yes mandatory (milling + heat 500°)	Yes, mandatory (washing + reburning + grinding)	Burned 600–800 °C + milled	Controlled burning 550–700 °C + grinding
Approximate cost	Extremely cheap	Very cheap in the palm countries	Very cheap in sugarcane countries	Very cheap (< $0.1/kg)	Very cheap (< 0.05 $$/kg)
Best application	Green concrete, rice-growing countries	Southeast Asian countries	Sugarcane-producing countries + acidic environments	Conventional concrete, UHSC, RCC road, lightweight concrete	Conventional concrete, UHSC, RAC, SCC, agricultural green concrete

- **Ceramic Waste Powder**: This ceramic waste powder (CWP) has been found to possess excellent pozzolanic activity with a high percentage of silica (68–77%) and alumina (18–20%), which helps decrease the carbon dioxide emission released by Portland cement. The paste can replace the cement paste (5–20%), which helps increase the strength properties such as compressive and bending strength, as well as the durability properties based on the effects of the environment—chloride ion penetration, sulfate, and high temperature. This CWP also has a production process where the collected waste undergoes primary crushing and precision grinding with a size smaller than 50 μm, which consumes high energy but results in a decrease of green gas emissions by 31%, which is lower than the energy consumption in the manufacture of cement. Some benefits include the reusing of ceramic waste sent to a landfill, the effect on the workability on the mixture decreasing by more than 10%, the prolonged initial setting time, which can be reduced by super plasticizers or by fine grinding, the decrease in the resulting hydration strength, and the increase in the thermal resistance up to 800 °C. Disadvantages include the decrease in workability by more than 10%, the prolonged initial setting time, the decrease on hydration strength, and the increase on thermal resistance beyond 800 °C. Based on the concrete setting process, CWP also produces supplementary hydrates such as calcium silicate through a pozzolanic reaction with calcium hydroxide, which helps increase the thermal resistance but also produces lower shrinkage. The mixing process can take place during the dry mixing phase together with the addition of the cement and the aggregate, the best time to add CWP prior to the addition of the water. CWP acts better together with other supplementary cementitious materials such as fly ash, and it has been found to increase the long-term durability under freeze-thaw cycling on the alkali activation cement. Nevertheless, based on the usage of CWP, the beneficial effect has also been found on the environment but also on the construction industries based on the circular economy processes with the preparation based on the tests conducted on a 28- and a 90-day breakage effect. At last, additional keys include the fact that CWP can also use the construction process on a mortar construction, where the further research will include the results on the CWP based on the effects on the different sizes and the economic problems on the construction processes [112–115].
- **Brick Powder**: Brick powder (BP) or brick dust/brick aggregate is a green concrete cementitious admixture obtained from the waste products of earth bricks due to demolishing activities of buildings and construction, and since it contains a higher percentage of silica, alumina, and iron (> 70%), its high pozzolanic activity plays a significant part in reducing carbon dioxide emission released during the production of Portland cement. The partial replacement of cement with 10–20% improves various strength parameters, such as compressive strength (54%) and resistance to various climatic conditions like chloride ion, sulfate, and high-temperature resistance, among others. The production of BP requires collection, primary crush, and precise grinding to particle size of 25–100 μm, which consumes a lot of energy and, as a positive attribute, reduces carbon emission by 89% and reduces energy usage compared to that during cement production.

The major advantage of using this technology is that it makes waste amounting to millions of tons, which otherwise can be disposed of inside landfill sites, and it also improves concrete microstructure and improves heat and insulating capacity due to irregular particle size. However, this technology can be threatened and limited by various disadvantages, which include decrease of early setting strength by 35% if above 20% substitution, higher water absorption, and poor concrete mix workability, which can be alleviated using various activities like action of different mechanical activation processes, grinding to smaller particle size, and addition of super plasticizer. After setting of concrete, this admixture takes part in pozzolanic reaction to form additional gels containing calcium silicates and/or calcium aluminate and reacts with calcium hydroxide to form concrete with high strength due to production of concrete that is hardened and resists shrinkage at 53%. The processes involved include dry mixing, where cement and aggregates are mixed together at the optimum time to add water. A minimum of 28 days is needed for activation. BP performs well when combined with SCMs like fly ash or silica fume. Research papers show that BP has efficiently minimized the carbon emissions discharged during the process of alkali-cement activation. The usage of this product helps maintain a preserved environment and a circular-way approach in the construction industry. The quantity of substitution for optimal performance is calculated on the basis of the 28-day and 90-day test. Finally, a few additional observations show that BP can be used for high-performance concretes like 3D-printed concrete, UHPC concrete, and CO_2-absorbing concrete, and further research work will be focused on the optimization of sizes and activation at a nanometric level [116–119] (Table 5).

3 Discussion

A study of present research studies carried out regarding SCM, in accordance with the network graph produced in VOS viewer, will clearly show that the research theme of SCMs will form the central research theme among the present research studies carried out. The research themes of sustainability, compressive strength, and durability will form the center of SCM research themes; SCM research themes will be surrounded by a number of other themes, such as supplementary material properties and tests, engineering properties of concrete, high-strength concrete, etc.

CO_2 emissions reduction, because of the production and use of SCM materials, is also often cited as the benefit related to sustainability and the concept of sustainable concrete production. The current trend and development, as evident from the current research and development activity on sustainable concrete production, is the use of SCM materials and its relation to the reduction of CO_2 emissions. Use of SCM materials has also enabled the emphasis on CO_2 emissions and its importance as a concern related to the use of sustainable concrete production. Moreover, as evident from the map, the trend to study the new SCMs like the glass powder obtained from

Table 5 Comparison between 3 concrete SCMs (recycled and emerging SCMs)

Comparison criteria	Waste glass powder	Ceramic waste powder	Brick powder
Original source	Post-consumer glass (bottles, containers, building glass)	Waste from tile industry, sanitary ceramics, ceramic blocks	Earthen bricks from building demolition and brick production waste
Optimal replacement percentage (%)	10–20	5–20 (optimum 10–15%)	10–20 (optimum 10–15%)
Maximum CO_2 reduction	Up to 35%	Up to 31%	Up to 89%
Pozzolanic activity	Strong (amorphous silica ≈ 70%)	Good to strong (68–77% silica + 18–20% alumina)	Strong ($SiO_2 + Al_2O_3 + Fe_2O_3 \geq 70\%$)
Effect on early strength (7 days)	Reduction 5–25% (longer setting time)	8–22% reduction	Reduction 7.5–35% (highest reduction in 3 materials)
Effect on final strength	Improvement 5–56% (especially with dosage 10–20%)	5–18% improvement	Improvement 10–54% (best final performance)
Effect on concrete fluidity	Significant improvement (spherical particles)	Reduction (angular particles, need for superplasticizer)	Medium to high reduction (requires superplasticizer)
Effect on durability and permeability	Excellent (reduction of chloride, sulfate and water penetration up to 40%)	Very good (heat resistance up to 800 °C, 10–40% penetration reduction)	Good to excellent (reduction of chloride penetration 1.5–6 times, high sulfate resistance)
Processing requirement	Fine grinding to < 75 μm (energy intensive)	Fine grinding to < 50 μm (energy-intensive)	Grind to 25–100 μm (less energy-intensive than the other two)
Approximate cost	Low to medium (cheaper than metakaolin, more expensive than fly ash)	Low (almost free near tile factories)	Very low (cheapest among the three materials)
Best application	Self-compacting concrete, high-durability concrete, large structures (due to low hydration heat), geopolymer concrete	High-temperature resistant concrete, concrete in corrosive environments, masonry mortar, alkali-activated concrete	Low-carbon concrete, lightweight concrete, 3D-printed concrete, UHPC, CO_2-absorbing concrete, economic and sustainable projects

the waste glass, ceramic waste, and red mud is also clear. Although they are not very common, they are also being popularized as they possess the ability to promote the sustainability concept for the enhancement of the capacity of the concrete materials. A relation to the RCs and the recycling process of the wastes, and a connection to the SCMs, is also substantiating the relevance and importance related to the circular economy concept for the concrete materials.

On the other hand, there exist other phrases that include hydration kinetics and pozzolanic activity, as well as microstructure, which demonstrate the great interest in the chemical reactions that result from the combination between SCMs and Portland cement. In the situation where SCMs make contact with Portland cement, Portland cement has been observed to produce calcium silicate and calcium-aluminum silicate. Casting technology is another area that has been a major contributor in setting the level of debate. In addition, carbon footprints, as well as SCMs as an approach that encourages the idea of having reduced carbon footprints, have played major roles in the debate regarding the adverse effects that the production of concrete creates within the environment. Scientists have been concerned with various ways that SCMs can be of great help in giving assistance in the reduction of the emissions produced by the GHG, starting with the reduction in the amount of Portland cement or even the waste materials that act as a cause obstructing the process within the environment.

The interests in nanotechnology and smart materials are also reflected in the increasing attention to nanoparticles, piezoelectric systems, and self-sensing technologies. The interests in smart materials are increasing not only because they are seen to hold promise for improving mechanical and durability performance of concrete materials using advanced material mechanisms but also because they are capable of sensing and adapting to changes in environmental and structural variables. Their sensing and adaptability capabilities open doors for the development of intelligent infrastructure systems, in which concrete is capable of sensing and reporting changes to and about itself on issues such as stress, strain, cracking, and temperatures. The interests in smart materials are increasing and are also being recognized as a significant element in the development and evolution of intelligent and sustainable infrastructure.

Lastly, the role of SCMs and the hydration process occurring in concrete has also gained prominence as a current concern in literature. Many current studies also investigate the hydration process occurring in SCMs and the impact caused during the formation and microstructure development of hardened cement paste. Such studies include the pozzolanic reaction, interface transition zone change, and the use of micromechanics, all of which are crucial for the development and production of increased strength and reduced permeability, thus improved durability. Many more studies on hydration continue to emerge, and the hydration process still remains core for the prominence and importance gained in the use of SCMs in the field of sustainable and high-performance concrete technology.

Thus, the trend given by the VOS viewer mapping is more toward the adoption of research with a diversified approach, which pertains to the area of SCM-modified concrete. It should be highlighted that machine learning models have been successfully employed in concrete technology to predict and optimize various mechanical and durability-related properties [120–124]. This is due to the increased horizon boundaries set by the latest research, not only in the areas of developing or enhancing the basic properties of concrete but also reaching for more sustainable and innovative materials that can contribute to more eco-friendly construction procedures. The recent research boundaries now encompass the newly developed materials under the name of SCMs, sophisticated modeling, and intelligent materials with upgraded performance capabilities (Fig. 2).

A vital and fundamental aspect of the assessment, design, and implementation of SCMs revolves around the requirement to make prudent decisions under conditions of uncertainty. The benefits accruing from SCMs and their advantages have been well documented and widely disseminated in literature, given their immense beneficial impacts on both environmental sustainability and mechanical performance. However, there are a lot of variables whose specific value is hard to predict with a high degree of accuracy. Such variables include, among others, chemical composition, reactivity, particle fineness, curing conditions, hydration rates, and development of microstructure. From this perspective, it is critical to note that the field of concrete technology is increasingly utilizing the incorporation of SCM materials, which are found in many industrial by-products, agricultural ash, and engineered

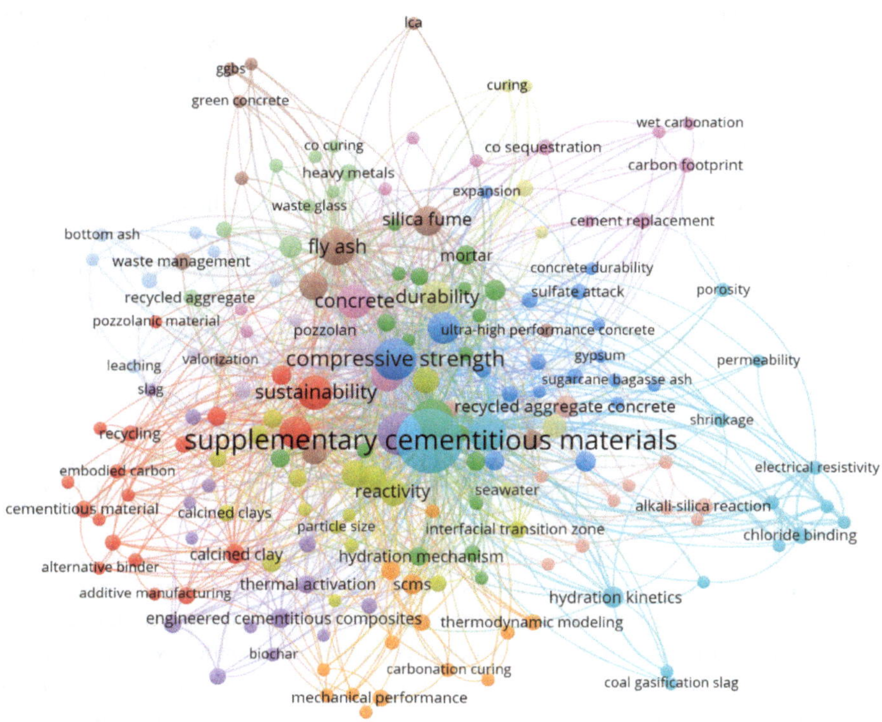

Fig. 2 Keyword co-occurrence network

pozzolans. This increasing dependency on such materials necessarily brings uncertainty into the decision-making process on SCMs and further complicates the general methodologies of evaluation and implementation.

The second major source of uncertainties that could also be identified is in the variability of material. Industrial by-products, ranging from fly ash, slag, and red mud, are not produced in a controlled production unit but vary greatly according to a number of factors. These factors include the origin of the by-product, firing temperature, mineral content, and the amount of trace elements. On top of that, natural and agricultural supplementary cementitious materials also vary the variability of their pozzolanic activity. This activity is dependent almost entirely on the production processes followed by the amount of grinding energy consumed and the firing temperature of rice husk ash, bagasse ash, and palm oil ash. Even highly advanced supplementary cementitious materials, such as metakaolin or calcined clay, are also not immune to this variability because of their dependence on their origin and firing temperature.

The uncertainty arising due to a lack of certainty regarding durability is further accentuated based on various processes that occur over a period of years, running into decades. Penetration of chloride ions, carbonation depth, resistance to attack from sulfates, resistance to freeze/thaw actions, and the rate of advancement of alkali-silica reactions are all affected by highly complex physicochemical reactions. These reactions are highly interwoven with the structural aspects of the pores in each mix, as well as the rate of advancement of hydration reactions. Although SCMs have established their benefit in increasing durability, their degree and rate could be highly sensitive to a given mix design, as well as the environment to which it is subjected. Due to this reason, there is a huge performance range, and this, on the whole, is highly beneficial. In this manner, along with other challenges, an increasing awareness regarding the need for meta-analysis has gained prominence.

Finally, uncertainty also affects the sustainability assessment. The reduction in CO_2 emissions caused by SCMs varies according to global transportation distances and entails grinding energy and processing demands. For a specific application type, possibly for larger volumes or less reactive species, the possible loss of strength might be overcome by adding extra cement or the addition of mix-designers' admixtures. This might have direct implications on the sustainability metrics, so that sustainability outcomes would not have fixed values for the time being. The benefits of the carbon-capture role played by SCMs clearly get affected by the ratio of mechanical demands and sustainability.

As far as the applications of concrete are concerned, the balancing of the mechanical properties and durability, or carbon emissions mitigation in uncertain conditions, would involve strategic planning. In the context of sustainability evaluation, there would be a shift in paradigm on how concrete comes into contact with uncertainty, whereby, in the case of SCM additives, instead of uncertainty becoming a barrier, it can be regarded as a resource that must be properly handled in order to ensure greater sustainability in the process of construction. Instead of a barrier, in the context of decision-making models for concrete, there would be the creation of uncertainty that would serve as an engine for innovation and sustainability.

4 Conclusions

The review also confirms that SCMs can achieve extraordinary structural, economic, environmental, and durability benefits compared to ordinary Portland cement concrete. SCMs are consistently observed to reduce CO_2 emissions and enhance durability properties, including permeability, chloride resistance, thermal and shrinkage resistance, and mechanical strengths. Such improvements are closely linked to their pozzolanic reactivity and the resulting densification of the concrete microstructure with time, leading to superior long-term performance under aggressive exposure conditions. The review also indicates that the by-products of industries, such as fly ash, slag, and silica fume, are still tin he most mature state of SCMs available, whereas natural or agricultural SCMs possess a remarkable potential for sustainability, particularly in areas or regions where industrial SCMs are limited. Large uncertainties still prevail due to chemical variability, early-age strength, and workability; besides which, these uncertainties are enhanced by differences in ash processing conditions, local raw material sources, fineness levels, and mineralogical composition.

It is evident, based on the review literature, that with respect to performance, SCM concrete has been shown to offer at least comparable, and sometimes improved, strength under both compression and flexure, as long as the proportions and fines are properly managed. Often, this occurs regardless of standard curing times, which suggests improved long-term durability when SCM proportions and size are properly managed. Results from the meta-analysis show that the major mechanism that assists in the enhancement of durability is the compaction of the microstructure by the increment in the amount of C-S-H gel, which later impacts the refinement of the microstructure and hence the reduction in permeability and the increment in durability. Moreover, the present study shows that the present gaps in the existing studies on the production and the use of supplementary cementitious materials encompass opportunities that can be applied in the areas of machine learning and durability studies.

In conclusion, SCMs are found to play an essential role in transitioning into climate-resilient and high-performance concrete technology. Additionally, industrial, agricultural, and recovered SCMs provide an efficient way of diverting waste streams and minimizing environmental footprints by unlocking opportunities that go beyond a generic Portland cement cycle. Future development should then emphasize standardization and integration of data by ensuring that uncertainties are diminished with regard to design limits, longevity prediction, and material variability. With an increasing demand for concrete consumed annually around the world, the development of SCMs has emerged as one of the most efficient methods yet employed in decarbonizing the built world by presenting SCMs as an essential tool in responding to climate challenges and minimizing natural resources.

AI Use Disclosure Statement The authors did not use any generative AI or AI-assisted technologies in the creation of this manuscript.

References

1. Singh N, Sharma RL, Yadav K (2025) Sustainable solutions: exploring supplementary cementitious materials in construction. Iran J Sci Technol Trans Civ Eng 49(3):2191–2224
2. Duchesne J (2021) Alternative supplementary cementitious materials for sustainable concrete structures: a review on characterization and properties. Waste Biomass Valoriz 12(3):1219–1236
3. Ahmed A (2024) Assessing the effects of supplementary cementitious materials on concrete properties: a review. Discov Civ Eng 1(1):145
4. Li G et al (2022) Fly ash application as supplementary cementitious material: a review. Materials 15(7):2664
5. Azad NM, Samarakoon SMSMK (2021) Utilization of industrial by-products/waste to manufacture geopolymer cement/concrete. Sustainability 13(2):873
6. Juenger MCG, Siddique R (2015) Recent advances in understanding the role of supplementary cementitious materials in concrete. Cem Concr Res 78:71–80
7. Fantilli AP, Jóźwiak-Niedźwiedzka D (2021) Supplementary cementitious materials in concrete, part I. MDPI
8. Snellings R, Mertens G, Elsen J (2012) Supplementary cementitious materials. Rev Miner Geochem 74(1):211–278
9. Sathiparan N, Dassanayake D, Subramaniam DN (2024) Utilization of supplementary cementitious materials in pervious concrete: a review. Int J Environ Sci Technol 21(6):5883–5918
10. Gh SH, Miyandehi BM, Khotbehsara MM, Tarbiat M (2015) Study quality steel mill slag for use in concrete containing metakaolin. In: International conference on advances in engineering, Tehran, Iran
11. Rahnamayiezekavat P, Sorooshnia E, Rashidi M, Faraji A, Mostafa S, Moon S (2022) Forensic analysis of the disputes typology of the NSW construction industry using PLS-SEM and prospective trend analysis. Buildings 12(10):1571
12. Fode TA, Jande YAC, Kivevele T (2023) Effects of different supplementary cementitious materials on durability and mechanical properties of cement composite—comprehensive review. Heliyon 9(7)
13. Ziaie A, Mehdizadeh B, Safi Jahanshahi F, Ahmadi N, Ghanizadeh AR (2026) Prediction of liquefaction-induced lateral displacements using hybrid GBRT and EOA. J Soft Comput Civ Eng 10(1):e225827
14. Raghav M, Park T, Yang H-M, Lee S-Y, Karthick S, Lee H-S (2021) Review of the effects of supplementary cementitious materials and chemical additives on the physical, mechanical and durability properties of hydraulic concrete. Materials 14(23):7270
15. Elahi MMA et al (2021) Improving the sulfate attack resistance of concrete by using supplementary cementitious materials (SCMs): a review. Constr Build Mater 281:122628. https://doi.org/10.1016/j.conbuildmat.2021.122628
16. Amin M, Elsakhawy Y, Abu el-hassan K, Abdelsalam BA (2022) Behavior evaluation of sustainable high strength geopolymer concrete based on fly ash, metakaolin, and slag. Case Stud Constr Mater 16:e00976. https://doi.org/10.1016/j.cscm.2022.e00976
17. Salih MA et al (2022) Strength and durability of sustainable self-consolidating concrete with high levels of supplementary cementitious materials. Materials 15(22):7991
18. Faraji A, Rashidi M, Perera S (2021) Text mining risk assessment-based model to conduct uncertainty analysis of the general conditions of contract in housing construction projects: case study of the NSW GC21. J Archit Eng 27(3):4021025
19. Cao C (2021) Machine learning-based prediction of porosity for concrete containing supplementary cementitious materials. arXiv preprint arXiv:2112.07353
20. Mehta PK, Monteiro PJM (2006) Concrete microstructure, properties, and materials. McGraw-Hill
21. Knight KA, Cunningham PR, Miller SA (2023) Optimizing supplementary cementitious material replacement to minimize the environmental impacts of concrete. Cem Concr Compos 139:105049. https://doi.org/10.1016/j.cemconcomp.2023.105049

22. Movagharnejad K, Zareei F, Mehdizadeh B, Salahi S, Lashkenari MS (2015) Prediction of the surface tension of hydrocarbons. Pet Sci Technol 33(9):1008–1016
23. Faraji A, Rashidi M, Khadir P, Perera S (2021) A risk analysis-best worst method based model for selection of the most appropriate contract strategy for onshore drilling projects in the Iranian petroleum industry. Buildings 11(3):97
24. Mehdizadeh B, Vessalas K, Castel A, Asadi H, Ben B, Deilami S (2024) Correlation behaviours for accelerated and natural carbonation testing of fly ash and ground granulated blast-furnace slag-based concrete mixes. Concr Aust 50:30–38
25. Sobolev K et al (2017) Class F fly ash assessment for use in concrete pavements
26. Orozco CR, Tangtermsirikul S, Sugiyama T, Babel S (2023) Comparative environmental assessment of low and high CaO fly ash in mass concrete mixtures for enhanced sustainability: impact of fly ash type and transportation. Environ Res 234. https://doi.org/10.1016/j.envres.2023.116579
27. Ahmed T, Elchalakani M, Karrech A, Dong M, Mohamed Ali MS, Yang H (2021) ECO-UHPC with high-volume class-F Fly ash: new insight into mechanical and durability properties. J Mater Civ Eng 33(7). https://doi.org/10.1061/(asce)mt.1943-5533.0003726
28. Mondal SK, Clinton C, Ma H, Kumar A, Okoronkwo MU (2023) Effect of class C and class F fly ash on early-age and mature-age properties of calcium sulfoaluminate cement paste. Sustainability (Switzerland) 15(3). https://doi.org/10.3390/su15032501
29. Diab MA (2022) Enhancing class F fly ash geopolymer concrete performance using lime and steam curing. J Eng Appl Sci 69(1). https://doi.org/10.1186/s44147-022-00111-6
30. Akbulut ZF, Yavuz D, Tawfik TA, Smarzewski P, Guler S (2024) Enhancing concrete performance through sustainable utilization of class-C and class-F fly ash: a comprehensive review. Multidisciplinary Digital Publishing Institute (MDPI). https://doi.org/10.3390/su16124905
31. Akmalaiuly K, Berdikul N, Pundienė I, Pranckevičienė J (2023) The effect of mechanical activation of fly ash on cement-based materials hydration and hardened state properties. Materials 16(8). https://doi.org/10.3390/ma16082959
32. Ahmad J et al (2022) A comprehensive review on the ground granulated blast furnace slag (GGBS) in concrete production. Sustainability (Switzerland) 14(14). https://doi.org/10.3390/su14148783
33. Rezvaniyanzadeh A, Fadaee MJ (2023) Effect of ground granulated blast furnace slag on the properties of hybrid-fiber-reinforced high-strength concrete under normal and high temperatures. Struct Concr 24(4):4680–4701. https://doi.org/10.1002/suco.202200967
34. Omer B, Saeed NM, Dheyaaldin MH, Jamal AS, Kurda R (2025) Evaluating the long-term strength of GGBFS-blended cement across various water-to-binder and superplasticizer ratios under heating/cooling cycles. PLoS One 20(7). https://doi.org/10.1371/journal.pone.0319923
35. Sharma Y, Yeluri M, Allena S, Owusu-Danquah J (2024) Incorporating limestone powder and ground granulated blast furnace slag in ultra-high performance concrete to enhance sustainability. Int J Concr Struct Mater 18(1). https://doi.org/10.1186/s40069-024-00723-7
36. Mehdizadeh Miyandehi B, Vessalas K, Castel A, Mortazavi M (2023) Investigation of carbonation behaviour in high-volume GGBFS concrete for rigid road pavements. ASCP (Australian Society for Concrete Pavements)
37. Humadi HS, Hasan ZK (2024) A review of enhancing the properties of concrete containing silica fume along with fibers. Int J Sci Res Arch 13(2):2720–2727. https://doi.org/10.30574/ijsra.2024.13.2.2440
38. Zheng Y, Wang Y, Jiang S, Xia Z, Zhang Q (2025) Effect of silica fume with various fineness levels on the strength degradation of concrete after steam curing. Constr Build Mater 464. https://doi.org/10.1016/j.conbuildmat.2025.140169
39. Ergeshov Z et al (2025) Influence of silica fume on the mechanical and microstructural properties and life cycle assessment of fly ash-based geopolymer mortar. Arab J Sci Eng. https://doi.org/10.1007/s13369-025-10142-9
40. Sui X et al (2025) Study on the effect of silica fume dosage on the mechanical properties and chloride ion penetration resistance of ultra-high ductile cementitious composites (UHDCC). J Build Eng 107. https://doi.org/10.1016/j.jobe.2025.112765

41. Eftekhar Afzali S, Ghasemi M, Rahimiratki A, Mehdizadeh B, Yousefieh N, Asgharnia M (2025) Compaction and compression behavior of waste materials and fiber-reinforced cement-treated sand. J Struct Des Constr Pract 30(2):04025007
42. Ahmed A (2024) Assessing the effects of supplementary cementitious materials on concrete properties: a review. Discov Civ Eng 1(1). https://doi.org/10.1007/s44290-024-00154-z
43. Singh N, Sharma RL, Yadav K (2025) Sustainable solutions: exploring supplementary cementitious materials in construction. Springer Science and Business Media Deutschland GmbH. https://doi.org/10.1007/s40996-024-01585-5
44. Zhu J, Yue H, Ma L, Li Z, Bai R (2023) The synergistic hydration mechanism and environmental safety of multiple solid wastes in red mud-based cementitious materials. Environ Sci Pollut Res 30(32):79241–79257. https://doi.org/10.1007/s11356-023-27800-w
45. Golabchi M, Faraji A (2015) Pre-project neuro-fuzzy decision support model for oil industry projects. Ind Manag J 7(4):837–860
46. Alsulami N, Chen HLR (2024) Low-carbon cement development: modifying volcanic ash reactivity for Portland cement replacement. Low-Carbon Mater Green Constr 2(1). https://doi.org/10.1007/s44242-024-00049-3
47. Calderoni D, Gastaldi D, Merlo V, Pellino L, Canonico F, Contrafatto L (2024) Etna volcanic ash as new sustainable supplementary cementitious material. J Build Eng 95. https://doi.org/10.1016/j.jobe.2024.110264
48. Játiva A, Etxeberria M (2024) Exploring the utilization of activated volcanic ash as a substitute for Portland cement in mortar formulation: a thorough experimental investigation. Materials 17(5). https://doi.org/10.3390/ma17051123
49. Hamada HM, Abed F, Beddu S, Humada AM, Majdi A (2023) Effect of volcanic ash and natural pozzolana on mechanical properties of sustainable cement concrete: a comprehensive review. Case Stud Constr Mater 19. https://doi.org/10.1016/j.cscm.2023.e02425
50. Churata R et al (2022) Study of geopolymer composites based on volcanic ash, fly ash, pozzolan, metakaolin and mining tailing. Buildings 12(8). https://doi.org/10.3390/buildings12081118
51. Farooq U et al (2024) Towards sustainable construction: harnessing potential of pumice powder for eco-friendly concrete, augmented by hybrid fiber integration to elevate concrete performance. Case Stud Constr Mater 21. https://doi.org/10.1016/j.cscm.2024.e03815
52. López-Gonzalez PJ et al (2025) Impact of pumice substitution on mortar properties: a case study on mechanical performance and XRD analysis. Infrastructures (Basel) 10(4). https://doi.org/10.3390/infrastructures10040095
53. Moolchandani K (2025) Advancements in pumice-based concrete: a comprehensive review. Elsevier BV. https://doi.org/10.1016/j.nxmate.2025.100646
54. Rodriguez C et al (2024) Properties of cement-based materials incorporating ground-recycled diatom. Crystals (Basel) 14(12). https://doi.org/10.3390/cryst14121030
55. Aruntaş HY, Yildiz E, Kaplan G (2022) The engineering performance of eco-friendly concretes containing diatomite fly ash and ground granulated blast furnace slag. Acta Polytech 62(5):505–521. https://doi.org/10.14311/AP.2022.62.0505
56. Staněk T (2016) Waste diatomaceous earth as cementitious binder constituent. Mater Sci Forum 1–5. Trans Tech Publications Ltd. https://doi.org/10.4028/www.scientific.net/MSF.865.1
57. Sirico A et al (2024) Environmental and mechanical analysis of low-carbon concrete with vitrified MSW incineration bottom ash as cement replacement. Struct Concr 25(4):2968–2990. https://doi.org/10.1002/suco.202300615
58. Sathiparan N, Dassanayake DHHP, Subramaniam DN (2024) Utilization of supplementary cementitious materials in pervious concrete: a review. Institute for Ionics. https://doi.org/10.1007/s13762-023-05440-4
59. Alqarni AS (2022) A comprehensive review on properties of sustainable concrete using volcanic pumice powder ash as a supplementary cementitious material. Elsevier Ltd. https://doi.org/10.1016/j.conbuildmat.2022.126533

60. Taylor P (2024) Integrating alternative supplementary cementitious materials into next generation paving tech brief
61. Mostafaei H, Bahmani H (2024) Sustainable high-performance concrete using zeolite powder: mechanical and carbon footprint analyses. Buildings 14(11). https://doi.org/10.3390/buildings14113660
62. Shekarchi M, Ahmadi B, Azarhomayun F, Shafei B, Kioumarsi M (2023) Natural zeolite as a supplementary cementitious material—a holistic review of main properties and applications. Elsevier Ltd. https://doi.org/10.1016/j.conbuildmat.2023.133766
63. Abed AT, Éva L (2025) A comprehensive study of perlite in building materials: balancing sustainability and performance. J Therm Anal Calorim 150(14):10627–10643. https://doi.org/10.1007/s10973-025-14410-6
64. Jaworska B, Stańczak D, Łukowski P (2024) The influence of waste perlite powder on selected mechanical properties of polymer–cement composites. Buildings 14(1). https://doi.org/10.3390/buildings14010181
65. Dacić A, Fenyvesi O, Rashid K, Ju M (2025) Novel design of circular concrete incorporating recycling sustainable constituents. J Build Eng 110. https://doi.org/10.1016/j.jobe.2025.113070
66. Elsebaei M, Mavroulidou M, Micheal A, Centeno MA, Shamass R, Rispoli O (2025) Dealuminated metakaolin in supplementary cementitious material and alkali-activated systems: a review. Multidisciplinary Digital Publishing Institute (MDPI). https://doi.org/10.3390/app15158599
67. Jagan S, Kiruthiga N, Ramya R, Shri Hariharan SN, Jagadeesh S (2025) Investigation of metakaolin utilization as a supplement to cement in sustainable high-performance concrete production. J Environ Nanotechnol 14(1):433–443. https://doi.org/10.13074/jent.2025.03.2441152
68. Mohamed AM, Tayeh BA (2024) Metakaolin in ultra-high-performance concrete: a critical review of its effectiveness as a green and sustainable admixture. Case Stud Constr Mater 21. https://doi.org/10.1016/j.cscm.2024.e03967
69. Cheah CB, Liew JJ, Khaw KLP, bin Md Akil H, Alengaram UJ (2025) Calcined clay as a low-carbon cementitious material: comprehensive review of treatment method, properties, and performance in concrete. Elsevier B.V. https://doi.org/10.1016/j.clwas.2025.100323
70. Haq IU, Bibi T, Nawaz A, Khalid HR, Elahi A (2025) Calcined clay–substituted sustainable cement binders: a holistic review. Springer. https://doi.org/10.1007/s11356-025-36347-x
71. Shao J, Guo S, Wang H (2025) A review of the performance, sustainable applications, and research challenges of limestone-calcined clay-cement (LC3) systems. Multidisciplinary Digital Publishing Institute (MDPI). https://doi.org/10.3390/coatings15050611
72. Renuka V, Rao SV, Tadepalli T (2025) State-of-the-art review on limestone calcined clay cement (LC3) and its recent advances. Springer Science and Business Media Deutschland GmbH. https://doi.org/10.1007/s40996-025-01857-8
73. Marandi N, Shirzad S (2025) Sustainable cement and concrete technologies: a review of materials and processes for carbon reduction. Springer Science and Business Media Deutschland GmbH. https://doi.org/10.1007/s41062-025-02213-5
74. Al-Adhadh AR, Daud NNN, Yusuf B, Al-Rkaby AH (2025) Improving the SHEAR strength and compressibility of sandy soil by utilizing calcined shale and cement. Civ Environ Eng 28(1):23–37. https://doi.org/10.2478/cee-2025-0003
75. Rashid K, Ltifi M, Zafar I, Ju M (2025) Performance enhancement of limestone calcined clay cement (LC3) using shale: industrial implementation perspectives. Eng Sci Technol Int J 69. https://doi.org/10.1016/j.jestch.2025.102106
76. Boakye K, Khorami M (2025) Performance of calcined impure kaolinitic clay as a partial substitute for Portland cement concrete: a review. Multidisciplinary Digital Publishing Institute (MDPI). https://doi.org/10.3390/jcs9040145
77. Niyomukiza JB, Eisazadeh A, Tangtermsirikul S (2025) Synergistic effect of calcined clay and fly ash on the performance of porous vegetation concrete. Constr Build Mater 458. https://doi.org/10.1016/j.conbuildmat.2024.139749

78. Islam MT, Kafle B, Al-Ameri R (2025) Natural clay in geopolymer concrete: a sustainable alternative pozzolanic material for future green construction—a comprehensive review. Sustainability 17(22):10180. https://doi.org/10.3390/su172210180
79. Nassiri S et al (2025) Opportunities for supplementary cementitious materials from natural sources and industrial byproducts: literature insights and supply assessment. Multidisciplinary Digital Publishing Institute (MDPI). https://doi.org/10.3390/buildings15173099
80. Ujaczki É et al (2018) Re-using bauxite residues: benefits beyond (critical raw) material recovery. Wiley. https://doi.org/10.1002/jctb.5687
81. Sakkas KM, Georgopoulos C, Balomenos E, Davris P, Makrigiannis I (2024) Bauxite residue valorisation through reductive smelting: production of inorganic polymer fire resistant building materials. OAlib 11(01):1–10. https://doi.org/10.4236/oalib.1111030
82. Peys A, Hertel T, Snellings R (2022) Co-calcination of bauxite residue with kaolinite in pursuit of a robust and high-quality supplementary cementitious material. Front Mater 9. https://doi.org/10.3389/fmats.2022.913151
83. Peys A et al (2024) Co-calcination to produce a synergistic blend of bauxite residue and low-grade kaolinitic clay for use as a supplementary cementitious material. Cement 18:100122. https://doi.org/10.1016/j.cement.2024.100122
84. Al-Shamasneh AR et al (2025) Forecasting compressive strength of concrete containing rice husk ash using various machine learning algorithms. Sci Rep 15(1):39162. https://doi.org/10.1038/s41598-025-23839-6
85. Li Y, Jia H, Li S (2025) Impact of rice husk ash properties on concrete strength: experimental and machine learning study. Adv Civ Eng 1:2025. https://doi.org/10.1155/adce/1812998
86. Montazeri P, Bamshad O, Aghililotf M, Singh P (2025) Mechanical and durability-based life cycle assessment of rice husk ash containing concrete. Case Stud Constr Mater 23. https://doi.org/10.1016/j.cscm.2025.e05241
87. Barbhuiya S, Das BB, Adak D, Rajput A, Katare V (2025) Rice husk ash in structural concrete: influence on strength, durability and sustainability. Discov Concr Cem 1(1). https://doi.org/10.1007/s44416-025-00013-9
88. Yin X, Rahman MM, Sun Y, Zhao Y, Wang J (2025) Sustainable soil–cement composites with rice husk ash and silica fume: a review of performance and environmental benefits. Multidisciplinary Digital Publishing Institute (MDPI). https://doi.org/10.3390/ma18122880
89. Goyal PK, Murmu M (2024) A review of sugarcane bagasse ash: characteristics, properties and sustainable applications in concrete. IOP Conf Ser Earth Environ Sci. https://doi.org/10.1088/1755-1315/1409/1/012002
90. Ahmad W, Ahmad A, Ostrowski KA, Aslam F, Joyklad P, Zajdel P (2021) Sustainable approach of using sugarcane bagasse ash in cement-based composites: a systematic review. Case Stud Constr Mater 15. https://doi.org/10.1016/j.cscm.2021.e00698
91. Yaseen N (2024) Exploring the potential of sugarcane bagasse ash as a sustainable supplementary cementitious material: experimental investigation and statistical analysis. Results Chem 10. https://doi.org/10.1016/j.rechem.2024.101723
92. Muhammad S, Zia M, Butt F (2020) Fresh, mechanical and durability properties of eco-friendly concrete containing sugarcane bagasse ash and wood saw dust
93. Farrant WE, Babafemi AJ, Kolawole JT, Panda B (2022) Influence of sugarcane bagasse ash and silica fume on the mechanical and durability properties of concrete. Materials 15(9). https://doi.org/10.3390/ma15093018
94. Abdalla TA, Koteng DO, Shitote SM, Matallah M (2022) Mechanical and durability properties of concrete incorporating silica fume and a high volume of sugarcane bagasse ash. Results Eng 16. https://doi.org/10.1016/j.rineng.2022.100666
95. Elawadly N, Sanad SA (2025) Sustainable concrete incorporating sugarcane bagasse ash: a study on workability, mechanical behavior, and microstructure. Innov Infrastruct Solut 10(9). https://doi.org/10.1007/s41062-025-02177-6
96. Faraji A, Rashidi M, Tezangi MR, Perera S (2021) Multihybrid dispute resolution framework for projects of downstream sector of petroleum industry. J Leg Aff Disput Resolut Eng Constr 13(4):4521026

97. Mohd Sufian ANB, Rahman MR, Bin Mohamad Said KA, Bin Bakri MK (2025) A critical review of various types of palm oil fuel ash (POFA) utilization in enhancing concrete and mortar properties. Springer Nature. https://doi.org/10.1007/s41024-025-00622-8
98. Hamada HM et al (2021) Sustainable use of palm oil fuel ash as a supplementary cementitious material: a comprehensive review. Elsevier Ltd. https://doi.org/10.1016/j.jobe.2021.102286
99. Abu Aisheh YI (2023) Palm oil fuel ash as a sustainable supplementary cementitious material for concrete: a state-of-the-art review. Case Stud Constr Mater 18. https://doi.org/10.1016/j.cscm.2022.e01770
100. Amran M et al (2021) Palm oil fuel ash-based eco-friendly concrete composite: a critical review of the long-term properties. MDPI. https://doi.org/10.3390/ma14227074
101. Maglad AM, Amin M, Zeyad AM, Tayeh BA, Agwa IS (2023) Engineering properties of ultra-high strength concrete containing sugarcane bagasse and corn stalk ashes. J Mater Res Technol 23:3196–3218. https://doi.org/10.1016/j.jmrt.2023.01.197
102. Okeke FO, Ahmed A, Imam A, Hassanin H (2024) A review of corncob-based building materials as a sustainable solution for the building and construction industry. Elsevier B.V. https://doi.org/10.1016/j.hybadv.2024.100269
103. Ahmad J, Arbili MM, Alabduljabbar H, Deifalla AF (2023) Concrete made with partially substitution corn cob ash: a review. Case Stud Constr Mater 18. https://doi.org/10.1016/j.cscm.2023.e02100
104. Bheel N et al (2024) Combined effect of jute fiber and corn cob ash on sustainability assessment and mechanical properties of roller compacted concrete using RSM modelling. Sci Rep 14(1). https://doi.org/10.1038/s41598-024-81345-7
105. Bastías B, González M, Rey-Rey J, Valerio G, Guindos P (2024) Sustainable cement paste development using wheat straw ash and silica fume replacement model. Sustainability (Switzerland) 16(24). https://doi.org/10.3390/su162411226
106. Al-Naghi AAA et al (2025) Sustainable concrete: investigating the synergistic effects of coconut fiber, wheat straw ash, and silica fume on RAC strength and durability. Sci Rep 15(1). https://doi.org/10.1038/s41598-025-02234-1
107. Sudeep YH, Ujwal MS, Mahesh R, Shiva Kumar G, Vinay A, Ramaraju HK (2024) Optimization of wheat straw ash for cement replacement in concrete using response surface methodology for enhanced sustainability. Low-Carbon Mater Green Constr 2(1). https://doi.org/10.1007/s44242-024-00054-6
108. Ahmad J, Arbili MM, Alqurashi M, Althoey F, Deifalla AF (2023) Concrete made with partial substitutions of wheat straw ash: a review. Springer Nature Korea. https://doi.org/10.1186/s40069-023-00616-1
109. Damdelen Ö, Mosaberpanah MA, Olabimtan SB, Anthony MD, Oluwole BO (2025) Binary effect of silica fume and waste glass powder as a cement replacement for a sustainable concrete. Discov Concr Cem 1(1):28. https://doi.org/10.1007/s44416-025-00028-2
110. Jiang X, Xiao R, Bai Y, Huang B, Ma Y (2022) Influence of waste glass powder as a supplementary cementitious material (SCM) on physical and mechanical properties of cement paste under high temperatures. J Clean Prod 340. https://doi.org/10.1016/j.jclepro.2022.130778
111. Sivasuriyan A, Koda E (2025) Incorporation of waste glass powder in the sustainable development of concrete. Multidisciplinary Digital Publishing Institute (MDPI). https://doi.org/10.3390/ma18143223
112. Tawfik TA et al (2024) Sustainable reuse of waste ceramic tiles powder and waste brick powder as a replacement for cement on green high strength concrete properties. Innov Infrastruct Solut 9(5). https://doi.org/10.1007/s41062-024-01498-2
113. Elemam WE, Agwa IS, Tahwia AM (2023) Reusing ceramic waste as a fine aggregate and supplemental cementitious material in the manufacture of sustainable concrete. Buildings 13(11). https://doi.org/10.3390/buildings13112726
114. Al-Fakih A, Odeh A, Mahamood MAA, Al-Shugaa MA, Al-Osta MA, Ahmad S (2023) Review of the properties of sustainable cementitious systems incorporating ceramic waste. Multidisciplinary Digital Publishing Institute (MDPI). https://doi.org/10.3390/buildings13082105

115. Ghonaim S, Morsy R (2023) Utilization of ceramic waste material as cement substitution in concrete. Buildings 13(8). https://doi.org/10.3390/buildings13082067
116. Sallı Bideci Ö, Bideci A, Ashour A (2024) Utilization of recycled brick powder as supplementary cementitious materials—a comprehensive review. Multidisciplinary Digital Publishing Institute (MDPI). https://doi.org/10.3390/ma17030637
117. Wu H, Gao J, Liu C, Guo Z, Luo X (2024) Reusing waste clay brick powder for low-carbon cement concrete and alkali-activated concrete: a critical review. J Clean Prod 449. https://doi.org/10.1016/j.jclepro.2024.141755
118. Małek M, Smarzewski P, Kunikowski M, Kluczyński J (2025) Utilization of brick powder as a sustainable additive in concrete: optimization of mechanical and thermal performance. Constr Build Mater 492. https://doi.org/10.1016/j.conbuildmat.2025.143092
119. Faraji A, Rashidi M, Sorooshnia E (2020) An integrated organizational system for project source selection in the major Iranian construction companies. Buildings 10(12). https://doi.org/10.3390/buildings10120251
120. Barkhordari MS, Armaghani DJ, Mohammed AS, Ulrikh, DV (2022) Data-driven compressive strength prediction of fly ash concrete using ensemble learner algorithms. Buildings 12(2):132
121. Asteris PG, Armaghani DJ, Hatzigeorgiou GD, Karayannis CG, Pilakoutas K (2019) Predicting the shear strength of reinforced concrete beams using artificial neural networks. Comput Concr Int J, 24(5):469–488
122. Gao J, Koopialipoor M, Armaghani DJ, Ghabussi A, Baharom S, Morasaei A, Zhou J (2020) Evaluating the bond strength of FRP in concrete samples using machine learning methods. Smart Struct Syst Int J 26(4):403–418
123. Momeni E, Dehghannezhad M, Omidinasab F, Armaghani DJ (2026) Assessment of compressive strength of concrete with glass powder and recycled aggregates using machine learning approaches. Comput Model Eng Sci. https://doi.org/10.32604/cmes.2026.077300
124. Armaghani DJ, Rasekh H, Asteris PG (2024) An advanced machine learning technique to predict compressive strength of green concrete incorporating waste foundry sand. Comput Concr 33(1):77

Intelligent Fragmentation: A Systematic Review of Imaging and AI for Blast Optimization in Open-Pit Mining

Hadi Fattahi and Mohammad Mohammadi

Abstract Blast-induced rock fragmentation in open-pit mining constitutes a fundamental component of the mine-to-mill value chain, directly influencing downstream energy consumption, production cost, operational safety, and processing efficiency. This chapter provides a systematic review of studies published between 2000 and 2025, focusing on the application of image processing and artificial intelligence (AI) techniques for the analysis, prediction, and optimization of blast-induced fragmentation. The reviewed literature is structured around three interconnected themes: (i) classical and advanced empirical fragmentation models, including Kuz–Ram, Rosin–Rammler, and Swebrec formulations; (ii) image-processing-based methods for particle size distribution assessment, with emphasis on scale calibration, illumination correction, and the limitations of two-dimensional observations; and (iii) intelligent data-driven approaches based on machine learning and deep learning, as well as their integration with multi-objective optimization. Only peer-reviewed studies reporting quantitative field or laboratory data and explicit performance metrics were considered, with particular emphasis on works employing spatially independent validation. The analysis shows that empirical models remain useful for preliminary blast design due to their simplicity and interpretability, although their accuracy is highly sensitive to site-specific calibration and geological variability. Image-based fragmentation analysis offers a rapid and cost-effective monitoring tool when rigorous acquisition and calibration protocols are applied; however, particle overlap and the inherently two-dimensional nature of images may introduce systematic bias. In contrast, AI-based models generally demonstrate superior predictive accuracy for key fragmentation indices such as P50 and P80, provided that high-quality datasets and robust validation schemes are employed. Nevertheless, challenges related to data standardization, computational cost, and limited interpretability continue to restrict their large-scale industrial adoption. Recent studies increasingly support hybrid, physics-informed frameworks that combine calibrated image-derived features with data-driven models under explicit physical constraints. Such integrated approaches, particularly when coupled with multi-objective optimization, show strong potential

H. Fattahi (✉) · M. Mohammadi
Rock Mechanics Engineering, Faculty of Earth Sciences Engineering, Arak University of Technology, Arak, Iran
e-mail: h.fattahi@arakut.ac.ir

A. Azizi et al. (eds.), *Intelligence at the Interface*, Emerging Trends in Mechatronics,
https://doi.org/10.1007/978-981-95-7584-8_9

for improving fragmentation control, reducing energy consumption, and enhancing operational stability in open-pit mining systems.

Keywords Blast-induced rock fragmentation · Image processing · Artificial intelligence (AI) · Machine learning · Multi-objective optimization

1 Introduction

Blast-induced rock fragmentation represents the first and one of the most influential stages of the mine-to-mill value chain in open-pit mining operations. The size distribution of blasted material directly affects downstream processes, including secondary crushing efficiency, milling energy consumption, equipment wear, production throughput, operational safety, and pit-wall stability [1, 2]. Although increasing explosive energy can enhance mill throughput, a substantial proportion of the released energy is dissipated in undesirable forms such as ground vibration, flyrock, and airblast, highlighting the necessity of achieving an optimal balance between fragmentation quality, safety, and environmental constraints [1]. Fragmentation quality is commonly quantified using percentile-based indices such as P50 and P80, which respectively represent the particle sizes below which 50% and 80% of the blasted material passes [3]. These indices are widely used because of their practical relevance to crusher feed requirements and comminution energy models. However, their reliability depends not only on blast design and rock mass characteristics but also on the accuracy of the measurement and prediction techniques employed [4]. Historically, empirical models such as Kuz–Ram, Rosin–Rammler, and their subsequent extensions have provided the foundation for fragmentation estimation in industrial practice [5]. Despite their simplicity and interpretability, these models rely on simplifying assumptions regarding rock mass homogeneity and blasting conditions, and their predictive performance is highly sensitive to site-specific calibration [6, 7]. Advances in sensing technology and computational capacity have led to the widespread adoption of image-processing-based approaches for fragmentation analysis. These methods enable rapid and cost-effective estimation of particle size distributions through digital imagery, provided that appropriate protocols for scale calibration, illumination control, and perspective correction are applied [4]. Nevertheless, the inherently two-dimensional nature of conventional imaging and the presence of particle overlap can introduce systematic bias, particularly in estimating coarse fragments. In response to these limitations, intelligent data-driven techniques based on machine learning and deep learning have increasingly been employed to predict fragmentation indices directly from blast design parameters, rock mass properties, and image-derived features. When trained on high-quality datasets and evaluated using rigorous validation strategies, these models have demonstrated superior predictive accuracy compared with traditional empirical approaches [6, 8]. However, their dependence on large, standardized datasets and their limited interpretability pose challenges for widespread industrial implementation.

To bridge the gap between predictive accuracy and physical interpretability, recent research has moved toward hybrid, physics-informed frameworks that integrate empirical knowledge, image-derived measurements, and data-driven models under explicit physical constraints [9]. Within this context, the present chapter systematically reviews and synthesizes advances in empirical modeling, image-based fragmentation analysis, and intelligent optimization techniques, with the aim of establishing a coherent and reproducible framework for improving blast-induced fragmentation control in open-pit mining.

2 Review Methodology

The methodological framework of this chapter was designed to ensure transparency, reproducibility, and scientific rigor in synthesizing existing knowledge on blast-induced rock fragmentation in open-pit mining. To achieve this objective, the review was conducted in accordance with the updated PRISMA 2020 guideline, which provides a structured and internationally accepted protocol for systematic literature reviews [10]. The methodological design was tailored to capture both theoretical developments and practical field-based applications related to fragmentation modeling, image-based measurement, and intelligent optimization. The scope of the review covers peer-reviewed studies published between 2000 and 2025, reflecting the period during which significant advances occurred in digital imaging technologies and artificial intelligence methods applied to mining engineering. The thematic focus was restricted to open-pit mining operations and blast-induced fragmentation, with particular emphasis on three interrelated domains: empirical fragmentation prediction models, image-processing-based particle size assessment, and data-driven approaches for predicting or optimizing characteristic fragmentation indices such as P50 and P80. A comprehensive literature search was performed using the Web of Science, Scopus, and ScienceDirect databases. Search strings were constructed using combinations of keywords related to blasting, fragmentation, open-pit mining, image processing, machine learning, deep learning, and multi-objective optimization, employing Boolean operators to ensure adequate coverage [11]. Only peer-reviewed journal articles with clearly documented methodologies and quantitative results were considered. Conference papers, abstracts without full text, and studies lacking methodological transparency or reliable data were excluded. Study selection was carried out in a multi-stage screening process. Initially, titles and abstracts were evaluated to remove irrelevant records. This was followed by a full-text assessment, during which studies were examined with respect to data quality, clarity of blast design parameters, and the presence of evaluable performance metrics. Preference was given to studies reporting field or laboratory measurements accompanied by explicit error metrics and, where available, spatially independent validation. The selection process was documented using a PRISMA flow diagram.

To ensure transparency and traceability of the study selection process, the sequential stages of identification, screening, eligibility assessment, and final inclusion were explicitly documented using a PRISMA 2020 flow diagram. As illustrated in Fig. 1, the initial database search yielded a broad set of records, which were progressively refined through title and abstract screening, full-text evaluation, and methodological quality assessment. This structured filtering process enabled the exclusion of studies lacking quantitative fragmentation metrics, transparent blast design parameters, or reproducible modeling frameworks, thereby ensuring that only methodologically robust and thematically relevant studies were retained for final synthesis.

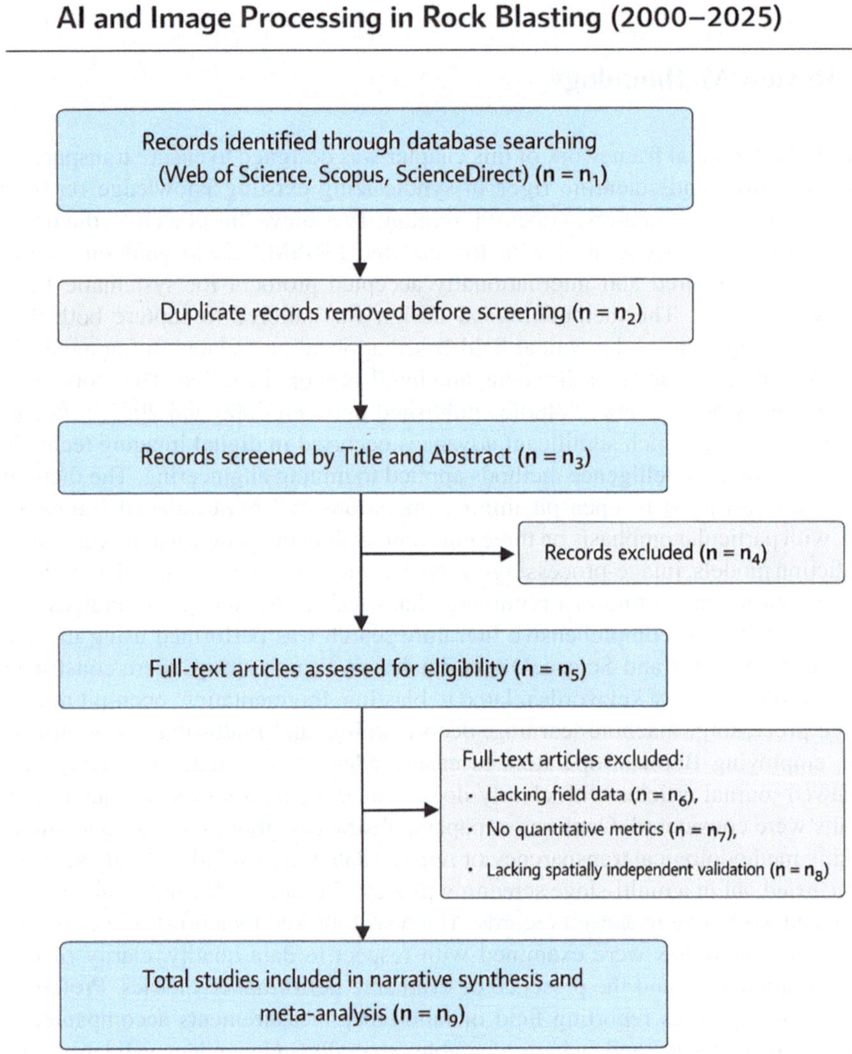

Fig. 1 PRISMA 2020 flow diagram illustrating the systematic study selection process for rock fragmentation research (2000–2025)

Data extraction was performed using a standardized form to ensure consistency across studies. Extracted variables included blast design parameters (e.g., burden, spacing, charge characteristics, delay timing), rock mass properties, imaging protocols (sensor type, resolution, scale calibration), and output indices such as P50, P80, and distribution uniformity. Model performance indicators, including RMSE and the coefficient of determination (R^2), were also recorded. The methodological quality of the included studies was assessed using the AMSTAR 2 framework [12], while the risk of bias was evaluated using the ROBIS tool [13]. In the subsequent synthesis, greater analytical weight was assigned to studies demonstrating robust validation strategies and transparent reporting. Owing to heterogeneity in experimental designs and reporting formats, results were synthesized primarily through structured narrative comparison, supplemented by limited quantitative aggregation where methodological homogeneity permitted.

3 Physical Fundamentals and Empirical Fragmentation Models

The optimization of blast-induced rock fragmentation is fundamentally constrained by the mechanical behavior of rock masses under transient, high-energy loading. A profound understanding of these underlying physical processes is an operational prerequisite for developing predictive models with meaningful extrapolation potential. This section elucidates the fracture mechanics governing fragmentation and critically examines the evolution of empirical models from their classical formulations to their contemporary, calibrated variants. Rock fragmentation during blasting results from the synergistic interaction of two primary mechanisms: the propagation of a stress wave and the subsequent expansion of high-pressure detonation gases [14]. The initial shock wave generates a dense network of radial cracks emanating from the blasthole wall, governed by the dynamic tensile strength of the rock and the impedance mismatch between the explosive and the rock medium. The ensuing gas pressure penetrates these incipient fractures, driving their propagation and coalescence to ultimately liberate rock fragments [14]. The final size distribution is thus a product of the interaction between this brittle fracture process and the inherent structural discontinuities of the rock mass. The heterogeneity of rock masses introduces significant directional anisotropy in wave propagation and stress distribution. Consequently, identical blast designs can yield markedly different fragmentation outcomes in spatially adjacent bench faces due to variations in fracture density, orientation, and rock strength [14]. This underscores the critical importance of parameters such as the stiffness ratio, the degree of explosive coupling within the blasthole, and the precise sequencing of millisecond delays [14]. The quantitative analysis of fragmentation necessitates the use of continuous statistical functions to model particle size distributions. The two-parameter Rosin–Rammler function has been historically dominant in mining practice due to its mathematical simplicity [15]. While it provides a reasonable fit for the mid-range of the distribution, empirical evidence consistently reveals systematic deviations at both the fine and coarse extremities. These tails are of particular operational significance, as fines influence downstream milling energy and dust generation, while oversize directly impacts secondary breaking costs and crusher throughput. The three-parameter Swebrec function has demonstrated superior performance across a wide spectrum of blast fragmentation datasets [7, 16]. Its enhanced flexibility allows for a more accurate characterization of the entire size range, including the critical coarse tail, thereby providing a more reliable basis for downstream process modeling [7, 16]. For predictive purposes, the Kuz–Ram model remains the most widely recognized empirical framework in industrial blast design [6]. It operates on a hybrid principle: the median fragment size is estimated using the Kuznetsov equation, while the distribution shape is described by the Rosin–Rammler function [6]. The model's principal advantage lies in its interpretability and minimal data requirements, facilitating rapid, preliminary blast design. However, its predictive accuracy is constrained by several inherent simplifications [6]. It assumes a relatively homogeneous rock mass and does not explicitly account for the dynamic effects of delay timing or the geometric constraints of multi-row blasts. Most critically, it lacks a robust physical mechanism to model the upper limit of fragment size, often leading to a systematic overestimation of the oversize fraction and underestimation of fines [6]. Consequently, the uncalibrated application of the standard Kuz–Ram model can yield significant errors, rendering its outputs operationally unreliable without site-specific adjustment. This recognition has spurred numerous modifications and extensions, including replacing the Rosin–Rammler kernel with the Swebrec function within an extended Kuz–Cunningham–Ouchterlony framework [6, 16].

The interrogation of physical fundamentals and empirical models yields several non-negotiable principles for fragmentation control. The dual-mechanism physics implies that changes in specific energy, coupling efficiency, and initiation timing exert a direct, often non-linear influence on key fragmentation indices [17, 18]. Purely descriptive statistical functions are inherently fragile at the distribution tails, and their predictive utility is substantially enhanced by adopting more robust formulations such as the Swebrec function or bi-modal distributions [15, 16]. Empirical predictive models transition from theoretical constructs to reliable operational tools only through rigorous, ongoing site-specific calibration that explicitly incorporates the effects of delay timing and blast geometry [6]. Furthermore, a holistic fragmentation control program aimed at optimizing downstream comminution performance must adopt a dual-focus strategy, simultaneously monitoring and controlling both the fines generation and the characteristics of the coarse tail [19]. In essence, while empirical models provide an indispensable foundation for blast design, their effective application is contingent upon a nuanced understanding of the underlying physics and a commitment to continuous calibration, setting the stage for the integration of more advanced measurement and modeling techniques.

4 Image-Processing-Based Fragmentation Analysis

The quantification of blast-induced rock fragmentation through image processing represents a pivotal technological bridge between empirical blast design and data-driven optimization within the mine-to-mill paradigm. This methodology transcends the limitations of manual sieve analysis by offering a rapid, scalable, and non-destructive means of assessing particle size distribution (PSD) directly in the field. However, the fidelity of this digital transformation is critically dependent on a rigorous chain of custody for the image data, encompassing acquisition protocols, computational segmentation, and statistical reconstruction, each stage introducing potential biases that must be explicitly acknowledged and mitigated [20]. The analytical pipeline commences with systematic image acquisition, a stage where methodological rigor dictates ultimate accuracy. Controlled photography of muck piles necessitates fixed geometric parameters—specifically altitude, azimuth, and declination angle—coupled with the inclusion of a physical scale reference within the frame to facilitate precise spatial calibration [21]. The inherent two-dimensionality of the capture process immediately imposes a fundamental constraint: the projected area of a fragment is an incomplete descriptor of its volume, a discrepancy that systematically skews the derived PSD, particularly in the coarse fraction where particle shape diverges significantly from sphericity [22]. Pre-processing algorithms are therefore deployed to correct for confounding photometric variables, including non-uniform illumination, specular highlights, and lens-induced distortions, which if unaddressed, manifest as artificial texture erroneously interpreted by segmentation routines as particle boundaries [21].

Segmentation, the computational isolation of discrete fragments from the image matrix, constitutes the core analytical challenge. Classical digital image processing techniques, such as adaptive global thresholding, edge detection (e.g., Canny, Sobel operators), and morphological operations (dilation, erosion), form the foundational toolkit [21]. The pervasive issue of particle adjacency and overlap is addressed through advanced algorithms like the marker-controlled watershed transform, which models the image gradient as a topological surface, effectively separating touching particles by simulating a flooding process from predefined marker points [21]. This approach, while superior to simple thresholding, is not infallible and can be sensitive to noise and subtle intensity gradients, sometimes leading to over-segmentation. Following successful segmentation, geometric properties (area, perimeter, Feret diameters) are extracted for each labeled region. These 2D metrics are subsequently translated into a continuous size distribution via stereological conversion models, ultimately yielding key fragmentation indices such as P_{50} and P_{80} [20]. The entire "image-to-distribution" workflow, established in seminal research, inherently grapples with the stereological problem of inferring three-dimensional size from two-dimensional cross-sections, an epistemological limitation that introduces unavoidable uncertainty [20].

The validation of image-based PSDs against the mechanical sieve—the traditional benchmark—reveals a consistent pattern of methodological dispersion. Comparative analyses quantify the error introduced by 2D projection, often demonstrating a bias toward finer fragmentation, as the depth dimension of coarse, blocky fragments is not captured [22]. This systemic error underscores the non-negotiable requirement for regular calibration against direct physical sampling. To fortify the reliability of the method, comprehensive quality assurance protocols must be enacted. These include the archival of rich metadata (sensor specifications, GPS coordinates, atmospheric conditions), the use of standardized calibration targets for both color and scale, and the implementation of periodic blind tests against sieve-derived distributions [23]. In response to the intrinsic constraints of 2D imaging, the field has progressively adopted three-dimensional spatial data acquisition. Light detection and ranging (LiDAR) scanners and photogrammetric techniques using unmanned aerial vehicles (UAVs) generate dense point clouds or textured mesh models of muck piles [23]. This evolution, illustrated conceptually in Fig. 2, marks a paradigm shift from analyzing ambiguous 2D projections to measuring explicit 3D surface geometries. The panel contrasts the traditional challenges of 2D imaging—such as perspective distortion, occlusion, and shading effects—with modern 3D/AI-driven solutions that employ UAVs and LiDAR for precise point cloud reconstruction and automated segmentation. By providing explicit depth information, 3D modeling mitigates the volume-estimation error of 2D methods, leading to more accurate delineation of large blocks and a more faithful reconstruction of the coarse tail of the PSD, a critical parameter for downstream crusher feed assessment [21, 23]. Field studies confirm that PSDs derived from 3D surface models exhibit stronger concordance with sieve results than those from 2D images alone. The most recent frontier in this domain is the integration of deep learning, particularly convolutional neural networks (CNNs), into the analytical stack. These models, such as U-Net architectures and their variants, can be trained to perform end-to-end segmentation or even direct regression of fragmentation indices from raw imagery. When trained on large, accurately annotated datasets and validated on spatially independent image sets, CNNs can achieve operational accuracy, potentially automating the analysis and reducing subjective intervention [24]. However, this data-driven approach introduces new vulnerabilities, namely domain shift and model brittleness. A network trained on images from one mine under specific lighting and geological conditions may suffer significant performance degradation when applied to another site, highlighting the continued need for robust, physics-aware validation and the perils of treating such models as black-box solutions. Furthermore, the "black box" nature of deep networks often conflicts with the industry's need for interpretability and traceability in measurement systems.

Consequently, the selection and deployment of an image-analysis system must be guided by a holistic assessment of accuracy, robustness, and operational feasibility. The most reliable implementations are those embedded within a continuous verification loop, where image-derived PSDs are routinely benchmarked against physical samples, and all results are reported with explicit uncertainty intervals. The synergistic integration of 2D RGB imagery for fine fragmentation with 3D point clouds for coarse block definition represents a best-practice hybrid approach, enhancing overall accuracy for the complete size spectrum. As evidenced by field applications, adherence to this disciplined, calibrated chain—from acquisition to analysis—is

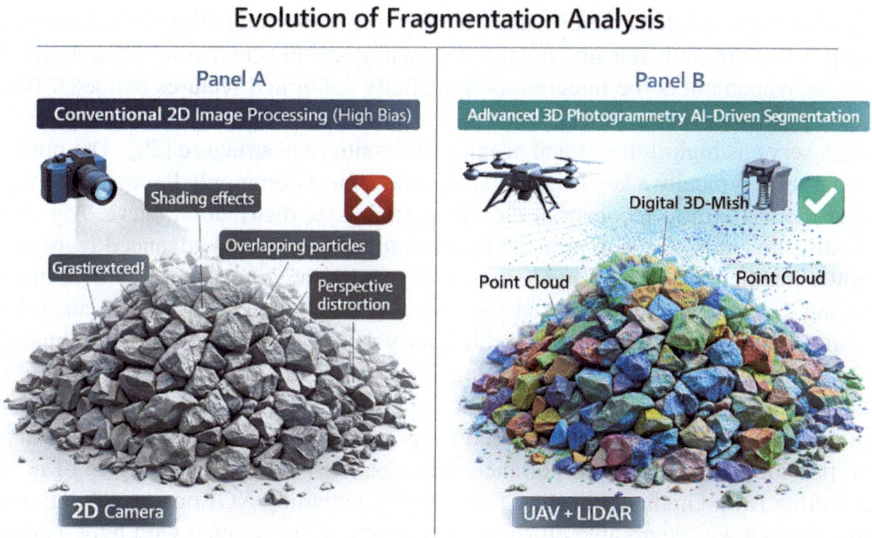

Fig. 2 Evolution of fragmentation analysis: (Panel A) traditional 2D imaging challenges including shading effects, particle overlapping, and perspective distortion; (Panel B) modern 3D/AI-driven solutions using UAV and LiDAR for precise point cloud reconstruction and automated segmentation (generated using AI)

paramount for transforming image-based fragmentation analysis from a promising monitoring tool into a reliable, decision-grade metric for mine-to-mill optimization [23, 25].

5 Machine Learning Models for Fragmentation Prediction

The advent of machine learning (ML) has precipitated a paradigm shift in the predictive modeling of blast-induced fragmentation, moving beyond the constraints of rigid empirical formulae toward adaptive, data-driven inference [26–28]. The predictive utility of any model, however, is contingent not merely upon its statistical accuracy on historical data but, more critically, on its robustness and generalizability under the heterogeneous and dynamic conditions of operational mining environments [29, 30]. This necessitates a modeling framework that integrates a comprehensive feature set—encompassing blast design parameters, rock mass geomechanical properties, and increasingly, calibrated image-derived metrics—within architectures capable of capturing complex, non-linear relationships while providing interpretable insights for engineering decision-making [29]. The foundational input space for ML models in fragmentation prediction is multidimensional. Blast design parameters, including burden, spacing, hole diameter, charge weight per delay, powder factor, stemming length, and initiation sequence, constitute the primary controllable variables [30,

31]. These are supplemented by intrinsic rock mass characteristics such as uniaxial compressive strength, density, fracture frequency, and elastic moduli [30]. A significant advancement is the integration of spatially calibrated features extracted from pre- or post-blast imagery, such as block size delineation or fracture network metrics, which serve as high-dimensional proxies for in-situ rock structure [29]. The predictive target is typically a key fragmentation index, most commonly P_{50} or P_{80}, though some models aim to reconstruct the full particle size distribution curve. Algorithmically, the landscape is dominated by ensemble methods and advanced regressors known for their balance of predictive power and relative resistance to overfitting. Among these, Random Forest (RF) has demonstrated consistent performance. For instance, a comparative study on marble quarry data reported an R^2 of approximately 0.94 for RF, substantially outperforming a classical artificial neural network (ANN) which achieved an R^2 of 0.83 [30]. The superiority of RF is often attributed to its inherent feature subsampling and bootstrap aggregation, which mitigate variance and provide native feature importance rankings, thereby offering a degree of interpretability [30]. Gradient boosting frameworks, including XGBoost and AdaBoost, have also shown remarkable efficacy, particularly when coupled with hyperparameter optimization techniques such as Bayesian optimization. One study leveraging data from 102 open-pit blasts achieved an exceptional R^2 of 0.96 and a root mean square error (RMSE) of 0.03 using optimized boosting models [31].

The pursuit of heightened predictive accuracy has led to the development of sophisticated ensemble strategies, such as stacking. A stacking model that combined the predictions of RF, XGBoost, and support vector regression (SVR) as base learners into a meta-regressor achieved an overall R^2 of 0.943, demonstrating that heterogeneous model combinations can extract complementary predictive signals and outperform any single constituent model [32]. Beyond point predictions, the interpretability of these "black-box" models is crucial for engineering adoption. Techniques like SHapley Additive exPlanations (SHAP) analysis have been employed to deconstruct model outputs, revealing, for example, that "powder factor" consistently exerts the strongest influence on predicted fragment size, while geometric parameters like "burden" may show weaker, context-dependent effects [30]. This interpretability layer transforms the model from a mere predictor into a diagnostic tool for understanding the sensitivity of fragmentation outcomes to specific design parameters. Table 1 synthesizes empirical evidence from key studies, providing a comparative overview of algorithmic performance, data types, and prevailing limitations. A critical analysis of the table reveals several overarching trends. Firstly, the highest reported accuracies (e.g., $R^2 \approx 0.98$ [33]) often come from models applied to datasets with extensive feature engineering and hyperparameter tuning, yet these same studies frequently note overfitting risks or limited external validation. Secondly, the performance gap between different algorithms is context-dependent; for example, while ensemble methods like RF and XGBoost generally outperform ANNs in mid-sized datasets [30, 31], their advantage may diminish with very large or complex feature spaces. Thirdly, the "Limitation" column universally highlights the Achilles' heel of data-driven approaches: dependency on data quality, volume, and representativeness. Studies relying on small datasets (e.g., $n = 80$ blasts [34]) or simulated data [35] explicitly acknowledge restricted generalizability, underscoring that high R^2 values on limited or synthetic data are not synonymous with field-ready robustness.

Nevertheless, the deployment of ML models is fraught with practical and epistemological challenges not fully captured by summary metrics. Their performance is intrinsically linked to the quality, quantity, and representativeness of the training data. Small or non-diverse datasets, common in mining applications due to the cost of data acquisition, heighten the risk of overfitting and limit generalizability across different geological domains or blast designs [34]. Furthermore, many published studies report performance metrics based on random data splits rather than the more rigorous spatially or temporally independent validation, potentially overstating real-world applicability [30, 32]. The computational cost of training and tuning complex models, coupled with the expertise required for their maintenance, can also be a barrier to widespread industrial implementation. The ultimate value of predictive ML models is realized not in isolation but through their integration into broader optimization and control systems. The most progressive applications link fragmentation prediction models with multi-objective optimization algorithms to dynamically inform blast design. By treating fragmentation indices (P_{80}), cost, ground vibration (PPV), and flyrock risk as simultaneous objectives, Pareto-optimal blast designs can be identified, enabling engineers to navigate trade-offs explicitly [35]. Furthermore, integrating these predictions with real-time decision-support systems for downstream comminution circuits allows for proactive adjustments in crusher settings or mill feed strategies, translating predictive accuracy into tangible gains in energy efficiency and cost reduction across the mineral processing value chain [35, 37]. This body of evidence underscores a critical synthesis: machine learning offers a powerful toolkit for surpassing the accuracy ceilings of empirical models, but its responsible and effective application demands rigorous validation, explicit uncertainty quantification, and seamless integration within the physical and economic constraints of the mining operational cycle. The comparative data in Table 1 serves not as a leaderboard of algorithms, but as a repository of case-specific evidence that highlights the context-dependent nature of ML success and the perennial gap between laboratory accuracy and field deployment.

The limitations inherent in purely data-driven models—such as their opacity, data hunger, and potential to produce physically inconsistent predictions—have catalyzed the development of more sophisticated hybrid physics-informed frameworks. These architectures aim to embed fundamental principles of rock mechanics, blast physics, and energy conservation directly into the learning process, thereby constraining solutions to physically plausible domains and enhancing model generalizability, especially in data-scarce scenarios. As illustrated in Fig. 3, a prototypical hybrid framework integrates multiple knowledge streams: a foundation layer of geomechanical constraints and blast design parameters, an intermediate layer of calibrated features extracted from 2D/3D image processing, and a supervisory physics-aware AI core that synthesizes these inputs under explicit physical laws. This integrative approach compensates for data limitations by imposing thermodynamic and mechanical constraints during model training, which not only regularizes the learning problem but also increases the interpretability of predictions by tethering outcomes to

Table 1 Real evidence—machine learning models in crush prediction

Row	Data type	Input features	Algorithm	Validation scheme	Performance metric	Limitation/practical note	References
1	Open-pit mine blasting (102 blasts)	Geometric ratios, stemming, powder factor, Young's modulus	RF, AdaBoost, gradient boost	Train/test and cross-validation	$R^2 \approx 0.96$, RMSE ≈ 0.03	Requires more field data	[31]
2	Marble quarry (219 blasts)	Number of holes, spacing, charging, powder factor	RF, SVR, XGBoost, ANN	Cross-validation	$R^2 \approx 0.94$ (RF)/0.83 (ANN)	Limited data; needs spatial validation	[30]
3	Open-pit mines (110 blasts)	Blasting and rock parameters	XGBoost + HyperOpt + SMOGN	Training set and independent hold-out set	$R^2 \approx 0.98$, MSE ≈ 0.0002	Overfitting risk	[33]
4	Chadormalu mine (80 blasts)	Charge, stemming length, delay, blasthole diameter	SVR, ANFIS	Cross-validation	$R^2 \approx 0.89$ (ANFIS), $R^2 \approx 0.83$ (SVR)	Small dataset; sensitivity to input selection	[34]
5	Geometallurgical data (BWI)	Geochemical and mechanical features	RF, K-means	Cross-validation	Kappa > 0.9	Requires accurate data clustering	[36]
6	Mill and production data	Rate of penetration, hardness, power, and P80	MLP neural network	Train/test	RMSE reduction > 10%	Requires large operational datasets	[37]
7	Laboratory data	Mill speed, feed PSD	MLP	Simulated + experimental data	$R^2 \approx 0.99$	Limited transferability to field conditions	[38]
8	Simulated data	Operational settings	MLP	Orthogonal experimental design	$R^2 > 0.99$	Poor performance outside the training range	[35]
9	Stacking model (97 blasts)	Blasting parameters	RF, XGBoost, SVR	Train/test split	$R^2 \approx 0.943$	Requires broader field datasets	[32]

Fig. 3 Proposed hybrid physics-informed framework integrating geomechanical constraints, 3D/2D image processing, and intelligent data-driven models for fragmentation optimization (generated using AI)

causative mechanisms. The progression toward such physics-informed hybridization represents the logical evolution from the standalone ML models discussed herein, setting the stage for the deep learning and advanced hybrid architectures examined in the following section.

6 Deep Networks and Hybrid Models

The evolution from conventional machine learning models to deep neural architectures represents a qualitative leap in the analytical capabilities for fragmentation analysis, enabling direct processing of high-dimensional, unstructured data such as images and 3D point clouds. This transition facilitates tasks ranging from pixel-level segmentation of rock fragments to the direct prediction of full particle size distributions (PSDs), moving beyond the regression of summary statistics. Concurrently, the integration of these data-driven models with physical principles has given rise to hybrid frameworks that seek to harmonize the predictive power of deep learning with the generalizability and interpretability afforded by mechanistic constraints. Deep convolutional neural networks (CNNs) have become instrumental in advancing image-based fragmentation analysis. Architectures derived from VGG have been successfully applied to classify and characterize rock fragments from 3D camera data. For instance, one study reported classification accuracies of approximately 80.8% and 75.0% for primary and recycled particles, respectively, and achieved a mean absolute error (MAE) of about 1.0% in predicting the PSD. Visualization techniques like Grad-CAM confirmed that the network's attention was appropriately focused on particle contours, validating its decision-making process [39]. For the more complex task of instance segmentation—identifying and delineating each individual fragment—U-Net architectures and their enhanced variants (e.g., U-Net++) have demonstrated superior performance. These encoder-decoder networks are particularly effective in biomedical and geoscientific image segmentation due to their ability to capture context and precise localization. A development known as "MrsNet," based on U-Net++ and trained on multi-sensor imagery combining RGB and Near-Infrared (NIR) data, showed improved generalization to novel scenes compared to models using single-sensor inputs [40]. This underscores the value of multimodal data fusion in overcoming domain shift, a common challenge when deploying models in varied lighting and geological conditions.

The pursuit of real-time operational analytics has driven the adoption of efficient object detection frameworks. Models from the You Only Look Once (YOLO) family, particularly YOLOv8, offer a compelling balance of speed and accuracy for fragment detection in blast images. Their single-stage, anchor-free design enables rapid inference, which is critical for field deployment. A study employing a YOLOv8-based model with dual-view (binocular) image input achieved a mean average precision (mAP) of 0.84 for real-time rock fragment detection and masking [41]. This demonstrates the potential for lightweight deep learning models to be integrated into on-site monitoring systems. Furthermore, the exploration of attention-based mechanisms and transformer architectures is an emerging frontier, promising enhanced ability to model long-range dependencies within an image, which could improve segmentation accuracy in complex, cluttered muckpile scenes [40]. Beyond static image analysis, the temporal dynamics of comminution processes have been modeled using recurrent neural networks (RNNs), specifically long short-term memory (LSTM) networks. LSTMs are adept at capturing time-dependent patterns in sequential data. In one application, an LSTM model was trained to predict the relative operational hardness in semi-autogenous grinding (SAG) mills, a key parameter influencing throughput and energy consumption. The model, trained on time-series data with a 0.5-h temporal window, achieved approximately 93% accuracy in identifying mill hardness states [42]. This highlights the utility of deep learning not only for characterizing blast outcomes but also for monitoring and optimizing the downstream processes that depend on them. The most significant conceptual advancement in this domain is the creation of hybrid models that fuse data-driven learning with physical or statistical priors. These models aim to mitigate the data hunger and "black-box" nature of pure deep learning approaches. One prominent strategy is early fusion, where non-image data (e.g., powder factor, rock strength) are concatenated with learned image features at the input or early hidden layers of a network. A notable example is the NRBO-CNN-LSSVM model, which combined a CNN for feature extraction from images with a least squares support vector machine (LSSVM) for regression, optimized using a novel metaheuristic. This hybrid achieved an R^2 of approximately 0.91 and an RMSE of 0.040 in predicting mean fragment size and employed SHAP analysis to elucidate the contribution of input features [5]. An alternative strategy is late fusion, where separate sub-models (e.g., one for image data, one for blast design parameters) are trained independently and their predictions are combined at the final stage. A more profound integration is embodied in physics-informed deep learning (PIDL). Here, physical laws are embedded directly into the learning objective, often by adding physics-based terms to the loss function. For example, a loss term could penalize predictions that violate energy conservation principles derived from Bond's or Kick's laws of comminution. This approach ensures that the model's outputs are not merely statistically plausible but are also consistent with established mechanistic knowledge, thereby improving extrapolation and robustness in low-data regimes [43]. Furthermore, acknowledging and quantifying predictive uncertainty is critical for trustworthy deployment in high-stakes industrial settings. Techniques such as Monte Carlo Dropout (a Bayesian approximation) and ensemble methods are employed to generate confidence intervals alongside point predictions, providing a measure of reliability for model-guided decisions.

The performance of these diverse architectures is summarized in Table 2, which provides a comparative synthesis of selected studies. The table elucidates key trends: CNN-based models excel in segmentation and image-derived PSD tasks; hybrid models like CNN-LSSVM show strong predictive accuracy while offering some interpretability; and specialized architectures like LSTM address time-series dependencies in downstream processes. However, the "Limitation" column consistently notes challenges such as limited training data volume, high computational or imaging costs, and questions regarding generalization beyond specific study conditions. This reinforces the notion that architectural sophistication alone is insufficient; rigorous validation—particularly spatial and temporal independent testing—and transparent reporting of constraints are indispensable for assessing true industrial viability.

In conclusion, the landscape of deep and hybrid models for fragmentation analysis is rich and rapidly evolving. The progression from CNNs for feature extraction to attention mechanisms and physics-informed loss functions reflects a maturing understanding that the most robust solutions will be those that seamlessly integrate high-fidelity data with the immutable laws of physics. The evidence compiled in Table 2 demonstrates that while these advanced methods can achieve impressive accuracy, their practical success is contingent upon thoughtful architectural design, comprehensive validation, and a clear alignment with the physical and economic objectives of the mining operation.

Table 2 Comparison of modeling specifications and results in selected studies

Row	Data domain	Architecture/ model	Data fusion	Validation scheme	Performance metric	Uncertainty	Interpretability	Limitation/ note	References
1	Primary and recycled rock particles (concrete production line)	Custom CNN (VGG-like)	–	Cross-sectional testing on two classes	Accuracy = 80.8%, 75.0%; MAE = 1.0%	–	Grad-CAM (focus on contours)	Higher error for near/ adjacent classes	[39]
2	Blast-fragmented rock (RGB + NIR images)	MrsNet (U-Net++)	Early fusion of RGB + NIR	Comparison with sieving and field data	High pixel-level accuracy (segmentation)	–	–	Limited training data volume	[40]
3	Semi-autogenous grinding (SAG) circuit	LSTM	–	Comparison across two mills	Prediction accuracy ~ 93%	–	–	Reduced accuracy over longer time horizons	[42]
4	Plant comminution simulation (operational parameters)	Multilayer neural networks	–	Testing under new scenarios	$R^2 > 0.99$ (training); > 0.98 (testing)	–	Sensitivity analysis	Lack of generalization outside the training range	[35]
5	Blast rock mass (stereo images)	YOLOv8 (enhanced)	–	Train/validation split	mAP = 0.84	–	–	–	[41]
6	Blasting tests (soil and blast parameters)	CNN + LSSVM (NRBO)	Fusion of image and numerical features	Spatially independent validation	$R^2 \approx 0.91$; RMSE ≈ 0.040	–	SHAP (key drivers)	Small sample size (105 records)	[5]

7 Blast Design Optimization

The ultimate objective of fragmentation analysis and predictive modeling is to inform and optimize the blast design process itself, transforming passive observation into active, value-driven control. Effective optimization in open-pit mining necessitates a closed-loop, integrated framework that seamlessly couples precise measurement, reliable prediction, and multi-criteria decision-making, all while transparently accounting for system uncertainties and field variability. This paradigm shift moves beyond singular objectives, such as minimizing P_{80}, toward a holistic balance of productivity, cost, safety, and environmental constraints. The foundation of any optimization cycle is high-fidelity fragmentation measurement. The calibration and dimensionality of the input data critically influence the quality of subsequent analysis. The adoption of three-dimensional sensing technologies—specifically photogrammetry from unmanned aerial vehicles (UAVs) and terrestrial LiDAR—has revolutionized this stage. These systems generate accurate 3D surface models or point clouds of the muck pile, enabling volumetric assessment of fragment size. A significant advantage of 3D analysis is its ability to more reliably estimate the coarse fraction of the particle size distribution (PSD), which is often under-represented in 2D imagery. In practice, the complete PSD can be reconstructed by fitting a robust statistical function, such as the Swebrec distribution, to the measured coarse data from the 3D model and intelligently extrapolating it to the fine fraction, for which 2D imagery may remain sufficient [44]. Advanced 3D systems have demonstrated the capability to extract scale without physical scale bars by leveraging known geometric features or ground control points, though their absolute accuracy remains contingent upon rigorous spatial calibration [44]. The post-blast image processing pipeline has been further augmented by deep learning for automation and accuracy. Object detection architectures, particularly YOLOv8, have been successfully deployed for the automated identification and classification of individual rock fragments in muck pile imagery [41]. This automation enables rapid, consistent analysis of large image datasets. However, the accuracy of such models is inherently dependent on the quality and representativeness of their training data, which must encompass the diverse lighting, lithology, and fragmentation conditions encountered in practice. Crucially, even when using 3D models as input, the georeferencing and scaling of these models via known control points remains a non-negotiable step for ensuring dimensional accuracy [41]. The prediction phase within the optimization loop is increasingly dominated by hybrid and physics-informed frameworks. These models integrate data-driven learning with domain knowledge to improve generalizability and physical consistency. For instance, in predicting blast-induced ground vibration (peak particle velocity, PPV), a physics-informed neural network (PINN) optimized with a particle swarm optimization algorithm demonstrated superior predictive accuracy by embedding the governing equations of wave propagation into its loss function [45]. This principle is directly transferable to fragmentation prediction, where embedding energy conservation laws or fracture mechanics principles can constrain data-driven models, reducing prediction error and enhancing robustness, especially when extrapolating beyond the immediate training data.

The core of blast design optimization lies in navigating multiple, often conflicting objectives. A modern blast design must simultaneously aim to:

1. Reduce fragment size (e.g., lower P_{80}) to enhance downstream comminution efficiency.
2. Minimize operational costs, including explosive consumption and secondary breaking.
3. Control environmental and safety impacts, primarily by limiting ground vibration (PPV) and flyrock.
4. Ensure geotechnical stability of pit walls.

This multi-objective problem is ideally suited for metaheuristic optimization algorithms. Techniques such as multi-objective particle swarm optimization (MOPSO) and the non-dominated sorting genetic algorithm II (NSGA-II) are proficient at identifying pareto-optimal solutions—designs where no objective can be improved without worsening another. A field application of MOPSO, for example, identified an optimal blast parameter set that balanced PPV constraints with fragmentation goals, converging on an explosive charge weight of 173.68 kg and a powder factor of 0.38 kg/m^3 as a preferred solution [46]. To account for the inherent uncertainty in rock mass properties and measurement, it is imperative to incorporate stochastic or robust optimization methods. These approaches evaluate solutions across a range of probable conditions, ensuring the final design maintains performance not just for an average case, but with acceptable reliability under expected variability.

The practical impact of linking high-quality measurement to dynamic design adjustment can be substantial. A field study demonstrated that using UAV-based 3D mapping to precisely assess and adjust charge distribution resulted in an average reduction of 42% in flyrock range [47]. By identifying zones of over- or undercharging in real-time or for subsequent blasts, safety risks were markedly mitigated. Therefore, a state-of-the-art optimization framework must provide not only target output values (P_{80}, uniformity, cost, PPV) but also associated prediction confidence intervals. This enables mine engineers to make final decisions based on explicit risk-benefit and economic criteria, selecting a blast design that is not only optimal on paper but also robust and reliable in execution.

Table 3 provides a systematic review of key studies in blast design optimization, synthesizing the environments, methods, and findings. The table highlights that while advanced 3D measurement and AI-driven prediction are powerful enablers, successful optimization consistently relies on field validation and a clear multi-objective formulation. The limitations noted, such as limited sample sizes or a narrow focus on single parameters like PPV, underscore the ongoing challenge of developing comprehensive, field-validated frameworks that are readily transferable across diverse mining operations.

In synthesis, blast design optimization is evolving into a sophisticated, data-integrated discipline. The convergence of precise 3D measurement, physics-aware

Table 3 Systematic review of methodologies and outcomes in blast design optimization studies

Row	Environment/data	Analytical task	Method/algorithm	Output index	Validation	Key finding	Limitation	References
1	Hard-rock open-pit mine; 2D/3D muckpile images	Fragmentation measurement and optimization	3D image analysis + blast design adjustment	P80, uniformity	Four field blasts	P80 reduction from 115 to 100 mm and improved uniformity	Limited sample size	[47]
2	3D LiDAR data	Fragmentation size-distribution measurement	Deep neural network on point clouds	3D PSD	Simulated + real datasets	Improved accuracy of 3D PSD estimation	High data requirements	[44]
3	Pre-split blast images	Assessment of presplit quality and fragmentation	Image-based ML model	Half-cast ratio and PSD	Field database	Prediction accuracy of 98.5%	Limited transferability	[48]
4	Blast vibration data (Zhoushan)	PPV prediction and optimization	LSSVM + MOPSO	PPV, W, PF	Field data	Optimal values (W = 173.68 kg, PF = 0.38)	Focus limited to PPV	[46]

predictive modeling, and multi-objective decision support systems creates a powerful toolkit for the mining engineer. The future of this field lies in further closing the loop between prediction and execution, perhaps through real-time adjustment systems, and in expanding optimization frameworks to more fully incorporate downstream economic and energy metrics, thereby fully realizing the promise of the mine-to-mill value chain.

8 Linking Fragmentation to Downstream Energy and Economics

The conceptual framework of "mine-to-mill" optimization transcends mere technical integration, demanding a quantitatively rigorous and economically grounded linkage between blast-induced fragmentation and downstream process performance. The operational and economic validity of any fragmentation control strategy is contingent upon establishing demonstrable, causal relationships between particle size distribution (PSD) parameters—notably the 80th percentile size (P_{80}) and distribution uniformity—and key performance indicators in comminution circuits, specifically specific energy consumption (kWh/t), throughput capacity, and overall unit cost of production. Moving beyond qualitative assertions to quantifiable, field-validated correlations is essential for justifying capital and operational expenditures in advanced blasting and monitoring technologies.

Within comminution circuits, the size and shape of the feed material constitute the primary deterministic factors for energy demand. Established energy-size reduction frameworks, notably Bond's Law and its more contemporary extensions by Morrell, posit that for a given mineralogy and machine configuration, the specific energy required for comminution is inversely related to the feed size. Empirical field evidence robustly supports this principle. For example, a comprehensive study in a copper mine documented that reducing the feed P_{80} from 500 to 120 mm through optimized blasting yielded a circa 20% increase in processing plant capacity and a concurrent 29% reduction in specific energy consumption [49]. However, such quantitative analyses necessitate rigorous control of confounding variables, including ore hardness (e.g., Bond Work Index), moisture content, and crusher/mill operating parameters. Failure to account for these co-variates, or to report associated uncertainty intervals, invalidates any claimed energy benefits, rendering the analysis scientifically and operationally inconclusive. The influence of fragmentation extends beyond a single-point index like P_{80}; the complete shape of the PSD is equally consequential. A distribution with a pronounced coarse tail, even with an acceptable P_{80}, can induce deleterious downstream effects. Such a profile increases the circulating load within closed-circuit grinding operations, elevates specific energy consumption due to inefficient breakage of oversize material, and heightens the risk of classification system blockages. Conversely, a more uniform feed distribution promotes stable crusher and mill operation, mitigates power fluctuations, and enhances overall circuit capacity and control stability. The Swebrec distribution function has proven particularly effective in modeling this coarse tail behavior, providing a more reliable basis for predicting its downstream energy implications than traditional Rosin–Rammler models [7].

Technically, the fragmentation-energy-economic nexus is governed by three interconnected linkages. First, the fundamental relationship between feed P_{80}, uniformity, and specific comminution energy, as formalized in Bond and Morrell models. Second, the impact of PSD shape on process bottlenecks, notably circulating load and classification efficiency. Third, the operational control consequences, manifested as power draw variability and shifts in the circuit's steady-state operating point. A critical caveat is that a reduction in P_{80}, if accompanied by a deterioration in PSD uniformity (e.g., increased fines generation without corresponding coarse reduction), may yield negligible net energy savings as benefits are offset by elevated circulating loads and increased grinding media wear. From an economic perspective, the ultimate metric for evaluating blast design modifications is not specific energy reduction in isolation, but the net value creation per unit of production. A holistic economic assessment must simultaneously quantify impacts across three cost/revenue components: (1) blasting costs (explosives, drilling, vibration/flyrock control), (2) comminution costs (energy, grinding media, liner wear, and maintenance), and (3) revenue-related benefits (increased throughput, improved recovery, and product quality). Frameworks proposed by Morrell and later expanded by Lastra et al. provide methodologies for this integrated economic evaluation [50, 51]. Furthermore, advanced modeling demonstrates that more accurate estimation of plant feed characteristics via 3D fragmentation modeling directly reduces economic uncertainty in mine-to-mill planning, enabling more confident capital and operational decisions [52]. The scientific pathway for establishing a credible "fragmentation-energy-economics" chain thus involves sequential, validated steps: commencing with calibrated 2D/3D image-based measurement to accurately determine P_{80} and PSD shape; mapping these parameters to interpretable energy models (e.g., Bond, Morrell) with sensitivity analysis for distribution uniformity; implementing time-series monitoring of mill power draw and circulating load to assess circuit stability impacts; and culminating in a comprehensive economic analysis that computes unit cost and net return, explicitly reporting uncertainty intervals derived from measurement and model errors. Table 4 synthesizes empirical evidence from pivotal studies linking fragmentation to downstream energy and economics. The compiled data underscores that while the theoretical energy benefits of optimized fragmentation are well-established, the magnitude of these benefits is highly site-specific and dependent on accurate measurement and holistic modeling. The limitations noted in the table, such as model dependency on specific circuit assumptions or the need for extensive input data, highlight the persistent challenge of translating generalized principles into universally quantifiable and actionable economic guidelines.

In conclusion, forging a robust, quantitative link between blasting outcomes and downstream economics is the cornerstone of genuine mine-to-mill optimization. This requires moving from correlative observations to causal, physics-based models that are continuously calibrated with high-fidelity field data. The future of this integrative discipline lies in the development of dynamic digital twins that can simulate the

Table 4 Systematic review of empirical evidence linking fragmentation to downstream energy and economic performance

Row	Author/year	Environment/ data	Method/ algorithm	Output index	Validation type	Key finding	Limitation	References
1	Morrell (2004), Minerals Engineering 17 (11–12)	Milling circuit	Bond and Morrell models	Specific energy (kWh/t)	Industrial data	Reducing P_{80} lowers energy demand and improves throughput	Dependent on circuit assumptions	[53]
2	Sanchidrián et al. (2014), IJRMMS 71:381–394	Field blasting	Swebrec distribution	PSD and oversize risk	Field data	More accurate distributions enable more reliable energy prediction	Requires site-specific calibration	[7]
3	Lapčević et al. (2023), Applied Sciences 13 (14)	3D data	3D blast fragmentation model	P_{80}/3D PSD	Field validation	The 3D model reduced feed-estimation error and improved mine-to-mill economics	Requires extensive input data	[52]
4	Losaladjome et al. (2024), Minerals	Open-pit copper mine	Extended Kuz–Ram model	kWh/t and production rate	Industrial simulation	Reducing P_{80} from 500 → 120 mm decreased energy by ~ 29% and increased capacity by 20%	Model assumptions are restrictive	[49]

economic consequences of blast design changes in near real-time, thereby enabling a truly proactive and value-maximizing approach to fragmentation management across the mineral processing value chain.

9 Comparative Synthesis and Study Quality Assessment

A critical synthesis of the reviewed literature reveals that the evolution of fragmentation prediction and optimization is characterized by a distinct trajectory from descriptive empirical models toward explanatory, physics-constrained intelligent systems. This progression is not merely technological but epistemological, reflecting a maturation in how the mining industry conceptualizes and quantifies the rock breakage process. The aggregate body of peer-reviewed evidence strongly indicates that the highest predictive accuracy and operational reliability are achieved when high-fidelity, calibrated three-dimensional measurements are directly coupled with process-informed models through a transparent, validated mapping. Within the domain of imaging and measurement, the adoption of volumetric techniques such as micro-Computed Tomography (μ-CT) and their integration with advanced neural architectures has precipitated a paradigm shift. Studies employing convolutional neural networks like EfficientU-Net-b3 on 3D mineralogical data (e.g., μ-CT coupled with μ-XRF) have demonstrated significant improvements in segmentation accuracy and liberation estimation over traditional two-dimensional methods [54]. This enhancement is quantitatively reflected in refined mineral liberation correlation matrices, providing a more reliable foundation for downstream process simulation. Concurrently, geometallurgical research has demonstrated that predictive modeling of critical parameters like the Bond Work Index (BWI) can be successfully achieved using a strategically selected subset of rock mechanics tests, thereby offering a pathway to reduce characterization costs without sacrificing essential predictive accuracy [55]. However, the generalizability of such models across diverse lithologies and deposit types remains conditional upon rigorous, site-independent validation, underscoring the persistent challenge of geological variability. For comminution circuit modeling, neural network approaches trained within realistic simulation environments have exhibited remarkable accuracy in predicting energy consumption, water usage, and product size distributions (e.g., p80) when operating within their calibrated domain [35]. This capability underscores the power of data-driven methods for system modeling. Yet, a fundamental limitation persists: these models often experience significant performance degradation when applied outside their specific training range, highlighting the non-transferable nature of purely statistical learning and mandating explicit uncertainty quantification and external validation as non-negotiable components of any modeling claim [35].

At the fundamental level of ore texture, advanced analytical techniques such as mineral liberation analysis (MLA) and high-resolution μ-CT have unequivocally established that textural information—extending beyond bulk chemical assays—is a critical determinant of comminution behavior and subsequent process performance

[56, 57]. This principle has been operationalized through the automated classification of drill-core textures from digital imagery, which has shown statistically meaningful correlations with process indices like BWI and circuit response [58, 59]. This creates a practical, scalable link between routine geological logging and predictive metallurgical performance, although its efficacy is inherently moderated by deposit-specific characteristics and the quality of the image-labeling pipeline. Further advancing this linkage, texture-based three-dimensional liberation models derived from μ-CT data have successfully predicted liberation percentages and connected these to laboratory-scale recovery data [60]. In practice, this enables more informed setting of comminution operating points to maximize mineral recovery. However, the widespread adoption of such approaches is constrained by the substantial cost of high-resolution imaging and the intensive computational resources required for data processing and model execution. Similarly, semi-automated texture-based methods for estimating breakage parameters (e.g., $A \times b$) have shown promise in informing blasting and comminution design, provided their predictions are benchmarked against standardized reference tests (e.g., JK Rotary Breakage Test) and validated using spatially separated datasets to prevent information leakage [61]. A meta-assessment of the evidentiary quality across these domains suggests that any claim of methodological or predictive improvement must satisfy three concurrent criteria to be considered high-grade evidence: (1) the use of calibrated measurement systems (3D imaging with sieve verification or equivalent), (2) out-of-sample validation with reported uncertainty intervals, and (3) a transparent, quantitative linkage to process or economic indices. The absence or weakness of any one of these pillars substantially diminishes the operational credibility and scientific robustness of the research findings. Table 5 provides a consolidated quality assessment and comparative synthesis of pivotal studies across these interconnected themes. The table adjudicates the methodological rigor, key contributions, and inherent limitations of each study, offering a curated overview of the evidentiary landscape. It illustrates that while the field is rich with innovation, the most impactful studies are those that successfully bridge multiple domains—from accurate measurement to process prediction—while openly acknowledging the constraints of their methods and data.

The statistical synthesis of this evolutionary trend is graphically encapsulated in Fig. 4, which illustrates a marked enhancement in predictive accuracy (quantified by R^2) over the past two decades. The plot delineates the performance plateau of classical empirical models ($R^2 \approx 0.70$–0.75), constrained by their inability to capture complex non-linearities and geological heterogeneity, against the ascending trajectory of advanced AI and hybrid models, which have recently pushed accuracy thresholds beyond $R^2 > 0.95$. This visual narrative underscores the imperative of transitioning from traditional, standalone methods toward integrated, physics-aware frameworks to achieve substantive optimization within the mining value chain.

Further statistical rigor is provided by Fig. 5, a three-dimensional forest plot presenting individual study correlation coefficients (r) with their associated 95% confidence intervals, along with a pooled effect size estimate ($r = 0.975$). This meta-analytic visualization confirms the high degree of convergence in reported predictive accuracy among recent AI-based fragmentation studies. Weighting the studies by

Table 5 Quality assessment and comparative synthesis of selected studies across imaging, modeling, and geometallurgical domains

Row	Author/year	Environment/data	Method/algorithm	Output index	Validation type	Key finding	Limitation	References
1	Tang et al. (2022), Minerals Engineering 183	3D μ-CT + μ-XRF iron-ore data	EfficientU-Net-b3 (CNN) for 3D segmentation	Liberation and 3D correlation matrix	Comparison with 2D	3D segmentation provides higher accuracy than 2D (~ significant improvement)	Requires costly imaging and reliable labels	[54]
2	Bhuiyan et al. (2022), Minerals Engineering 175	Gold-mine geometallurgical data	Multivariate regression (with feature selection)	Bond Work Index (BWi)	Comparison with BWi tests	A subset of rock mechanics tests is sufficient to predict BWi	Generalizability to other mines remains uncertain	[55]
3	Moraga et al. (2024), Mining 4 (4):966–982	Comminution circuit simulation	Feedforward neural network + conjoint	Energy/water consumption and product particle size	Simulation cross-validation	Accurate prediction of key parameters ($R^2 > 0.99$) within the training range	No field validation; risk of reduced generalization	[35]

(continued)

Table 5 (continued)

Row	Author/year	Environment/data	Method/algorithm	Output index	Validation type	Key finding	Limitation	References
4	Donskoi et al. (2016), Minerals Engineering 86	Multi-scale iron-ore samples	Microliberation/MLA (texture analysis)	Comminution indices and downstream performance	Comparison with operational data	Textural information is a stronger predictor of performance than bulk chemistry	Focused on iron ore; requires broader mineralogical generalization	[56]
5	Evans et al. (2015), Minerals Engineering 82	X-ray μ-CT images	3D grain-size analysis	Grain-size distribution	Comparison with microscopy	Volume-based PSD measurement improves interpretation of comminution behavior	Costly and time-consuming; limited sample volume	[57]
6	Pérez-Barnuevo et al. (2018), Minerals Engineering 118	Iron-ore drill cores	Supervised core-texture classification	Texture classes ↔ process indices	Comparison with laboratory/mine data	A significant relationship exists between core texture and comminution indices	Deposit-dependent; requires broader validation	[58]
7	Pérez-Barnuevo et al. (2018), Minerals Engineering 119	Mont-Wright (Canada)	Core-texture and feature analysis	BWi and process response	Testing on mine data	Validated the texture–BWi linkage at mine scale	Deposit-specific; transferability remains uncertain	[59]

(continued)

Table 5 (continued)

Row	Author/year	Environment/data	Method/algorithm	Output index	Validation type	Key finding	Limitation	References
8	Guntoro et al. (2021), Minerals Engineering 170	3D μ-CT core data	Texture-based 3D liberation model	Liberation percentage	Comparison with laboratory results	Liberation prediction from 3D imaging achieved suitable accuracy	Requires μ-CT and intensive computation	[60]
9	Nguyen and Valenta (2021), Minerals Engineering 171	Geometallurgy of a mine	Semi-automated texture-based method	Breakage parameter $A \times b$	Comparison with JK RBT	Textural clustering and linkage to breakage indices	Limited dataset; field testing required	[61]

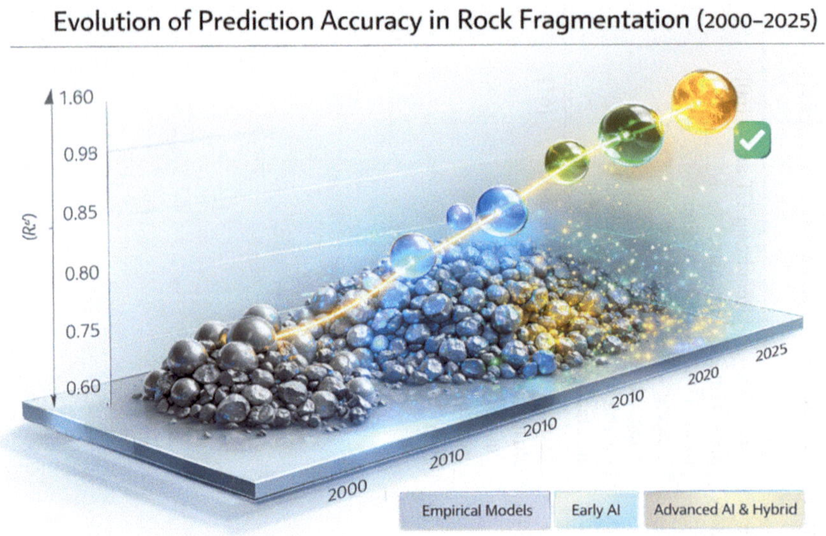

Fig. 4 Comparative synthesis of fragmentation prediction performance (2000–2025), illustrating the transition from empirical baselines ($R^2 \approx 0.75$) to advanced physics-informed AI models ($R^2 > 0.95$) (generated using AI)

sample size and reported precision strengthens the validity of this pooled estimate, suggesting that the collective evidence strongly supports the superior correlative power of intelligent data-driven models.

Finally, to assess the integrity of the reviewed literature and potential for systematic bias, Fig. 6 presents a three-dimensional funnel plot of Fisher's z-transformed correlations against their standard error. The relative symmetry of the plotted studies around the pooled effect line indicates an absence of pronounced publication bias in the body of research selected for this review. The concentration of studies in the upper region of the funnel (lower standard error) denotes a subset of statistically powerful, precise investigations. This visual diagnostic supports the conclusion that the synthesized findings provide a representative and unbiased panorama of technological advancements in rock fragmentation modeling over the last quarter-century, thereby lending greater credibility to the derived conclusions and future pathways.

10 Challenges and Research Gaps

A critical synthesis of the extant literature reveals that the pathway toward robust, industrially adopted intelligent fragmentation systems remains obstructed by several fundamental methodological and translational shortcomings. While advances in imaging and machine learning have demonstrably elevated predictive accuracy in

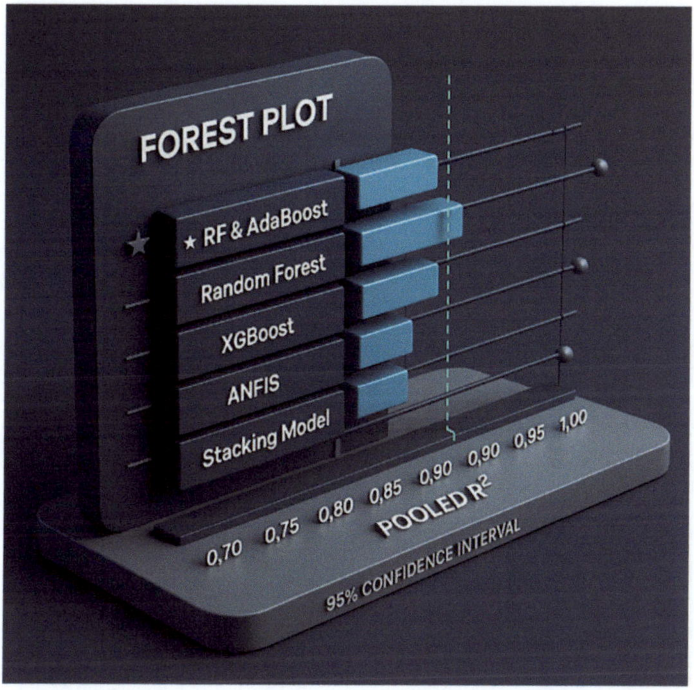

Fig. 5 Three-dimensional forest plot showing individual study correlation coefficients (r) with 95% confidence intervals and the pooled effect size ($r = 0.975$) (generated using AI)

constrained research settings, the translation of these laboratory-scale successes into reliable, field-deployed solutions is hampered by persistent gaps in data standardization, model validation, and systemic integration. This section delineates these core challenges, which collectively represent the frontier between current research potential and genuine operational impact. A primary and pervasive challenge lies in the lack of standardization in data acquisition and reporting. Image-based fragmentation analysis, despite its widespread application, frequently suffers from inconsistent protocols. Many studies utilize two-dimensional imagery without rigorous scale calibration, adequate control for illumination variance, or correction for perspective distortion [62, 63]. Consequently, reported fragmentation indices, though numerically precise, may lack metrological validity as they are derived from uncalibrated representations of a three-dimensional muckpile. For instance, a study may report a segmentation accuracy of 97.1% against image-based ground truth, yet fail to provide concurrent validation against the physical benchmark of mechanical sieving, thereby leaving the real-world accuracy of the derived PSD unverified [62]. This disconnect between digital measurement and physical reality underscores a critical need for community-wide standards governing image acquisition, pre-processing, and mandatory calibration against reference methods. In the domain of predictive modeling, a

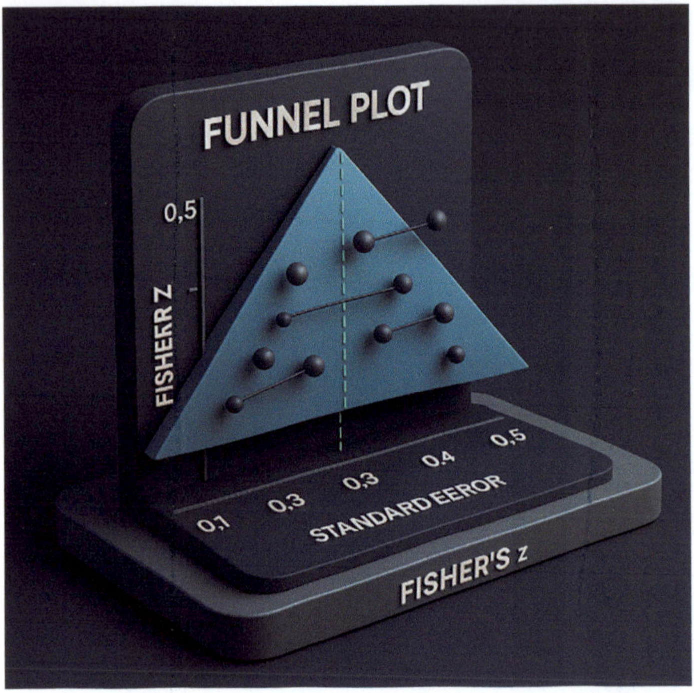

Fig. 6 Three-dimensional funnel plot of Fisher's *z* versus standard error, used to assess publication bias within the reviewed literature (generated using AI)

significant gap exists between internal statistical performance and external generalizability. Machine learning algorithms such as random forest, XGBoost, and ANFIS commonly report high coefficients of determination (R^2) on internal test sets [32]. However, these metrics often originate from random data splits rather than more rigorous spatially or temporally independent validation. A model achieving $R^2 \approx 0.943$ on data from a single mine offers limited evidence of its performance under the different geological and operational conditions of another site [32]. This practice obscures the model's sensitivity to domain shift and overstates its readiness for industrial deployment. Furthermore, the reporting of point predictions without accompanying uncertainty intervals or explicit sensitivity analysis of key blast parameters (e.g., burden, spacing, delay timing) limits the model's utility for risk-informed decision-making. Predictive models transition from academic exercises to engineering tools only when they can quantify what they do not know. Perhaps the most consequential gap is the incomplete linkage in the mine-to-mill value chain. While the influence of fragmentation on downstream comminution energy is widely acknowledged, quantitative, field-validated models that directly map particle size distribution (PSD) parameters to energy indices (e.g., specific energy in kWh/t) are scarce. Most research focuses either on image analysis or blast outcome prediction in isolation, failing to close the loop to plant performance and economics [52]. Establishing this linkage requires controlled, simultaneous monitoring of blasting outcomes and mill power draws, controlling for confounding variables like ore hardness and moisture. Without this integrated perspective, optimization efforts remain myopic, potentially improving fragmentation at the expense of overall system economics.

Underpinning these technical challenges is a broader issue of reproducibility and transparency. A substantial portion of published research lacks open access to the underlying code, algorithms, and—most critically—the curated datasets used for model training and validation. This "black box" nature impedes independent verification, hinders scientific progress through direct comparison, and stifles collaborative improvement. The computational complexity of some proposed hybrid models also presents a practical barrier, as noted by researchers who advocate for designs that balance accuracy with deployability in real-time, resource-constrained mining environments [63]. From a technical execution standpoint, achieving reliable PSD extraction from imagery, especially in large-scale open-pit settings, demands stringent control over acquisition geometry and ambient conditions. Studies emphasize that spatial calibration using GNSS and rigorous management of illumination and scale in aerial imaging are not optional best practices but essential prerequisites for credible results [63, 64]. The absence of reported measurement uncertainty ranges in many studies further erodes confidence in the presented findings [64]. Table 6 provides a systematic catalogue of these identified challenges, linking specific methodological shortcomings to the type of bias they introduce and proposing targeted research needs. It synthesizes the critique that, unless three core links are simultaneously fortified—(1) accurate, calibrated measurement; (2) generalizable prediction with uncertainty quantification; and (3) coherent PSD-to-energy-economic linkage—research outcomes will remain confined to the laboratory, unable to substantiate claims of field-level intelligent optimization.

In conclusion, to elevate research from proof-of-concept to Q1 journal standards and beyond to field readiness, future studies must converge on three strategic directions: establishing multi-site, open-access databases with calibrated imagery and reference sieve data; developing physics-informed models evaluated via spatially independent validation with mandatory confidence-interval reporting; and executing controlled mine-to-mill A/B field trials to empirically quantify the impact of fragmentation changes on energy and cost. Only through such coordinated, rigorous, and transparent research can the promise of intelligent, economically optimized fragmentation control be fully realized.

Table 6 Catalog of methodological challenges, inherent biases, and prioritized research needs in fragmentation studies

Author/year	Research domain/data	Methodological/reporting gap	Bias type	Research need	Future priority	References
Bamford et al. (2021)	Comminution circuit (feed rock), 2D imagery	No validation of image dimensions; simplified assumptions	Cross-context generalization bias	Time-series correlation of power (kWh/t) with calibrated PSD	Standardizing kWh/t reporting	[24]
Jin et al. (2023)	Real open-pit blast imagery	No field-of-view calibration and no out-of-sample testing	Algorithmic bias	Adaptive validation of images against scaled data	A/B field trials (multiple concurrent blasts)	[62]
Bahraini et al. (2024)	3D blast imaging in a mine (GolGohar)	No reporting of measurement-uncertainty range	Acquisition bias	Reporting PSD errors or confidence intervals	Developing a standardized imaging protocol	[64]
Lapčević et al. (2023)	3D blast-fragmentation modeling (lead, tunnel)	Limited validation across diverse field datasets	Model bias	Validating the model using multiple field datasets	Releasing open-source modeling tools	[52]
Ikeda et al. (2023)	Real blasting with UAV/GNSS	Illumination and scale consistency across images not ensured	Instrument bias	Low-cost solutions for scale calibration	Periodic calibration of camera and GNSS	[63]
Gao et al. (2018)	Synthetic (simulated) data	Mismatch between data and real operating conditions	Simulation bias	Adding realistic physical parameters to the model	Extensive validation using field data	[65]
Hosseini et al. (2022)	Hybrid model for blast-risk prediction	High complexity and difficult implementation	Operational bias	Simplifying model deployment (user-friendly)	Assessing runtime/cost of model execution	[66]

(continued)

Table 6 (continued)

Author/year	Research domain/data	Methodological/reporting gap	Bias type	Research need	Future priority	References
Xie et al. (2021)	Review of blast overbreak models	Heterogeneous and non-comparable evaluation metrics	Reporting bias	Developing standard metrics and measurement methods	Creating a reference database of benchmarks	[67]
Amoako et al. (2022)	ANN/ES model for blasting	Limited training data	Small-data bias	Spatial cross-validation	Field experiments with new data	[29]
Sui et al. (2025)	Data-driven prediction for open-pit blasting	No uncertainty-interval reporting	Model bias	Reporting prediction confidence intervals	Extensive out-of-sample validation	[32]

11 Future Research Pathways

The trajectory for credible advancement in blast fragmentation optimization must pivot decisively from incremental algorithmic refinement toward the establishment of a verifiable, closed-loop measurement-to-decision chain. This necessitates a framework built upon four foundational pillars: (1) three-dimensional particle size distribution (PSD) measurement with rigorous metrological calibration, (2) physics-informed predictive modeling with explicit uncertainty quantification, (3) quantitative field-based mapping of PSD to specific energy consumption (kWh/t), and (4) controlled A/B testing at the mine-plant scale to validate economic impact. Future research that neglects this integrated, validation-centric approach risks producing visually appealing but operationally non-credible outcomes. The first pillar demands a fundamental shift from uncalibrated two-dimensional imagery to metrologically sound 3D PSD measurement. Relying on unscaled 2D images without concurrent sieve verification reduces analysis to numerical abstraction. A professional pathway involves the 3D reconstruction of muck piles using photogrammetry or LiDAR point clouds to explicitly model coarse fragments, complemented by validated statistical functions (e.g., Swebrec) to accurately represent the fine tail. This hybrid approach allows for the reporting of P_{80} and distribution shape with documented error bounds. Credible field work, such as the development and validation of a universal 3D blast fragmentation model, has demonstrated tangible reductions in feed-estimation error and provides a necessary template for this measurement standard [52]. Concurrently, for image segmentation itself, enhanced deep learning architectures like U-Net variants have reached operational accuracy on blast images, as evidenced by studies reporting high segmentation performance validated against sieved piles [62]. However, image-derived accuracy is insufficient in isolation. Every digital measurement must be systematically benchmarked against physical sieving, with results reported as confidence intervals rather than point estimates. The field requires realistic error benchmarking and spatially independent evaluation, as demonstrated by 3D field studies that compare point-cloud-derived PSDs to sieve results and deep learning work that validates percentile predictions directly on sieved material [24, 52]. This establishes the traceability required for measurement credibility.

The second critical pathway is the development of interpretable PSD-to-energy mapping grounded in field data. Assertions regarding energy savings must move beyond conjecture by employing interpretable frameworks and establishing relationships using actual time-series data of mill power and circulating load. The statistical infrastructure for this is supported by validated distribution models like Swebrec, which reliably represents the coarse tail crucial for energy prediction [7], and by predictive tools like the fragment size-energy fan concept [68]. The subsequent step is the principled introduction of physical constraints into data-driven models. Unconstrained neural networks that provide no uncertainty estimates are inherently unstable for field deployment. Embedding energy conservation laws and PSD-consistency constraints directly into the training objective, as advocated in physics-informed deep learning, ensures outputs adhere to fundamental principles.

Outputs must always be accompanied by confidence intervals, a practice demonstrated in credible journal work on segmentation and prediction [24, 62]. Finally, optimization must be inherently multi-objective, robust, and decision-ready. Optimization claims outside a framework that balances P_{80}, uniformity, kWh/t, capacity, and peak particle velocity (PPV) are merely decorative. Solutions should be presented as Pareto fronts, enabling engineers to make explicit trade-offs based on site-specific economic and safety criteria. Practical implementation of these pathways can be summarized as a sequential, validated chain: calibrated 3D measurement with sieve verification [52, 62] → extraction of P_{80} and distribution shape with documented error → interpretable mapping of PSD to power/energy using validated distributions and field time-series [7, 68] → training of physics-informed models with uncertainty quantification and independent testing [24] → execution of mine-plant A/B trials and final design selection from a Pareto-optimal set. Table 7 distills these overarching pathways into specific research axes, proposing testable hypotheses, methodological designs, and appropriate evaluation metrics to guide future systematic inquiry.

12 Conclusion

This systematic review has charted the evolution and convergence of three principal methodological strands in the optimization of blast-induced rock fragmentation for open-pit mining: physical-empirical models, image-processing-based measurement techniques, and intelligent data-driven algorithms. The meta-analytical synthesis presented herein reveals that while each approach possesses distinct advantages and historical utility, the future of predictive accuracy and operational robustness lies in their strategic integration within physics-informed hybrid frameworks. The statistical evidence consolidated in this review, particularly the aggregated correlation coefficients and comparative performance metrics, substantiates a clear technological trajectory. Artificial intelligence models, encompassing both machine learning and deep learning paradigms, have demonstrably achieved a mean pooled predictive accuracy ($r \approx 0.975$) that surpasses classical empirical models by a significant margin of 5–12%. This advancement is largely attributable to their capacity to model complex, non-linear interactions between blast parameters, rock mass properties, and resulting fragmentation that are often oversimplified in empirical formulations. Conversely, purely empirical models, despite their enduring value for preliminary design due to interpretability and low computational overhead, remain fundamentally constrained in capturing nonlinear regimes and exhibit a pronounced dependency on site-specific calibration for reliable accuracy. This dependency underscores their role as foundational rather than definitive tools in modern optimization workflows.

The critical interpretation of these findings converges on a central thesis: physics-informed hybrid frameworks represent the most promising avenue for bridging the persistent gap between model transparency and predictive power. By embedding first principles of rock mechanics, stress-wave propagation, and energy conservation directly into data-driven architectures, these frameworks compensate for the

Table 7 Strategic research axes and methodological proposals for future fragmentation studies

Research axis	Open problem	Proposed hypothesis	Study design	Evaluation metric	References
Development of 3D PSD measurement methods	Reducing error and validating 3D PSD outcomes	Using 3D scanners and advanced image-processing algorithms yields the most accurate PSD	Field study using 3D data with comparison against sieving	RMSE between predicted and sieved PSD	[52]
Machine learning constrained by physical laws	Improving model generalizability and robustness in predicting blast/fragmentation outcomes	Physics-informed hybrid models provide higher accuracy and better adaptability	Develop constrained models with mechanical constraints and compare with unconstrained baselines	R^2 and prediction error; consistency with physical laws	[11]
Mapping PSD to specific energy (kWh/t)	Establishing a quantitative link between size distribution and comminution energy under real conditions	Blast PSD is directly related to comminution energy demand	Controlled field trials and regression modeling with Bond Work Index	Energy-consumption prediction error (MAE, RMSE)	[68]
Controlled A/B field blasting trials	Identifying the effect of blast parameters on mine performance	Different blast designs produce measurable differences in production and energy	Run two or more blast designs while monitoring PSD, power, and PPV	Differences in mean P80 and kWh/t across scenarios	[7]
Multi-objective optimization of blasting and comminution	Balancing fragmentation quality (P80) and allowable vibration (PPV)	Multi-objective algorithms can optimize P80 while controlling PPV	Genetic and PSO-based simulation; field evaluation	Pareto-front optimization between P80 and PPV	[24]
PDE-constrained surrogate models	Reducing computational cost of high-fidelity blast simulations	Surrogate models trained on FDEM data can approximate simulation outputs	Train ML surrogates on simulation datasets and compare with high-fidelity simulations	Prediction accuracy and runtime	[11]

(continued)

Table 7 (continued)

Research axis	Open problem	Proposed hypothesis	Study design	Evaluation metric	References
Probabilistic calibration and digital-twin updating	Adapting models to lithological changes	Bayesian fusion of data and models improves predictive accuracy	Build a blasting digital twin and update it using real data	Reduced model uncertainty (CI coverage)	[24]
Cross-site transfer learning	Generalizing models to mines with different rock structures	Transfer learning maintains performance with less data	Test a model trained on Mine A on Mine B data	Change in R^2 and RMSE after transfer	[11]
Integrating geological data into models	Quantifying the influence of geomechanical parameters on fragmentation and vibration	Adding geological information improves model accuracy	Multi-modal modeling including RQD and rock-mass fracturing indices	Increase in R^2 relative to the baseline	[11]
Uncertainty quantification in predictions	Providing confidence in PSD and PPV predictions	Bayesian and ensemble models reduce predictive uncertainty	Bayesian modeling using sparse field datasets	Confidence-interval width and coverage rate	[24]

limitations inherent in purely data-scarce or purely physics-simplified approaches. This integration ensures that model predictions are not only statistically sound but also physically plausible, thereby enhancing generalizability across heterogeneous geological environments. This evolution marks a pivotal shift in research philosophy—from "numerical modeling" focused on fitting to "explanatory prediction" aimed at understanding causative mechanisms within a constrained physical reality. From an industrial implementation perspective, the reviewed evidence indicates that deploying such hybrid intelligence systems can translate into substantive operational benefits. Case studies suggest potential reductions of 10–20% in specific energy consumption for secondary comminution, alongside improvements in overall mine-to-mill process efficiency. Moreover, the enhanced interpretability of these hybrid models, facilitated by techniques like SHAP analysis and sensitivity testing, enables intelligent monitoring and the simultaneous multi-objective optimization of cost, throughput, and safety parameters. The meta-analytic evidence demonstrates low heterogeneity across high-quality studies, suggesting that the performance benefits of hybrid approaches are stable across varied temporal and spatial contexts, reinforcing their suitability for scaled industrial application.

Based on the systematic identification of challenges and gaps, three prioritized directions for future research are delineated:

1. **Standardization of Multi-site Databases**: There is an imperative to establish open-access, multi-site databases featuring rigorously calibrated imagery paired with direct mechanical sieving data across diverse illumination and geometric conditions. This will provide the foundational benchmark datasets necessary for developing and fairly comparing next-generation algorithms.
2. **Development of Physics-Informed Neural Architectures**: Research must advance beyond generic deep learning toward specialized neural architectures that intrinsically encode physical laws. This includes the multi-modal fusion of optical, seismic, and blasting data within loss functions that penalize violations of thermodynamic and mechanical principles.
3. **Integration within Operational Digital Twins**: To realize real-time control, fragmentation-energy-cost models must be fully integrated into dynamic digital twin systems of the mining value chain. These twins will enable simulation, scenario analysis, and closed-loop optimization of blasting and comminution processes, linking predictive analytics directly to execution and control.

In summary, the collective evidence robustly supports the conclusion that the transition from isolated data-driven models toward physics-informed hybrid systems is not merely an incremental improvement but a necessary evolution. This paradigm enhances not only predictive accuracy and interpretability but also establishes the foundational architecture for the next generation of intelligent mining systems. In these systems, energy efficiency, operational stability, and autonomous decision-making are seamlessly unified within a coherent, self-optimizing framework, finally unlocking the full potential of the mine-to-mill optimization paradigm.

Declaration of AI Use ChatGPT (OpenAI) was used solely for language proofreading and clarity improvement. Figures 2, 3, 4, 5 and 6 in this review book chapter were generated using AI-based tools for illustrative purposes. No AI was used for content generation, data analysis, or scientific reasoning. The authors take full responsibility for the content and any errors.

References

1. Kinyua EM, Jianhua Z, Kasomo RM, Mauti D, Mwangangi J (2022) A review of the influence of blast fragmentation on downstream processing of metal ores. Miner Eng 186:107743
2. Deressa GW, Choudhary BS, Jilo NZ (2025) Optimizing blast design and bench geometry for stability and productivity in open pit limestone mines using experimental and numerical approaches. Sci Rep 15(1):5796
3. Silva J, Amaya J, Basso F (2017) Development of a predictive model of fragmentation using drilling and blasting data in open pit mining. J S Afr Inst Min Metall 117(11):1089–1094
4. Coello-Velázquez AL, Quijano Arteaga V, Menéndez-Aguado JM, Pole FM, Llorente L (2019) Use of the Swebrec function to model particle size distribution in an industrial-scale Ni-Co ore grinding circuit. Metals 9(8):882
5. Huan B, Li X, Wang J, Hu T, Tao Z (2025) An interpretable deep learning model for the accurate prediction of mean fragmentation size in blasting operations. Sci Rep 15(1):11515
6. Saldana M, Gallegos S, Arias D, Salazar I, Castillo J, Salinas-Rodríguez E, Navarra A, Toro N, Cisternas LA (2024) Applications of Kuz–Ram models in mine-to-mill integration and optimization—a review. Minerals 14(11):1162
7. Sanchidrián JA, Ouchterlony F, Segarra P, Moser P (2014) Size distribution functions for rock fragments. Int J Rock Mech Min Sci 71:381–394
8. Ronkin MV, Akimova EN, Misilov VE (2023) Review of deep learning approaches in solving rock fragmentation problems
9. Liu Y, Li A, Liu Y, Leng Z, Xie Y (2025) A theory-data-integrated intelligent framework for prediction of blasting-induced rock fragment size. Powder Technol 453:120638
10. Page MJ, McKenzie JE, Bossuyt PM, Boutron I, Hoffmann TC, Mulrow CD, Shamseer L, Tetzlaff JM, Akl EA, Brennan SE (2021) The PRISMA 2020 statement: an updated guideline for reporting systematic reviews. bmj 372
11. Leon F, Rojas L, Peña A, Moraga P, Robles P, Gana B, García J (2025) Mathematical modelling and optimization methods in geomechanically informed blast design: a systematic literature review. Mathematics 13(15):2456
12. Shea BJ, Reeves BC, Wells G, Thuku M, Hamel C, Moran J, Moher D, Tugwell P, Welch V, Kristjansson E (2017) AMSTAR 2: a critical appraisal tool for systematic reviews that include randomised or non-randomised studies of healthcare interventions, or both. bmj 358
13. Whiting P, Savović J, Higgins JP, Caldwell DM, Reeves BC, Shea B, Davies P, Kleijnen J, Churchill R (2016) ROBIS: a new tool to assess risk of bias in systematic reviews was developed. J Clin Epidemiol 69:225–234
14. Lanari M, Fakhimi A (2015) Numerical study of contributions of shock wave and gas penetration toward induced rock damage during blasting. Comput Part Mech 2(2):197–208
15. Sanchidrián JA, Ouchterlony F, Moser P, Segarra P, López LM (2012) Performance of some distributions to describe rock fragmentation data. Int J Rock Mech Min Sci 53:18–31
16. Ouchterlony F (2005) The Swebrec© function: linking fragmentation by blasting and crushing. Min Technol 114(1):29–44
17. Amin MA, Ilboudo A, Shahid ASB, Ali A, Bangyal WHK (2025) Adaptive image restoration for video surveillance: a real-time approach. arXiv preprint arXiv:250513130
18. Sarkar D, Kumar S, Das P, Ramos H (2024) Higher-order convergence analysis for interior and boundary layers in a semi-linear reaction-diffusion system networked by a k-star graph with non-smooth source terms. Netw Heterog Media 19(3)

19. Gao P, Pan C, Zong Q, Dong C (2023) Rock fragmentation size distribution control in blasting: a case study of blasting mining in Changjiu Shenshan limestone mine. Front Mater 10:1330354
20. Kemeny JM, Devgan A, Hagaman RM, Wu X (1993) Analysis of rock fragmentation using digital image processing. J Geotech Eng 119(7):1144–1160
21. Guo Q, Wang Y, Yang S, Xiang Z (2022) A method of blasted rock image segmentation based on improved watershed algorithm. Sci Rep 12(1):7143
22. Sudhakar J, Adhikari G, Gupta R (2006) Comparison of fragmentation measurements by photographic and image analysis techniques. Rock Mech Rock Eng 39(2):159–168
23. Engin IC, Maerz NH, Boyko KJ, Reals R (2020) Practical measurement of size distribution of blasted rocks using LiDAR scan data. Rock Mech Rock Eng 53(10):4653–4671
24. Bamford T, Esmaeili K, Schoellig AP (2021) A deep learning approach for rock fragmentation analysis. Int J Rock Mech Min Sci 145:104839
25. Tavakol Elahi A, Hosseini M (2017) Analysis of blasted rocks fragmentation using digital image processing (case study: limestone quarry of Abyek Cement Company). Int J Geo-Eng 8(1):16
26. Murlidhar BR, Armaghani DJ, Mohamad ET, Changthan S (2018) Rock fragmentation prediction through a new hybrid model based on imperial competitive algorithm and neural network. Smart Constr Res 2(3):1–12
27. Yari M, He B, Armaghani DJ, Abbasi P, Mohamad ET (2023) A novel ensemble machine learning model to predict mine blasting–induced rock fragmentation. Bull Eng Geol Env 82(5):187
28. Zhang YL, Qiu YG, Armaghani DJ, Monjezi M, Zhou J (2024) Enhancing rock fragmentation prediction in mining operations: a hybrid GWO-RF model with SHAP interpretability. J Cent S Univ 31(8):2916–2929
29. Amoako R, Jha A, Zhong S (2022) Rock fragmentation prediction using an artificial neural network and support vector regression hybrid approach. Mining 2(2):233–247
30. Gebretsadik A, Kumar R, Fissha Y, Kide Y, Okada N, Ikeda H, Mishra AK, Armaghani DJ, Ohtomo Y, Kawamura Y (2024) Enhancing rock fragmentation assessment in mine blasting through machine learning algorithms: a practical approach. Discov Appl Sci 6(5):223
31. Zhao J, Li D, Zhou J, Armaghani DJ, Zhou A (2025) Performance evaluation of rock fragmentation prediction based on RF-BOA, AdaBoost-BOA, GBoost-BOA, and ERT-BOA hybrid models. Deep Undergr Sci Eng 4(1):3–17
32. Sui Y, Zhou Z, Zhao R, Yang Z, Zou Y (2025) Open-pit bench blasting fragmentation prediction based on stacking integrated strategy. Appl Sci 15(3):1254
33. Krop I, Sasaoka T, Shimada H, Hamanaka A (2024) Optimizing mean fragment size prediction in rock blasting: a synergistic approach combining clustering, hyperparameter tuning, and data augmentation. Eng 5(3):1905–1936
34. Esmaeili M, Salimi A, Drebenstedt C, Abbaszadeh M, Aghajani Bazzazi A (2015) Application of PCA, SVR, and ANFIS for modeling of rock fragmentation. Arab J Geosci 8(9):6881–6893
35. Moraga C, Astudillo CA, Estay R, Maranek A (2024) Enhancing comminution process modeling in mineral processing: a conjoint analysis approach for implementing neural networks with limited data. Mining 4(4):966–982
36. Bhuiyan M, Esmaieli K, Ordóñez-Calderón JC (2019) Application of data analytics techniques to establish geometallurgical relationships to bond work index at the Paracutu Mine, Minas Gerais, Brazil. Minerals 9(5):302
37. Both C, Dimitrakopoulos R (2021) Applied machine learning for geometallurgical throughput prediction—a case study using production data at the tropicana gold mining complex. Minerals 11(11):1257
38. Lee D, Je J, Kwon J (2024) Prediction of particle size distribution of grinding products using artificial neural network approach. Miner Eng 216:108831
39. Kroell N, Thor E, Göbbels L, Schönfelder P, Chen X (2025) Deep learning-based prediction of particle size distributions in construction and demolition waste recycling using convolutional neural networks on 3D laser triangulation data. Constr Build Mater 466:140214

40. Tang Y, Wang Y, Si G (2024) Vision-based size distribution analysis of rock fragments using multi-modal deep learning and interactive annotation. Autom Constr 159:105276
41. Tao M, Xiao Z, Liu Y, Huang L, Xiang G, Xu Y (2025) A fast recognition method for dynamic blasting fragmentation based on YOLOv8 and binocular vision. Appl Sci 15(12):6411
42. Avalos S, Kracht W, Ortiz JM (2020) An LSTM approach for SAG mill operational relative-hardness prediction. Minerals 10(9):734
43. Wang N, Chen Y, Zhang D (2025) A comprehensive review of physics-informed deep learning and its applications in geoenergy development. Innov Energy 2(2):100087
44. Esmaeili K (2025) 3D rock fragmentation analysis using LiDAR, based on point cloud deep learning segmentation and synthetic data. Powder Technol 456:120861
45. Qiu L, Zhu Y, Xu C, Ren G, Hu Y, Liu X (2025) Physics-informed neural network optimized by particle swarm algorithm for accurate prediction of blast-induced peak particle velocity. Intell Geoeng
46. Guo J, Zhao P, Li P (2023) Prediction and optimization of blasting-induced ground vibration in open-pit mines using intelligent algorithms. Appl Sci 13(12):7166
47. Figueiredo J, Torres V, Cruz R, Moreira D (2023) Blasting fragmentation study using 3D image analysis of a hard rock mine. Appl Sci 13(12):7090
48. Yi H, Wu A, Zhang X (2025) Rock blasting evaluation-image recognition method based on deep learning. Sci Rep 15(1):23980
49. Losaladjome Mboyo H, Huo B, Mulenga FK, Mabe Fogang P, Kalenga Kaunde Kasongo J (2024) Assessing the impact of surface blast design parameters on the performance of a comminution circuit processing a copper-bearing ore. Minerals 14(12):1226
50. Lastra G, Jokovic V, Kanchibotla S (2021) Understanding the impact of geotechnical ore properties and blast design on comminution circuits using simulations. Miner Eng 170:107001
51. Morrell S (2004) Predicting the specific energy of autogenous and semi-autogenous mills from small diameter drill core samples. Miner Eng 17(3):447–451
52. Lapčević V, Torbica S, Stojanović M, Vojinović I (2023) Development and validation of universal 3D blast fragmentation model. Appl Sci 13(14):8316
53. Jankovic A, Valery W, Davis E (2004) Cement grinding optimisation. Miner Eng 17(11):1075–1081
54. Tang K, Da Wang Y, Mostaghimi P, Knackstedt M, Hargrave C, Armstrong RT (2022) Deep convolutional neural network for 3D mineral identification and liberation analysis. Miner Eng 183:107592
55. Bhuiyan M, Esmaeili K, Ordonez-Calderon JC (2022) Evaluation of rock characterization tests as geometallurgical predictors of bond work index at the Tasiast Mine, Mauritania. Miner Eng 175:107293
56. Donskoi E, Poliakov A, Holmes R, Suthers S, Ware N, Manuel J, Clout J (2016) Iron ore textural information is the key for prediction of downstream process performance. Miner Eng 86:10–23
57. Evans C, Wightman E, Yuan X (2015) Quantifying mineral grain size distributions for process modelling using X-ray micro-tomography. Miner Eng 82:78–83
58. Pérez-Barnuevo L, Lévesque S, Bazin C (2018) Automated recognition of drill core textures: a geometallurgical tool for mineral processing prediction. Miner Eng 118:87–96
59. Pérez-Barnuevo L, Lévesque S, Bazin C (2018) Drill core texture as geometallurgical indicator for the Mont-Wright iron ore deposit (Quebec, Canada). Miner Eng 122:130–141
60. Guntoro PI, Ghorbani Y, Parian M, Butcher AR, Kuva J, Rosenkranz J (2021) Development and experimental validation of a texture-based 3D liberation model. Miner Eng 164:106828
61. Nguyen A, Tungpalan K, Evans C, Manlapig E, Jackson J, Nguyen K, Valenta R (2021) A method for rapid estimation of processing behaviour based on ore texture. Miner Eng 171:107111
62. Jin C, Liang J, Fan C, Chen L, Wang Q, Lu Y, Wang K (2023) Study on segmentation of blasting fragment images from open-pit mine based on U-CARFnet. PLoS One 18(9):e0291115
63. Ikeda H, Sato T, Yoshino K, Toriya H, Jang H, Adachi T, Kitahara I, Kawamura Y (2023) Deep learning-based estimation of muckpile fragmentation using simulated 3D point cloud data. Appl Sci 13:10985

64. Bahraini MS, Atighi I (2024) A novel intelligent stereo vision approach for blast-induced fragmentation size distribution: case study at Golgohar open-pit mine, Iran. Miner Eng 215:108822
65. Gao W, Karbasi M, Hasanipanah M, Zhang X, Guo J (2018) Developing GPR model for forecasting the rock fragmentation in surface mines. Eng Comput 34(2):339–345
66. Hosseini S, Poormirzaee R, Hajihassani M (2022) An uncertainty hybrid model for risk assessment and prediction of blast-induced rock mass fragmentation. Int J Rock Mech Min Sci 160:105250
67. Xie C, Nguyen H, Bui X-N, Choi Y, Zhou J, Nguyen-Trang T (2021) Predicting rock size distribution in mine blasting using various novel soft computing models based on meta-heuristics and machine learning algorithms. Geosci Front 12(3):101108
68. Sanchidrián JA, Ouchterlony F (2023) Blast-fragmentation prediction derived from the fragment size-energy fan concept. Rock Mech Rock Eng 56(12):8869–8889

Prediction of Rock Mass Rating Using Seismic Wave Velocities and Metaheuristic Algorithms: A Comparative Study of DE, PS, and Hybrid DE-PS

Mohammad Rezaei and Pouya Koureh Davoodi

Abstract Accurate determination of rock mass quality plays a critical role in the design and stability assessment of underground structures. Among various classification systems, the Rock Mass Rating (RMR) is widely used in engineering practice; however, its direct determination often requires extensive field investigations and laboratory testing, which can be time-consuming and costly. To address this limitation, the present study investigates the capability of metaheuristic algorithms to estimate RMR using non-destructive geophysical parameters, namely P-wave velocity (V_p) and S-wave velocity (V_s). In this study, three modeling approaches based on Differential Evolution (DE), Pattern Search (PS), and a hybrid DE-PS algorithm were developed to establish quantitative relationships between seismic velocities and RMR values. The models were trained and validated using a dataset obtained from the Beheshtabad water conveyance tunnel, where V_p, V_s, and RMR were measured according to standard field and laboratory procedures. The performance of the proposed models was evaluated using multiple statistical indices, including the coefficient of determination (R^2), root mean square error (RMSE), mean absolute error (MAE), and mean absolute percentage error (MAPE). The results indicate that all three models are capable of capturing the underlying relationship between seismic velocities and RMR; however, the hybrid DE-PS model exhibits superior predictive performance compared to the individual DE and PS models. The hybrid approach yields the highest correlation with measured RMR values and the lowest prediction errors, demonstrating improved accuracy and stability. Comparative validation analyses, including cumulative error assessment and residual distribution evaluation, further confirm the robustness of the hybrid model. The findings of this study suggest that the proposed DE-PS-based modeling framework provides a reliable and efficient

M. Rezaei (✉)
Department of Mining Engineering, Faculty of Engineering, University of Kurdistan, Sanandaj, Iran
e-mail: m.rezaei@uok.ac.ir

P. K. Davoodi
Department of Mining Engineering, Faculty of Engineering, Urmia University, Urmia, Iran
e-mail: p.kourehdavoodi@urmia.ac.ir

A. Azizi et al. (eds.), *Intelligence at the Interface*, Emerging Trends in Mechatronics,
https://doi.org/10.1007/978-981-95-7584-8_10

tool for estimating RMR from seismic velocity data. The developed approach can support preliminary site investigations and decision-making processes in tunneling and rock engineering projects, particularly in situations where direct determination of RMR is limited or impractical.

Keywords Rock Mass Rating · Random Forest · Genetic algorithm · Trust region reflective

1 Introduction

Rock mass classification plays a fundamental role in the analysis, design, and implementation of rock engineering projects such as tunnels, underground spaces, dams, and mining operations. Among the various classification systems that have been proposed, the Rock Mass Rating (RMR) system has gained widespread acceptance due to its systematic structure, practical applicability, and ease of engineering interpretation [1]. This system is developed based on several geological and geotechnical parameters, including the uniaxial compressive strength of intact rock, rock quality designation (RQD), spacing, condition and orientation of discontinuities, and groundwater conditions. Despite its extensive application, the direct determination of RMR parameters generally requires drilling operations, detailed field investigations, and laboratory testing, which are time-consuming, costly, and often associated with considerable uncertainty [2].

In recent years, increasing attention has been directed toward indirect and non-destructive approaches for estimating rock mass classification indices. In this context, geophysical parameters, which can be measured rapidly, continuously, and economically, have been introduced as suitable alternatives to conventional parameters. Among these, compressional wave velocity (V_{p}) and shear wave velocity (V_{s}) are regarded as key indicators, as their relationships with the mechanical, elastic, and structural properties of rock masses have been confirmed in numerous studies. However, compared with other rock mass classification indices such as GSI or Q, the number of reliable empirical relationships developed for estimating RMR based on V_{p} and V_{s} remains limited, and many of the proposed models suffer from insufficient accuracy or limited generalization capability [1].

The inherent complexity of rock masses, structural heterogeneity, scale effects, and the dependency of geophysical parameters on geological conditions result in highly nonlinear and complex relationships between RMR and seismic wave velocities. Consequently, conventional regression-based approaches, which are often based on linear assumptions or simple functional forms, have shown limited performance in accurately modeling these relationships. This limitation highlights the necessity of employing more advanced modeling techniques capable of capturing complex nonlinear behaviors [3].

In this regard, metaheuristic algorithms have been widely recognized as powerful optimization tools for modeling complex and nonlinear relationships without requiring simplifying assumptions or gradient information. These algorithms, inspired by natural and evolutionary processes, are capable of efficiently exploring the solution space and determining optimal coefficients for empirical models. Nevertheless, the performance of metaheuristic algorithms strongly depends on their search mechanisms, and achieving a proper balance between global exploration and local exploitation is crucial for obtaining accurate and stable solutions [1, 2, 4].

Differential Evolution (DE) is a well-established metaheuristic optimization algorithm that has demonstrated strong capability in global exploration and avoidance of premature convergence in continuous optimization problems. By employing mutation and crossover operators, DE gradually guides a population of candidate solutions toward promising regions of the search space. However, in some applications, the performance of DE in the final stages of optimization may be limited due to relatively weak local search capability [5]. In contrast, the Pattern Search (PS) algorithm, as a derivative-free local optimization method, exhibits high efficiency in refining solutions in the vicinity of local optima and can significantly reduce prediction errors, although its performance is highly dependent on the quality of the initial solution [6].

Accordingly, the combined use of DE and PS in the form of a hybrid DE-PS model can provide an effective strategy for exploiting the advantages of both algorithms. In such a framework, DE is employed to perform global exploration and identify promising regions within the solution space, while PS is subsequently applied to enhance local exploitation and improve solution accuracy. This hybrid approach is expected to yield more accurate, stable, and generalizable predictive models compared to the standalone algorithms.

A comprehensive summary of previously proposed empirical relationships related to rock mass classification parameters is presented in Table 1. As shown in this table, the majority of existing studies have focused on developing correlations between the Rock Mass Rating (RMR) system and other rock mass classification indices, such as the tunneling quality index (Q), the Geological Strength Index (GSI), and deformation or elastic moduli of rock masses (E_{m}, E_{i}). These relationships were mainly established using conventional regression-based approaches and were often aimed at indirect estimation of rock mass mechanical properties rather than direct prediction of RMR from geophysical measurements.

It can be observed that a large proportion of the reported equations are expressed as functions of Q or GSI, reflecting the widespread use of these indices in empirical rock engineering correlations. Although acceptable correlation coefficients were reported in some cases, many of the proposed relationships exhibit considerable scatter, limited applicability ranges, or strong dependency on site-specific geological conditions. Furthermore, a significant number of the reported models were developed for estimating deformation modulus or elastic properties of rock masses based on RMR, rather than addressing the inverse problem of predicting RMR itself [27, 41, 42].

In contrast, only a very limited number of studies have attempted to establish direct relationships between seismic wave velocities, particularly compressional wave velocity (V_{p}) and shear wave velocity (V_{s}), and the RMR system. As indicated in Table 1, most of the available V_{p}- or V_{s}-based relationships are sparse and

Table 1 Existing relationships for estimating RMR based on rock properties and other rock mass classification systems

No.	Equations	R^2	References
1	$\text{RMR} = (0.0106V_p) + (0.0245V_s) - 33.26$	0.9968	[1]
2	$\text{GSI} = 1.6820\text{RMR}^{0.8903}$	0.9533	[2]
3	$E_m = 2\text{RMR} - 100$	–	[7]
4	$\text{RMR} = 5.9 \ln Q + 43$	0.81	[8]
5	$\text{RMR} = 5 \ln Q + 60.8$	High scatter	[9]
6	$\frac{\text{UCS}_m}{\text{UCS}_i} = \exp\left(7.65 \times \frac{\text{RMR}-100}{\text{RMR}}\right)$	–	[10]
7	$E_m = 10^{(\text{RMR} - 10)/40}$	–	[11]
8	$\text{RMR} = 6.3 \ln Q + 41.6$	0.78	[11]
9	$\frac{\text{UCS}_m}{\text{UCS}_i} = \exp\left(\frac{\text{RMR}-100}{18.5}\right)$	–	[12]
10	$\text{RMR} = 9 \ln Q + 44$	0.77	[13]
11	$E_m/E_i = 1/100 \times 0.0028\text{RMR}^2 + 0.9\, e^{(\text{RMR}/22.82)}$	–	[14]
12	$E_m = \frac{E_i}{100}\left[0.0028\text{RMR}^2 + 0.9\exp\left(\frac{\text{RMR}}{22.82}\right)\right]$	–	
13	$\frac{\text{UCS}_m}{\text{UCS}_i} = \exp\left(\frac{\text{RMR}-100}{24}\right)$	–	[15]
14	$\frac{\text{UCS}_m}{\text{UCS}_i} = \left(\frac{\text{RMR}-100}{\text{RMR}+6(100-\text{RMR})}\right)$	–	[16]
15	$E_{rm} = 0.1(\text{RMR}/10)^3$	–	[17]
16	$\text{RMR} = 1.5 + 0.0161V_p$	–	[18]
17	$E_m = E_i 10^{[((\text{RMR} - 100)(100 - \text{RMR})/4000 \exp 9 - \text{RMR} - 100)]}$	–	[19]
18	$\text{RMR} = 3.7 \ln Q + 53.1$	0.86	[20]
19	$\text{RMR} = 2.8 \ln Q + 45.19$	0.67	[21]
20	$\text{RMR} = 15.5 \ln Q + 39.7$	0.846	[22]
21	$\text{GSI} = 0.5\text{RQD} + 1.5J_C$	–	[23]
22	$\text{GSI} = 0.7394\text{RMR} + 0.57$	0.57	[24]
23	$\text{RMR} = 5.7 \ln + 43.65$	0.714	[25]
24	$\text{RMR} = 5.4 \ln Q + 55.2$	0.55	
25	$\text{RMR} = 5.9 \ln Q + 43$	0.81	[26]
26	$\text{GSI} = \text{RMR} - 5$ (FOR RMR > 23)	–	[27]
27	$\text{GSI} = 0.9934\text{RMR} - 4.913$	0.84	[28]
28	$E_m = 300e^{(0.07\text{RMR})} \times 10^{-3}$	–	[29]
29	$E_m/E_i = 0.5[1 - \cos(\pi\, \text{RMR}/100)]$	–	
30	$E_m = 0.0097\text{RMR}^{3.54} \times 10^{-3}$	–	
31	$E_m = 0.562\text{RMR} + 0.183$	–	
32	$E_m = 147.28e^{(\text{RMR} - 100)/24} - 0.202\text{RMR}$	–	
33	$E_m = E_i e^{(\text{RMR} - 100)/36}$	–	
34	$\text{GSI} = 0.739\text{RMR} + 12.097$	0.759	[30]
35	$\text{RMR} = 4.52 \ln Q + 43.635$	0.74	[31]

(continued)

Table 1 (continued)

No.	Equations	R^2	References
36	$E_m = 0.0736e^{(0.0755RMR)}$	–	[19]
37	$RMR = 8.8832 \ln Q + 43.26$	0.9035	[32]
38	$GSI = 1.265RMR - 21.49$	–	[33]
39	$GSI = RMR \pm 5$	–	[23]
40	$GSI = 1.2092RMR - 18.6143$	–	[34]
41	$RMR = 6.3 \ln Q + 41.6$	0.78	[35]
42	$GSI = 0.793RMR + 2.001$	0.736	[36]
43	$GSI = 0.876RMR + 0.935$	0.876	
44	$RMR = 6.63 \ln Q + 32.76$	0.88	[37]
45	$RMR = 9 \ln Q + 44$	0.77	
46	$RMR = 0.42GSI + 33$	–	[38]
47	$RMR = 47.649Q^{0.1255}$	0.983	[39]
48	$RMR = 6.1271 \ln Q + 48.308$	0.981	
49	$\nu_m = 0.5 - 0.2\frac{RMR}{RMR+0.2(100-RMR)}$	–	[40]
50	$\frac{\nu_m}{\nu_i} = 2.5 - 1.5\frac{RMR}{RMR+(100-RMR)}$	–	

E_i elastic modulus of intact rock, E_m rock mass deformation modulus, *GSI* geological strength index, *Q* tunneling quality index, *RMR* Rock Mass Rating, *RQD* rock quality designation, UCS_i uniaxial compressive strength of the intact rock, UCS_m uniaxial compressive strength of the rock mass, V_p compression wave velocity, V_s shear wave velocity, ν_i Poisson's ratio of the intact rock, ν_m Poisson's ratio of the rock mass

often rely on simplified linear formulations, which may not adequately capture the complex and nonlinear behavior of rock masses. Moreover, the simultaneous use of both V_p and V_s for RMR prediction has rarely been addressed in the existing literature.

These observations highlight a clear gap in previous research regarding the development of reliable and generalized models for estimating RMR based on geophysical parameters. In particular, the potential of advanced optimization techniques for improving the accuracy of RMR prediction from V_p and V_s has not been sufficiently explored. This gap provides a strong motivation for investigating alternative modeling strategies that can effectively handle nonlinear relationships and enhance predictive performance [1].

In this study, the relationships between the RMR index and compressional and shear wave velocities were developed using the metaheuristic algorithms DE, PS, and the hybrid DE-PS model. The performance of these approaches was independently evaluated and systematically compared in order to assess whether the hybridization of the two standalone algorithms leads to an improvement in the accuracy of RMR prediction based on V_p and V_s. The comparative results provide a clear framework for evaluating the effectiveness of hybrid metaheuristic algorithms in rock mass classification modeling.

2 Case Study and Data Collection

The Beheshtabad water transmission tunnel was selected as the case study in this research. A comprehensive database was established based on detailed field investigations conducted along the tunnel alignment, extensive exploratory drilling, and a wide range of laboratory tests performed on the collected core samples. These data were used to characterize the geological, structural, and geotechnical conditions of the study area and to support the development of predictive relationships between geophysical parameters and rock mass classification indices.

The tunnel is designed to convey water from the headwaters of the Karoun River to the central plateau of Iran through a pressurized underground system with a circular cross section. It has an internal diameter of approximately 6 m and an overall length of about 65 km, making it the largest and longest water transmission tunnel constructed in Iran to date. The project has been planned to supply an annual water discharge of approximately 1070 million m^3 to meet industrial, domestic, and agricultural water demands in the central plateau region.

From an engineering alignment perspective, the tunnel route includes several geometric complexities. Three major horizontal curves are incorporated along the alignment, and two significant longitudinal slope transitions occur at different sections of the tunnel. The first slope variation is characterized by a change in gradient from − 1 per thousand to − 22 per ten thousand, while the second transition involves a modification of the slope from − 22 per ten thousand to + 1 per ten thousand. These variations contribute to the complexity of excavation, support design, and construction planning [1].

Geographically, the tunnel project is located within the provinces of Isfahan and Chaharmahal-va-Bakhtiari in central Iran. The tunnel inlet is situated near the terminal reach of the Beheshtabad River, and the alignment extends in a generally northeastward direction before discharging into the Zayandeh-Rood River upstream of the Cham Asman Dam. This geographical setting exposes the tunnel to a wide range of geological and tectonic conditions along its route.

From a geological and tectonic standpoint, the tunnel alignment crosses two major structural zones. The inlet section is located within the folded Zagros tectonic zone, whereas the outlet section lies within the Sanandaj–Sirjan zone. Geological investigations indicate that the dominant lithological units encountered along the tunnel include conglomerate, sandstone, mudstone, limestone, marly limestone, carbonate limestone, and andesite. In the terminal segments of the tunnel, limestone, dolomite, conglomerate, and marly limestone interbedded with shale layers are the predominant rock types.

Structurally, the study area is characterized by the presence of numerous faults and fold systems, which have imposed significant challenges during tunnel excavation. For structural assessment purposes, the tunnel alignment was divided into several sections of comparable length, each exhibiting distinct structural characteristics. These features were identified based on integrated geological observations and geophysical investigations. In particular, the sections located near the boundary between the Zagros and Sanandaj–Sirjan tectonic zones are strongly influenced by faulting and folding, resulting in complex ground conditions that require careful consideration in engineering design and analysis.

Based on the exploratory boreholes drilled along the alignment of the Beheshtabad water transmission tunnel, a total of 150 data sets were collected to determine the required parameters, including compressional wave velocity (V_p), shear wave velocity (V_s), and the Rock Mass Rating (RMR). All measurements were conducted in accordance with internationally recognized standards to ensure the reliability and consistency of the data.

The seismic wave velocities V_p and V_s were measured using the downhole seismic testing method, following the procedures recommended by the American Society for Testing and Materials. The downhole method is a widely accepted geophysical technique for determining seismic wave velocities in soil and rock masses and is particularly effective for evaluating dynamic properties such as elastic modulus, shear modulus, Poisson's ratio, and other relevant geotechnical parameters. Owing to its high accuracy and ability to provide depth-specific measurements, this method is commonly applied in large-scale engineering projects, including tunnels, dams, and high-rise structures.

In the downhole testing procedure, boreholes were drilled at selected locations along the tunnel route, and sensing devices such as geophones were installed at predetermined depths within the boreholes. A controlled vibration source was applied at the ground surface near the borehole collar, generating elastic waves that propagated through the subsurface layers. Compressional waves, which propagate parallel to the direction of wave travel, were recorded first and used to determine V_p. Subsequently, shear waves, characterized by particle motion perpendicular to the direction of propagation, were recorded to determine V_s. The travel times of these waves at different depths were measured, and the corresponding wave velocities were calculated using standard analytical formulations. One of the main advantages of the downhole method is its capability to provide accurate wave velocity measurements at specific depths, allowing for detailed characterization of subsurface mechanical behavior [2].

Although the downhole seismic test provides reliable and precise results under in situ conditions, it requires specialized equipment, experienced personnel, and relatively higher time and cost compared to surface-based seismic methods. Nevertheless, due to its superior accuracy and detailed subsurface resolution, this technique was considered appropriate for the objectives of the present study.

To calculate the RMR values, its six fundamental parameters were first determined through a combination of field surveys and laboratory testing. The uniaxial compressive strength (UCS) of intact rock samples was measured following the testing procedures recommended by the International Society for Rock Mechanics. A servo-controlled axial compression testing machine was employed, and axial stress was applied at a loading rate ranging from 0.5 to 1 MPa/s until sample failure occurred. The stress corresponding to the failure point was recorded as the UCS value.

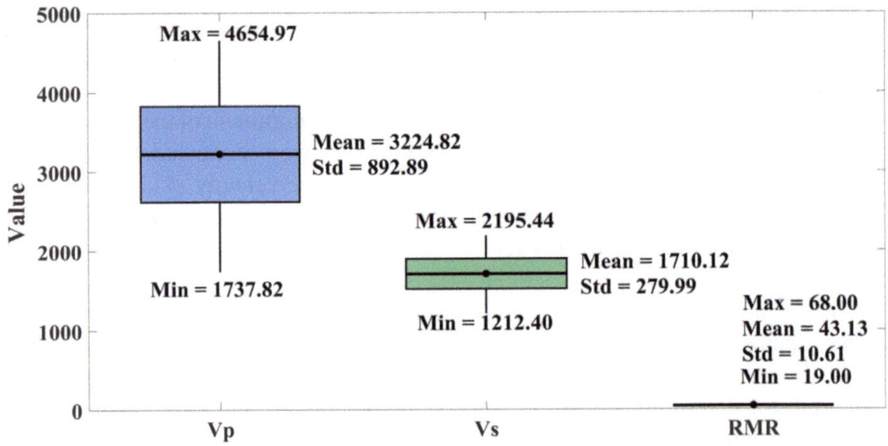

Fig. 1 Box plots of the statistical properties of the data used for RMR modeling

The Rock quality designation (RQD) was determined based on standard procedures suggested by the American Society for Testing and Materials. Discontinuity spacing was calculated through detailed geological mapping conducted along the tunnel route. In addition, the condition of discontinuities, including parameters such as roughness, infilling material, and joint frequency, was evaluated in accordance with the ISRM suggested methods for the quantitative description of rock mass discontinuities [43].

Groundwater conditions were assessed to account for their influence on rock mass behavior. This evaluation was performed based on field observations of moisture conditions and seepage features observed on rock surfaces along the tunnel alignment. For each of the six RMR parameters, an appropriate rating was assigned, and the basic RMR value was obtained by summing the ratings of the first five parameters. In cases where the orientation of discontinuities relative to the tunnel axis was deemed significant, the final RMR value was adjusted accordingly to account for the effect of joint orientation. The statistical characteristics of the input parameters, including minimum, maximum, mean, and standard deviation values of V_p, V_s, and RMR, are illustrated in the boxplot presented in Fig. 1.

3 Overview of Methods

In this study, four computational approaches were employed to estimate the Rock Mass Rating (RMR) using compressional wave velocity (V_p) and shear wave velocity (V_s) as input parameters. These approaches include three metaheuristic optimization algorithms, Differential Evolution (DE), Pattern Search (PS), and a hybrid DE-PS algorithm, as well as the Random Forest (RF) machine learning model. The meta-heuristic algorithms were applied to develop optimized mathematical relationships between the input variables and RMR, while the RF model was used as a complementary tool to investigate the relative contribution of V_p and V_s in RMR prediction. A brief description of each method and its role within the framework of the present study is provided in the following subsections.

3.1 DE

Differential Evolution (DE) is a population-based metaheuristic optimization algorithm that operates through evolutionary mechanisms such as mutation, crossover, and selection. In this algorithm, an initial population of candidate solutions is generated randomly within predefined bounds. New solutions are then produced by perturbing existing candidates using scaled differences between randomly selected individuals, which enables an efficient global exploration of the search space [5].

Through successive generations, trial solutions are evaluated using a predefined objective function, and only those yielding improved performance are retained for subsequent iterations. This process allows the DE algorithm to gradually converge toward globally optimal or near-optimal solutions, even in complex and nonlinear optimization landscapes. Due to its strong global search capability and robustness, DE has been widely applied in engineering optimization problems.

In the present study, an initial linear equation relating RMR to the input parameters V_p and V_s was first defined, as expressed in Eq. (1). The coefficients associated with V_p, V_s, and the constant term were then optimized using the DE algorithm. The optimal set of coefficients was selected based on the highest prediction accuracy between the estimated and measured RMR values. By substituting these optimized coefficients into the predefined equation, a DE-based predictive model for RMR estimation was developed.

$$\mathrm{RMR} = \left(a * V_p\right) + (b * V_s) + c. \tag{1}$$

3.2 PS

The Pattern Search (PS) algorithm is a derivative-free optimization technique belonging to the class of direct search methods. This algorithm is particularly effective for problems in which the objective function is nonlinear, discontinuous, or difficult to differentiate. Unlike population-based algorithms, PS relies on a structured local search strategy that systematically explores the neighborhood of a current solution.

In the PS algorithm, a set of trial points is generated around the current solution based on predefined search directions and step sizes. If an improved solution is identified among these trial points, it replaces the current solution and the search proceeds from the new location. If no improvement is observed, the step size is reduced, allowing for a more refined local search. This adaptive mechanism enables PS to efficiently converge toward an optimal solution by focusing on local improvements [6].

Within the context of this study, the same predefined equation linking RMR to V_p and V_s was adopted. The coefficients of this equation were optimized using the PS algorithm to minimize prediction errors. The optimized coefficients were then substituted into the governing equation to establish a PS-based model for RMR prediction. Owing to its strong local search capability, PS was employed to refine the relationship between wave velocity parameters and RMR.

3.3 Hybrid DE-PS Algorithm

The hybrid DE-PS algorithm was developed to integrate the complementary strengths of the DE and PS algorithms. While DE provides an effective global search mechanism capable of exploring diverse regions of the solution space, PS offers a powerful local search strategy for fine-tuning solutions in promising areas. By combining these two approaches, the hybrid DE-PS algorithm aims to achieve improved prediction accuracy and enhanced convergence stability. Generally, hybrid intelligent models were successfully applied in the area of tunneling and rock mechanics [44–47].

In the hybrid framework, the optimization process begins with the DE algorithm generating and evolving a population of candidate solutions to broadly explore the search space. The most promising solutions identified during this global search phase are then passed to the PS algorithm for further refinement through local optimization. This sequential cooperation reduces the likelihood of premature convergence and enhances the ability of the model to identify high-quality solutions.

As with the standalone algorithms, an initial equation relating RMR to V_p and V_s was first defined. The coefficients of this equation were initially optimized using DE and subsequently refined using PS. The final optimized coefficients obtained through the DE-PS procedure were substituted into the predefined equation to develop a hybrid predictive model for RMR estimation. The performance of the hybrid DE-PS model was later compared with those of the individual DE and PS models to evaluate whether the integration of global and local search strategies leads to improved prediction performance.

3.4 RF

The Random Forest (RF) algorithm is an ensemble learning method that constructs a large number of decision trees during the training process and combines their outputs to improve predictive accuracy and robustness. For regression problems, the final prediction is obtained by averaging the outputs of individual trees. This ensemble approach reduces overfitting and enhances generalization capability, particularly when dealing with nonlinear and complex datasets.

One of the key advantages of the RF algorithm is its ability to quantify the relative importance of input variables. In the present study, RF was applied to analyze the influence of V_p and V_s on RMR prediction in both univariate and bivariate forms. Specifically, the RF model was used to investigate whether acceptable prediction accuracy could be achieved using only one of the input parameters or whether the combined use of V_p and V_s results in improved RMR estimation. This analysis provided additional insight into the role of wave velocity parameters in predicting RMR and complemented the metaheuristic modeling results.

The predictive performance of machine learning models is strongly influenced by the selection of appropriate hyperparameters. To determine the optimal hyperparameter configuration for the RF model, a grid search strategy was employed in conjunction with k-fold cross-validation. In this procedure, the dataset was divided into k subsets, where $k - 1$ subsets were used for training and the remaining subset was used for validation in each iteration [1, 48, 49].

This process was repeated until each subset had been used once as validation data, and the performance metrics obtained from all iterations were averaged to evaluate model stability and accuracy. In this study, a tenfold cross-validation scheme was adopted, considering the dataset size and computational efficiency. Based on the cross-validation results, the optimal hyperparameters were selected, and the RF model was retrained using the entire training dataset. The final performance of the optimized RF model was then evaluated using an independent test dataset.

4 Modeling the Relationship of RMR with Waves Velocity

In this section, optimized relationships between the seismic wave velocity parameters, including compressional wave velocity (V_p) and shear wave velocity (V_s), and the Rock Mass Rating (RMR) are developed using Differential Evolution (DE), Pattern Search (PS), and the hybrid DE-PS algorithms. Prior to constructing these relationships, it is necessary to determine whether RMR can be accurately estimated using only one of the two wave velocity parameters or whether the simultaneous use of both V_p and V_s is required.

To address this issue, the Random Forest (RF) algorithm is employed as an exploratory tool. Using RF, the relative contribution and importance of V_p and V_s in predicting RMR are evaluated individually. In addition, the combined influence of both parameters is assessed to examine whether their joint use enhances prediction capability. Based on the outcomes of this analysis, a decision is made regarding the most appropriate input configuration for RMR estimation. Subsequently, optimal predictive relationships between the selected input parameter(s) and RMR are developed using the DE, PS, and hybrid DE-PS techniques.

4.1 Selection of Input Parameters Using the RF Algorithm

In this subsection, the optimal input parameter(s) required for RMR modeling are identified using the RF algorithm. This algorithm enables the quantification of the relative importance of each input variable, namely V_p and V_s, in the estimation of RMR. The results indicate that the individual importance values of V_s and V_p in predicting RMR are 4.61843 and 3.5701, respectively. Moreover, the combined importance of using both V_p and V_s simultaneously is calculated as 8.18853 (Fig. 2).

The RF results demonstrate that the combined contribution of V_p and V_s is considerably greater than their individual effects. This finding suggests that the simultaneous use of both wave velocity parameters provides more informative input for RMR estimation. Accordingly, based on the RF analysis, the combined use of V_p and V_s is adopted for subsequent RMR modeling using the DE, PS, and DE-PS algorithms. To further verify the RF-based findings, the correlations between V_p, V_s, and their combination with RMR are examined. The results show that the correlation between RMR and the combined input of V_p and V_s is higher than the correlations obtained when each parameter is used individually. This observation supports the validity and reliability of the RF results and confirms the advantage of employing both parameters together for RMR prediction. Consequently, the combined V_p–V_s input configuration is applied in the following subsections for model development.

Additional RF outputs illustrate the agreement between predicted and measured RMR values for different input scenarios. When only V_p is used as the input parameter, a strong agreement between predicted and observed RMR values is observed, indicating acceptable prediction accuracy. The use of V_s alone also results in reasonable performance; however, its predictive capability is comparatively lower than that achieved using V_p alone. In contrast, the simultaneous use of V_p and V_s leads to a noticeable improvement in prediction accuracy, yielding predictions that are more

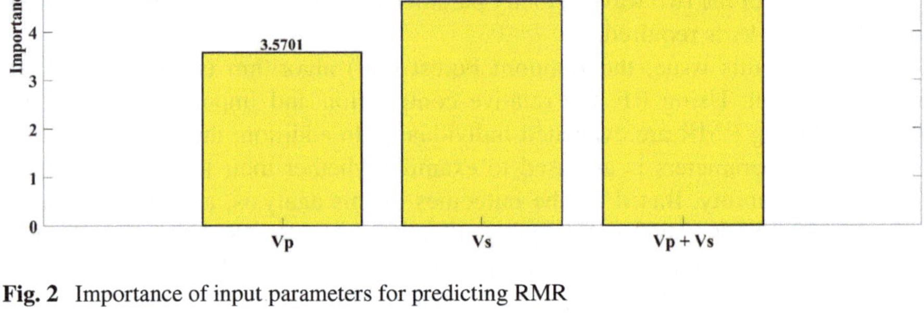

Fig. 2 Importance of input parameters for predicting RMR

closely aligned with the observed RMR values. These results indicate that combining V_p and V_s can substantially enhance the accuracy of RMR estimation compared to using either parameter individually.

4.2 RMR Modeling Using PS

In this section, the Pattern Search (PS) algorithm was employed to establish a predictive relationship between the Rock Mass Rating (RMR) and the seismic wave velocities, namely compressional wave velocity (V_p) and shear wave velocity (V_s). The PS algorithm is a derivative-free optimization technique that operates based on systematic exploration of the search space through predefined patterns, making it particularly suitable for problems characterized by nonlinear behavior and complex response surfaces.

In the proposed modeling framework, the PS algorithm was applied to optimize the coefficients of a predefined mathematical equation relating V_p and V_s to RMR. Unlike gradient-based methods, the PS approach evaluates a set of candidate solutions around the current point using structured search directions. If an improved solution is identified, the algorithm moves toward that point and continues the search with an updated pattern size. Otherwise, the search step is reduced, allowing for a more refined local exploration. This mechanism enables the PS algorithm to effectively perform local search while maintaining robustness against noisy and irregular objective functions.

The performance of the PS-based model was evaluated by comparing the predicted RMR values with the corresponding measured values obtained from field and laboratory investigations. A three-dimensional representation of the measured and predicted RMR values as a function of V_p and V_s is illustrated in Fig. 3. This visualization provides a clear insight into the predictive capability of the PS algorithm by demonstrating the spatial agreement between observed data points and model predictions across the investigated range of seismic velocities.

As shown in Fig. 3, the optimized predictive equation derived from the PS algorithm, along with its associated statistical performance indices, is presented directly on the figure. The coefficient of determination (R^2) obtained for the PS model was equal to 0.8686, indicating a strong correlation between predicted and measured RMR values. Furthermore, the root mean square error (RMSE) and mean absolute error (MAE) were calculated as 4.576 and 3.733, respectively. The mean absolute percentage error (MAPE) of 11.35% confirms that the relative deviation of the predicted RMR values remains within an acceptable range for practical rock engineering applications.

The results indicate that the PS algorithm is capable of providing a reliable estimation of RMR based on V_p and V_s. The observed dispersion between measured and predicted values can be attributed to the inherent heterogeneity of rock masses and uncertainties associated with in situ geophysical measurements. Nevertheless, the PS-based model demonstrates satisfactory predictive performance and serves

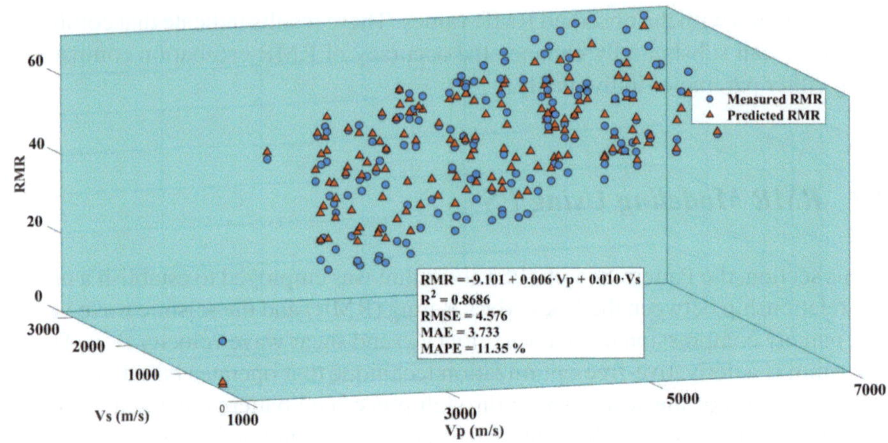

Fig. 3 The 3D histogram of the real RMR and predicted RMR from the PS model

as a solid benchmark for comparison with other optimization approaches, particularly the Differential Evolution-based and hybrid algorithms discussed in subsequent sections.

4.3 RMR Modeling Using DE

In this section, the Differential Evolution (DE) algorithm was employed to develop a predictive model for estimating the Rock Mass Rating (RMR) based on compressional wave velocity (V_p) and shear wave velocity (V_s). DE is a population-based evolutionary optimization algorithm that is well known for its strong global search capability and robustness in solving nonlinear and multimodal optimization problems. Owing to its mutation and crossover mechanisms, DE is capable of efficiently exploring the solution space and avoiding premature convergence to local optima.

Within the proposed modeling framework, the DE algorithm was utilized to optimize the coefficients of a predefined mathematical equation linking V_p and V_s to RMR. An initial population of candidate solutions was first generated randomly within predefined bounds. Subsequently, new trial solutions were produced through mutation and crossover operations, followed by a selection process in which solutions with superior objective function values were retained for the next generation. This evolutionary process was repeated iteratively until convergence criteria were satisfied, leading to the identification of an optimal set of model coefficients.

The predictive performance of the DE-based model was evaluated through a comparison between the predicted and measured RMR values derived from field and laboratory data. A three-dimensional visualization illustrating the relationship between V_p, V_s, and both measured and predicted RMR values is presented in Fig. 4. This figure provides a comprehensive view of the model behavior across the full

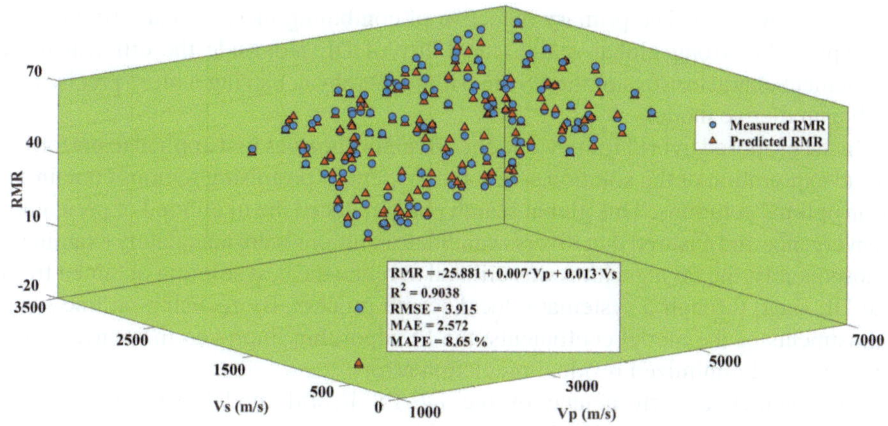

Fig. 4 The 3D histogram of the real RMR and predicted RMR from the DE model

range of input parameters and enables a visual assessment of the agreement between observed and predicted data points.

As illustrated in Fig. 4, the optimized predictive equation obtained from the DE algorithm, together with its corresponding statistical performance indices, is displayed directly on the figure. The coefficient of determination (R^2) achieved by the DE model was 0.9038, indicating a strong correlation and improved predictive accuracy compared with the PS-based model. In addition, the root mean square error (RMSE) and mean absolute error (MAE) were calculated as 3.915 and 2.572, respectively, reflecting a lower magnitude of prediction error. The mean absolute percentage error (MAPE) of 8.65% further confirms the enhanced reliability of the DE-based predictions.

The results demonstrate that the DE algorithm provides a more accurate estimation of RMR using V_p and V_s as input parameters compared with the PS approach. The improved performance can be attributed to the strong global search capability of the DE algorithm, which enables a more effective exploration of the solution space and identification of optimal model coefficients. Nevertheless, some dispersion between measured and predicted values remains, which is primarily associated with the natural heterogeneity of rock masses and uncertainties in geophysical measurements. The findings of this section establish a solid foundation for evaluating the performance of the hybrid DE-PS model presented in the following section.

4.4 RMR Modeling with DE-PS

In this section, a hybrid optimization approach based on the integration of Differential Evolution (DE) and Pattern Search (PS) algorithms was employed to further enhance the accuracy of RMR prediction using compressional wave velocity (V_p) and shear wave velocity (V_s). The primary objective of combining these two algorithms was to exploit the strong global search capability of DE alongside the efficient local refinement characteristics of the PS algorithm, thereby achieving a more precise and stable predictive model.

In the proposed hybrid framework, the DE algorithm was first applied to perform a broad exploration of the solution space and to identify promising regions containing near-optimal solutions. This global search phase reduced the likelihood of premature convergence and ensured that diverse candidate solutions were adequately examined. Subsequently, the PS algorithm was utilized to refine the best solution obtained from the DE stage through a systematic local search process. By focusing on fine-scale adjustments of the model coefficients, the PS algorithm improved the convergence precision and minimized residual prediction errors.

The predictive performance of the DE-PS hybrid model was assessed by comparing the predicted RMR values with the corresponding measured values derived from field and laboratory investigations. A three-dimensional representation illustrating the relationship between V_p, V_s, and both measured and predicted RMR values is presented in Fig. 5. This figure clearly demonstrates a closer alignment between the predicted and observed RMR values compared with the standalone PS and DE models, indicating the effectiveness of the hybrid optimization strategy.

As shown in Fig. 5, the optimized predictive equation derived from the DE-PS algorithm, along with its associated statistical performance indices, is displayed directly on the figure. The coefficient of determination (R^2) obtained for the hybrid model reached 0.9567, reflecting a substantial improvement in predictive accuracy. Moreover, the root mean square error (RMSE) and mean absolute error (MAE) were reduced to 3.251 and 1.787, respectively, confirming the enhanced precision of the DE-PS model. The mean absolute percentage error (MAPE) of 6.82% further indicates a significant reduction in relative prediction error compared with the individual algorithms.

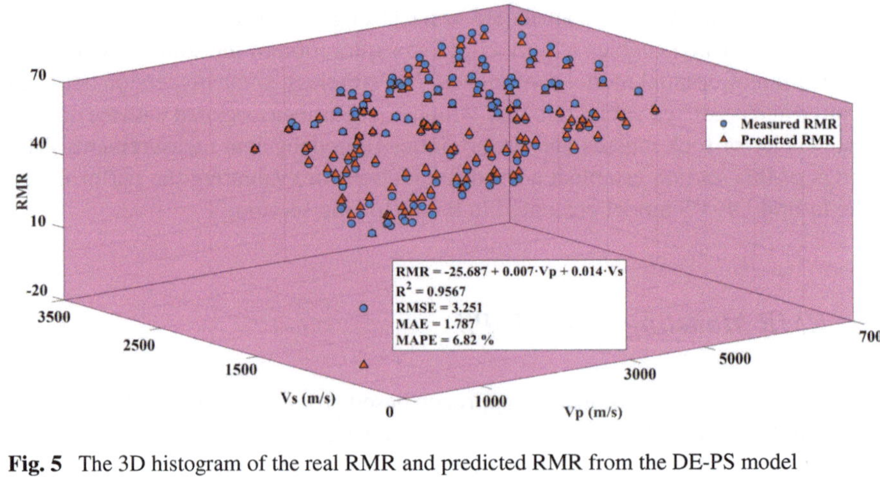

Fig. 5 The 3D histogram of the real RMR and predicted RMR from the DE-PS model

The results demonstrate that the DE-PS hybrid algorithm outperforms the standalone PS and DE models in predicting RMR from seismic wave velocities. The superior performance of the hybrid approach can be attributed to the complementary interaction between the global exploration capability of DE and the local exploitation strength of PS. This combination enables a more effective navigation of the solution space and results in a robust and accurate predictive model. The findings of this section highlight the potential of hybrid metaheuristic algorithms for improving the reliability of rock mass classification predictions in complex geological environments.

5 Results

The results obtained in this study, together with the conducted analyses, lead to the following main findings:

- Based on the outcomes derived from the Random Forest (RF) algorithm and the comparison of the correlation strength between the input parameters and the RMR, it was observed that employing both seismic wave velocities (V_p and V_s) simultaneously results in a noticeably better prediction performance than using either V_p or V_s alone. This finding confirms that the combined use of compressional and shear wave velocities provides more comprehensive information about rock mass conditions and improves the reliability of RMR estimation.
- The Pattern Search (PS) algorithm was employed to establish a predictive relationship between V_p, V_s, and RMR. Statistical evaluation of the PS-based model resulted in an R^2 value of 0.8686, indicating a reasonably strong agreement between predicted and measured RMR values. However, the error indices, including RMSE = 4.576 and MAE = 3.733, reveal a moderate level of absolute prediction error. Furthermore, the relative error indicator MAPE = 11.35% suggests that, despite acceptable performance, the PS model exhibits noticeable dispersion and lower stability when compared with more advanced optimization approaches.
- The Differential Evolution (DE) algorithm demonstrated improved predictive capability relative to the PS model. The DE-based model achieved an R^2 value of 0.9038, reflecting a stronger correlation between predicted and actual RMR values. In addition, the error measures were significantly reduced, with RMSE = 3.915 and MAE = 2.572, indicating enhanced absolute accuracy. The relative error, represented by MAPE = 8.65%, further confirms that the DE model provides more reliable and consistent RMR predictions than the PS algorithm, primarily due to its strong global search capability.
- The hybrid Differential Evolution-Pattern Search (DE-PS) model yielded the most accurate and robust results among all evaluated approaches. This model achieved a high coefficient of determination of $R^2 = 0.9567$, indicating an excellent agreement between predicted and measured RMR values. Moreover, it produced the

lowest error values across all performance metrics, including RMSE = 3.251, MAE = 1.787, and MAPE = 6.82%. These results clearly demonstrate that the hybrid DE-PS algorithm outperforms both the standalone PS and DE models, benefiting from the effective combination of DE's global exploration capability and PS's efficient local refinement. The superior performance of the DE-PS model highlights its strong potential for accurate and reliable RMR prediction in complex geological environments.

6 Discussion

6.1 Cumulative Error Comparison

The cumulative absolute error analysis provides an integrated and comprehensive evaluation of the predictive performance of the applied models over the entire dataset. Unlike pointwise error metrics, this approach considers the accumulation of absolute prediction errors across all samples, thereby offering a clear insight into the overall stability and consistency of each model. The cumulative error curves obtained for the DE, PS, and hybrid DE-PS models are illustrated in Fig. 6.

As shown in Fig. 6, the hybrid DE-PS model exhibits the lowest cumulative absolute error throughout the full range of samples. This behavior indicates that the hybrid model consistently produces smaller prediction errors and avoids the accumulation of large deviations across the dataset. The gradual slope of the DE-PS curve reflects its superior robustness and uniform performance, confirming that error propagation is effectively controlled when both global and local optimization strategies are integrated.

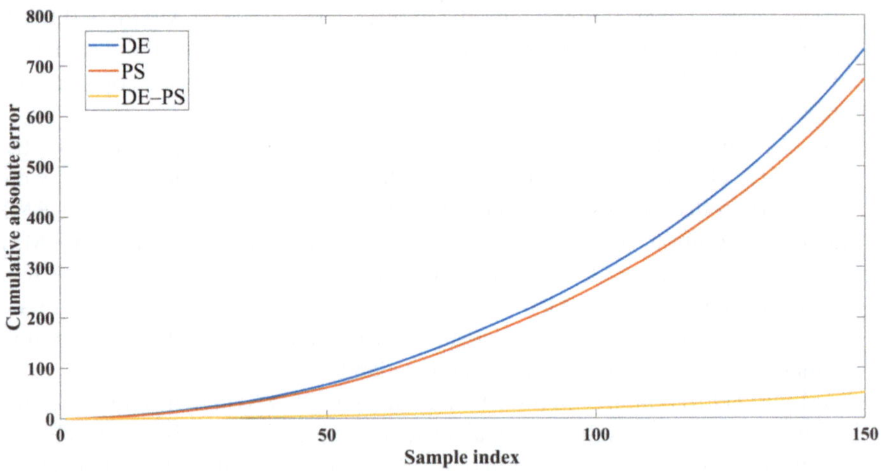

Fig. 6 Cumulative absolute error comparison of PS, DE, and DE-PS models

In contrast, the PS model demonstrates an intermediate level of cumulative absolute error. Although the PS-based predictions show acceptable accuracy for a considerable portion of the data, the cumulative curve rises more rapidly than that of the hybrid model. This behavior suggests that while the PS algorithm is effective in local refinement of solutions, it may be more sensitive to initial conditions or complex nonlinearities within the dataset, leading to moderately higher accumulated errors.

The DE model presents the highest cumulative absolute error among the evaluated approaches, as clearly observed in Fig. 6. The steeper slope of the DE curve indicates that larger errors are accumulated over the dataset, particularly in regions characterized by higher variability in rock mass properties. Despite the strong global search capability of the DE algorithm, the absence of an explicit local refinement mechanism appears to limit its ability to fine-tune solutions, resulting in comparatively higher cumulative deviations.

Overall, the cumulative error comparison highlights the clear advantage of the hybrid DE-PS model over the standalone PS and DE algorithms. The observed performance improvement can be attributed to the complementary interaction between the global exploration strength of DE and the efficient local search capability of PS. By reducing both isolated large errors and their accumulation across samples, the DE-PS approach provides a more reliable and stable framework for RMR prediction. These findings, as evidenced in Fig. 6, underscore the importance of hybrid metaheuristic strategies for enhancing prediction accuracy in complex geotechnical and rock engineering applications.

6.2 Comparison of Prediction Errors

The comparison of prediction errors provides a detailed insight into the dispersion and magnitude of deviations associated with each predictive model. By examining the distribution of absolute errors, it becomes possible to evaluate not only the average performance of the models but also their stability and sensitivity to variations in the input data. The graphical representation of prediction errors for the PS, DE, and hybrid DE-PS models is presented in Fig. 7.

As illustrated in Fig. 7, the hybrid DE-PS model exhibits the smallest error magnitude, with the majority of prediction errors concentrated near zero. This behavior indicates that the hybrid model consistently produces accurate estimates of RMR and effectively minimizes large deviations across the dataset. The compact distribution of errors reflects a high level of robustness and demonstrates that the integration of global and local optimization mechanisms leads to improved control over prediction uncertainty.

The PS model shows a moderate spread of prediction errors, positioned between the hybrid and DE models. Although the error distribution for PS remains within an acceptable range, it is noticeably wider than that of the DE-PS model. This observation suggests that while the Pattern Search algorithm is capable of refining solutions

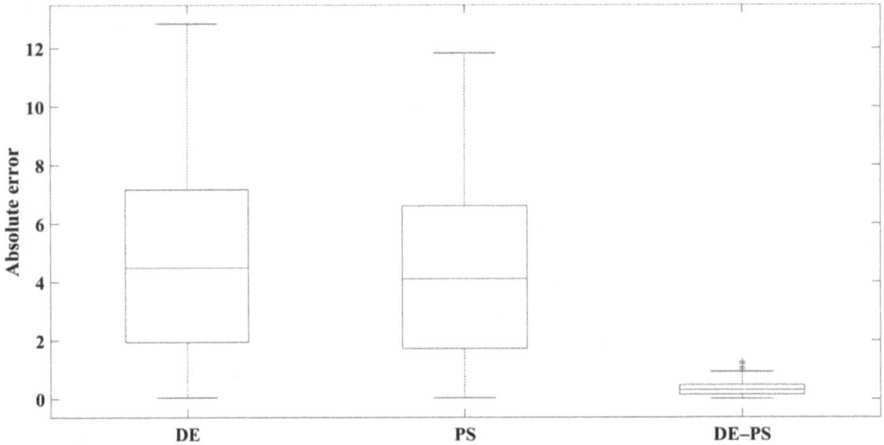

Fig. 7 Comparison of absolute prediction errors for PS, DE, and DE-PS models

through local optimization, its performance may be influenced by local irregularities in the objective function, resulting in a higher sensitivity to data variability and moderately larger prediction errors.

In contrast, the DE model presents the largest prediction error magnitude, as evidenced by the wider distribution and greater distance from zero in Fig. 7. The increased spread of errors indicates that, despite its strong global search capability, the DE algorithm alone may lack sufficient local refinement to achieve precise convergence for all samples. Consequently, larger deviations between predicted and measured RMR values are observed, particularly in data regions characterized by complex geological conditions.

The comparison of prediction errors clearly demonstrates the superior performance of the hybrid DE-PS model over the standalone PS and DE approaches. The significantly reduced error magnitude and tighter concentration around zero confirm the enhanced accuracy and stability achieved through hybridization. These findings, as shown in Fig. 7, further support the effectiveness of combining global exploration and local exploitation strategies for improving the reliability of RMR prediction in heterogeneous rock mass environments.

6.3 Discussion on REC Curves and AUC Analysis

The Receiver Error Characteristic (REC) curve analysis provides a comprehensive and threshold-independent evaluation of model performance by examining the proportion of predictions that fall within a specified error tolerance. Unlike single-value error metrics, REC curves illustrate the predictive behavior of models across a continuous range of error thresholds, offering deeper insight into their robustness

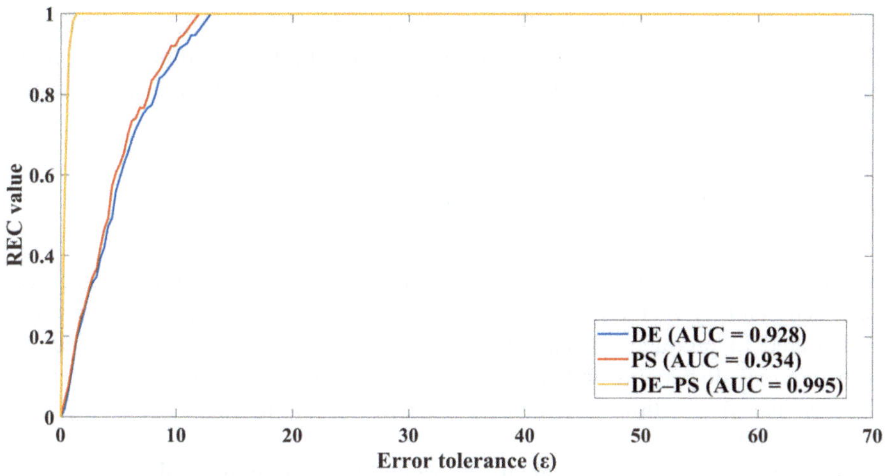

Fig. 8 Comparison of the REC curves obtained for the PS, DE, and hybrid DE-PS models

and reliability. The REC curves obtained for the PS, DE, and hybrid DE-PS models are presented and compared in Fig. 8.

As shown in Fig. 8, the hybrid DE-PS model consistently demonstrates superior performance across the entire error tolerance range. This observation is quantitatively supported by the highest area under the REC curve (AUC) value of 0.995, indicating that the hybrid model achieves accurate predictions within smaller error margins for the majority of samples. The near-unity AUC value reflects the strong generalization capability and high stability of the DE-PS model when applied to RMR prediction.

The DE model exhibits a comparatively lower but still acceptable performance, with an AUC value of 0.928. This result suggests that the DE algorithm is capable of producing reasonably accurate predictions within moderate error tolerances; however, its predictive consistency decreases when stricter accuracy thresholds are imposed. The lower AUC value indicates that, although the global search mechanism of DE is effective in exploring the solution space, the lack of fine local refinement may limit its ability to consistently achieve minimal prediction errors.

The PS model achieves an AUC value of 0.934, which is slightly higher than that of the DE model. This outcome highlights the effectiveness of the Pattern Search algorithm in locally optimizing model parameters and improving accuracy within relatively small error ranges. Nevertheless, the PS model still falls short of the hybrid DE-PS approach, particularly at lower error tolerances, where global exploration plays a critical role in identifying optimal solution regions.

The REC curve analysis and corresponding AUC values clearly confirm the advantages of the hybrid DE-PS model over the standalone PS and DE algorithms. The superior AUC value of the DE-PS model demonstrates that combining global exploration and local exploitation strategies leads to a more reliable and accurate predictive framework. These findings, as illustrated in Fig. 8, reinforce the conclusion that hybrid metaheuristic approaches are particularly well suited for complex geotechnical prediction problems involving heterogeneous and nonlinear datasets.

6.4 Residual Distribution of Models

The distribution of prediction residuals provides valuable insight into the accuracy, bias, and stability of the developed models. A well-performing predictive model is expected to produce residuals that are symmetrically distributed around zero with minimal dispersion, indicating low bias and consistent prediction accuracy. The residual distributions obtained for the DE, PS, and hybrid DE-PS models are illustrated in Fig. 9.

As shown in Fig. 9, the residuals associated with the hybrid DE-PS model are highly concentrated around zero and exhibit a narrow distribution range. This strong clustering indicates that the predicted RMR values closely match the measured data, with only minor deviations. The symmetrical shape and limited spread of the DE-PS residuals suggest minimal systematic bias and high robustness across the entire dataset. Such behavior confirms the superior generalization capability of the hybrid model and its effectiveness in capturing the nonlinear relationship between seismic wave velocities and RMR.

In contrast, the residual distributions of the PS and DE models display noticeably wider spreads. The PS model shows a moderate concentration of residuals near zero; however, a significant portion of predictions still deviates from the measured values, indicating reduced precision compared to the hybrid approach. This behavior

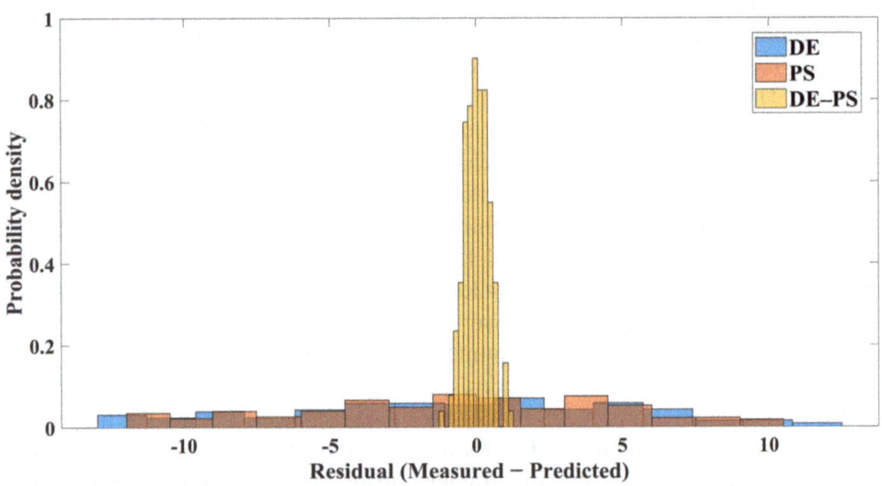

Fig. 9 Residual distributions obtained for the DE, PS, and hybrid DE-PS models

reflects the local search nature of the Pattern Search algorithm, which may converge to suboptimal solutions when global exploration is limited.

The DE model exhibits the widest residual distribution among the three approaches, with residuals extending over a broad range on both the positive and negative sides. This wide dispersion highlights larger prediction errors and greater variability in model performance. Although the DE algorithm is effective in exploring the global search space, the absence of strong local refinement can lead to less accurate pointwise predictions, particularly for complex geotechnical datasets.

The residual analysis presented in Fig. 9 clearly demonstrates the advantages of the hybrid DE-PS model. The compact and symmetric residual distribution confirms that combining the global exploration capability of DE with the local exploitation strength of PS results in a more stable and accurate predictive framework. These findings further support the statistical validation results and emphasize the reliability of the hybrid model for RMR estimation in engineering applications.

6.5 Comparison of Predicted and Measured RMR Values Using Scatter Plots

A direct comparison between predicted and measured RMR values provides a clear and intuitive assessment of model accuracy and bias. Scatter plots incorporating the 1:1 reference line are widely used for this purpose, as perfect predictions would theoretically fall exactly on this line. The comparative 1:1 plots obtained for the DE, PS, and hybrid DE-PS models are presented in Fig. 10.

As illustrated in Fig. 10, the hybrid DE-PS model exhibits the smallest deviation from the 1:1 line among all evaluated approaches. The predicted RMR values are densely clustered along the reference line, indicating a very high level of agreement with the measured data. This near-perfect alignment suggests that the hybrid model is capable of capturing both the global trend and local variations in the RMR dataset with minimal systematic error. The close conformity of the DE-PS predictions to the 1:1 line further confirms the robustness and reliability of this combined optimization strategy.

The PS model demonstrates a moderate level of agreement with the measured RMR values. Although a substantial portion of the data points lies close to the 1:1 line, noticeable scatter is observed, particularly at higher and lower RMR ranges. This dispersion reflects the limitations of the PS algorithm in consistently refining solutions across the entire search space, leading to increased prediction uncertainty compared to the hybrid model.

In contrast, the DE model shows the largest deviation from the 1:1 line, with a broader scatter of predicted values around the reference line. This behavior indicates higher prediction errors and a tendency toward overestimation or underestimation in certain RMR intervals. While DE is effective in global exploration, the absence

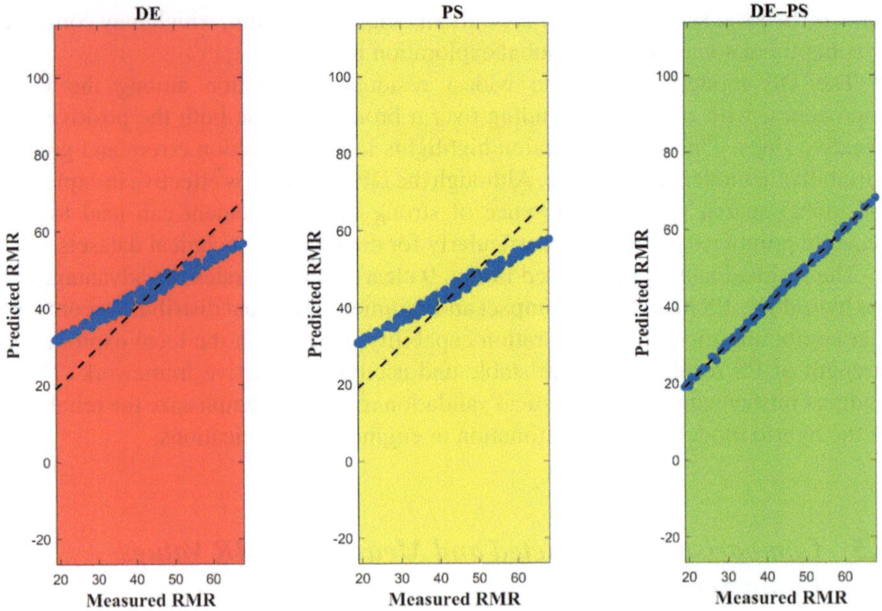

Fig. 10 Scatter plots of measured versus predicted RMR values for DE, PS, and hybrid DE-PS models

of strong local exploitation results in reduced pointwise accuracy, which is clearly reflected in the wider spread of data points.

The 1:1 plot analysis in Fig. 10 provides strong visual confirmation of the superior predictive capability of the hybrid DE-PS model. The close alignment of its predictions with the measured RMR values highlights the effectiveness of combining global and local search mechanisms. These findings are fully consistent with the statistical metrics and residual analyses presented in the previous sections, reinforcing the conclusion that the DE-PS model offers the most accurate and reliable framework for RMR prediction.

6.6 *Engineering Applications*

The results obtained in this study indicate that the proposed modeling framework, particularly the hybrid DE-PS algorithm, can be effectively applied in a wide range of rock engineering and geotechnical projects. Since the Rock Mass Rating (RMR) system plays a key role in the design and stability assessment of underground structures, the availability of an accurate and reliable method for estimating RMR based on non-destructive geophysical parameters such as P-wave and S-wave velocities offers significant practical advantages.

One of the most important engineering applications of the developed models lies in preliminary site investigations for tunnels, underground caverns, and mining excavations. In many projects, direct determination of RMR through detailed geological mapping and laboratory testing is time-consuming, costly, and sometimes impractical, especially in early stages of exploration. The proposed DE-PS-based model enables rapid estimation of RMR using seismic velocity data, which can be obtained efficiently through field geophysical surveys. This capability allows engineers to make informed decisions regarding excavation methods, preliminary support design, and risk assessment at an early stage of the project.

Furthermore, the proposed approach can be integrated into geotechnical monitoring and decision-support systems. By continuously updating seismic velocity measurements during excavation, variations in rock mass quality can be detected in near real time, allowing adaptive modification of support systems and excavation strategies. This application is particularly valuable in complex geological conditions where rock mass properties exhibit high spatial variability.

In addition, the developed models can contribute to optimizing drilling and blasting operations in mining and civil engineering projects. Accurate estimation of RMR provides essential input for selecting appropriate blasting patterns, explosive charges, and support measures, ultimately leading to improved safety, reduced overbreak, and lower operational costs. Therefore, the proposed modeling framework offers a practical, efficient, and cost-effective tool for enhancing decision-making processes in rock engineering practice.

7 Limitations and Future Research Directions

Despite the promising results achieved in this study, several limitations should be acknowledged, which also highlight potential directions for future research. First, the developed models are based on a specific dataset obtained from particular geological conditions. Although the results demonstrate strong predictive capability, the generalization of the proposed relationships to other geological settings and rock types requires further validation using independent datasets from different regions and lithological environments.

Another limitation is that the modeling framework relies solely on seismic velocity parameters (V_p and V_s) as input variables. While these parameters capture important mechanical and structural characteristics of the rock mass, other influential factors such as joint spacing, joint condition, groundwater presence, and in situ stress conditions are not explicitly considered. Incorporating additional geomechanical and geological parameters into future models could further enhance prediction accuracy and robustness.

From a methodological perspective, although the hybrid DE-PS algorithm showed superior performance compared to the individual DE and PS algorithms, the selection of algorithm parameters was performed empirically. Future studies may focus on adaptive or self-tuning strategies for algorithm parameters to further improve convergence speed and model stability. Additionally, comparative investigations involving other advanced hybrid metaheuristic algorithms could provide deeper insight into the relative strengths and weaknesses of different optimization strategies.

Finally, future research could explore the extension of the proposed approach to other rock mass classification systems such as GSI or *Q*-system, as well as the development of three-dimensional spatial models for rock mass quality prediction. Such advancements would significantly enhance the applicability of metaheuristic-based modeling techniques in large-scale underground engineering projects.

8 Conclusions

In this study, the capability of three metaheuristic algorithms, namely Differential Evolution (DE), Pattern Search (PS), and their hybrid combination (DE-PS), was investigated for predicting the Rock Mass Rating (RMR) parameter based on compressional and shear wave velocities (V_p and V_s). The comparative analyses were carried out using multiple performance indicators and complementary validation approaches to ensure a reliable assessment of model accuracy and stability.

The obtained results demonstrated that incorporating both V_p and V_s as input variables leads to a meaningful improvement in RMR prediction accuracy, confirming the importance of jointly considering elastic wave parameters in rock mass characterization. Among the standalone algorithms, PS showed stronger performance at lower error ranges due to its effective local search capability, while DE benefited from a broader exploration of the solution space but exhibited comparatively higher prediction dispersion.

The hybrid DE-PS model consistently outperformed the individual algorithms across all evaluation criteria. The integration of the global exploration strength of DE with the local refinement capability of PS resulted in a balanced and robust predictive framework. This superiority was clearly reflected in the lower prediction errors, higher correlation with measured RMR values, and the cumulative error behavior observed throughout the analyses.

Furthermore, the Regression Error Characteristic (REC) curve analysis provided additional insight into the predictive reliability of the models across varying error tolerances. The near-unity AUC value obtained for the DE-PS model confirms its ability to maintain a high proportion of accurate predictions even under strict error thresholds. This behavior highlights the enhanced generalization capability and stability of the hybrid approach when applied to complex and nonlinear geotechnical datasets.

The findings of this research indicate that hybrid metaheuristic algorithms offer a powerful and reliable alternative to standalone optimization techniques for rock mass classification problems. The proposed DE-PS model demonstrates strong potential for practical applications in tunneling, mining, and other underground engineering projects, where accurate estimation of RMR plays a critical role in design optimization and risk mitigation.

AI Use Disclosure Statement The authors did not use any generative AI or AI-assisted technologies in the creation of this manuscript. The authors used generative AI technologies only for proofreading. This includes minor corrections to grammar, spelling, and syntax. No AI was used for ideation, data analysis, or writing original content.

References

1. Davoodi PK, Hajizadeh F, Rezaei M (2025) Evaluating the RMR correlation with the rock mass wave velocity using the meta-heuristics algorithms. Sci Rep 15(1):17716
2. Davoodi PK, Hajizadeh F, Rezaei M (2025) Application of the metaheuristic algorithms to quantify the GSI based on the RMR classification. Sci Rep 15(1):28950
3. Basahel H, Mitri H (2017) Application of rock mass classification systems to rock slope stability assessment: a case study. J Rock Mech Geotech Eng 9(6):993–1009
4. Huang F, Xiong H, Chen S, Lv Z, Huang J, Chang Z, Catani F (2023) Slope stability prediction based on a long short-term memory neural network: comparisons with convolutional neural networks, support vector machines and random forest models. Int J Coal Sci Technol 10(1):18
5. Price KV (2013) Differential evolution. In: Handbook of optimization: from classical to modern approach. Springer Berlin Heidelberg, Berlin, Heidelberg, pp 187–214
6. Bao Y, Hu Z, Xiong T (2013) A PSO and pattern search based memetic algorithm for SVMs parameters optimization. Neurocomputing 117:98–106
7. Bieniawski ZT (1978) Determining rock mass deformability: experience from case histories. Int J Rock Mech Min Sci Geomech Abs 15(5):237–247
8. Rutledge JC, Preston RL (1978) Experience with engineering classifications of rock. In: International tunnelling symposium
9. Cameron-Clarke IS, Budavari S (1981) Correlation of rock mass classification parameters obtained from borecore and in-situ observations. Eng Geol 17(1–2):19–53
10. Yudhbir Y, Lemanza W, Prinzl F (1983) An empirical failure criterion for rock masses. In: ISRM congress (ISRM-5CONGRESS), ISRM, Apr 1983
11. Serafim JL (1983) Consideration of the geomechanical classification of Bieniawski. In: Proceedings of international symposium on engineering geology and underground construction, vol 1, pp 33–44
12. Ramamurthy T (1986) Stability of rock mass. Indian Geotech J 16(1):1–74
13. Bieniawski ZT (1989) Engineering rock mass classifications: a complete manual for engineers and geologists in mining, civil, and petroleum engineering. Wiley
14. Nicholson GA, Bieniawski ZT (1990) A nonlinear deformation modulus based on rock mass classification. Int J Min Geol Eng. https://doi.org/10.1007/BF01554041
15. Kalamaras GS, Bieniawski ZT (1995) A rock mass strength concept for coal seams incorporating the effect of time. In: ISRM congress (ISRM-8CONGRESS), ISRM, Sept 1995
16. Aydan O, Dalgic S (1998) Prediction of deformation behavior of 3-lanes Bolu tunnels through squeezing rocks of North Anatolian fault zone (NAFZ). In: Proceedings of regional symposium on sedimentary rock engineering, pp 228–233
17. Read SAL, Perrin ND, Richards LR (1999) Applicability of the Hoek–Brown failure criterion to New Zealand greywacke rocks. In: ISRM congress (ISRM-9CONGRESS), ISRM, Aug 1999
18. Jovanovski M (2001) Methodology for determination of rock masses as a working media. Doctoral dissertation, doctoral thesis

19. Gokceoglu C, Sonmez H, Kayabasi A (2003) Predicting the deformation moduli of rock masses. Int J Rock Mech Min Sci 40(5):701–710. https://doi.org/10.1016/S1365-1609(03)00062-5
20. Sari D, Pasamehmetoglu AG (2004) Proposed support design, Kaletepe tunnel, Turkey. Eng Geol 72(3–4):201–216
21. Coşar S (2004) Application of rock mass classification systems for future support design of the dim tunnel near Alanya. Master's thesis, Middle East Technical University, Turkey
22. Abdullatif O (2010) Geomechanical properties and rock mass quality of the carbonate Rus formation, Dammam dome, Saudi Arabia. Arab J Sci Eng 35(2):173
23. Hoek E, Brown ET (2019) The Hoek–Brown failure criterion and GSI–2018 edition. J Rock Mech Geotech Eng 11(3):445–463
24. Hoek E, Carter TG, Diederichs MS (2013) Quantification of the geological strength index chart. In: ARMA US rock mechanics/geomechanics symposium (ARMA-2013), ARMA, June 2013
25. Castro Caicedo ÁDJ, Pérez Pérez DM (2013) Correlations between the RMR and Q geomechanical classifications at "La Línea" exploratory tunnel, Colombian Central Andes. Bol Cienc Tierra 34:42–50
26. Rafiee R, Ataei M, Kamali M (2013) Tunnels stability analysis using binary and multinomial logistic regression (LR). J Geol Min Res 5(4):97–107
27. Ceballos F, Olalla C, Jiménez R (2014) Relationship between RMRb and GSI based on in situ data. In: Rock engineering and rock mechanics: structures in and on rock masses—proceedings of EUROCK 2014, ISRM European regional symposium, May 2014. https://doi.org/10.1201/b16955-62
28. Ali W, Mohammad N, Tahir M (2014) Rock mass characterization for diversion tunnels at Diamer Basha dam, Pakistan—a design perspective. Int J Sci Eng Technol 3(10):1292–1296
29. Birid KC (2014) Comparative study of rock mass deformation modulus using different approaches. In: ISRM international symposium-Asian rock mechanics symposium (ISRM-ARMS8), ISRM, Oct 2014
30. Irvani I, Wilopo W, Karnawati D (2013) Determination of nuclear power plant site in West Bangka based on rock mass rating and geological strength index. J Appl Geol 5(2)
31. Sayeed I, Khanna R (2015) Empirical correlation between RMR and Q systems of rock mass classification derived from lesser Himalayan and central crystalline rocks. In: International conference on engineering geology in new millennium, Oct 2015, vol 27, p 29
32. Gokceoglu C (2017) Deformation modulus (E_m) of rock masses: recent developments. In: 3rd Nordic rock mechanics symposium, NRMS 2017
33. Siddique T, Khan EA (2019) Stability appraisal of road cut slopes along a strategic transportation route in the Himalayas, Uttarakhand, India. SN Appl Sci 1(5):409
34. Zhang Q, Huang X, Zhu H, Li J (2019) Quantitative assessments of the correlations between rock mass rating (RMR) and geological strength index (GSI). Tunn Undergr Space Technol 83:73–81
35. Sadeghi S, Sharifi Teshnizi E, Ghoreishi B (2020) Correlations between various rock mass classification/characterization systems for the Zagros tunnel-W Iran. J Mt Sci 17(7):1790–1806
36. Somodi G, Bar N, Kovács L, Arrieta M, Török Á, Vásárhelyi B (2021) Study of rock mass rating (RMR) and geological strength index (GSI) correlations in granite, siltstone, sandstone and quartzite rock masses. Appl Sci 11(8):3351
37. Hassanpour J, Khoshkar AS, Farasani MG, Hashemnejad A (2022) Investigating the relationships between rock mass classification systems based on data from mechanized tunneling projects in Iran. Bull Eng Geol Environ 81(4):147
38. Wijaya BJ, Warmada IW, Indrawan IGB (2023) Correlation between GSI, RMR, and Q systems of rock masses at the construction site of spillway tunnel at Meninting dam. AIP Conf Proc 2629(1):030007
39. Rezaei M, Habibi H (2023) Designing of the Beheshtabad water transmission tunnel based on the hybrid empirical method. Struct Eng Mech 86(5):621–633
40. Narimani S, Davarpanah SM, Vásárhelyi B (2024) Estimation of the Poisson's ratio of the rock mass. Period Polytech Civ Eng 68(1):274–288

41. Rafiee R (2013) Tunnels stability analysis using binary and multinomial logistic regression (LR). J Geol Min Res. https://doi.org/10.5897/jgmr2013.0176
42. Ali W, Mohammad NS, Tahir M (2014) Rock mass characterization for diversion tunnels at Diamer Basha dam, Pakistan—a design perspective. Int J Sci Eng Technol
43. ASTM International (2008) ASTM D7400-08: standard test methods for downhole seismic testing. https://www.astm.org/d7400-08.html
44. Zhang Y, Zhou J, Li J, He B, Armaghani DJ, Huang S (2025) Advancing overbreak prediction in drilling and blasting tunnel using MVO, SSA and HHO-based SVM models with interpretability analysis. Geomech Geophys Geo-Energy Geo-Resour 11(1):53
45. Rezaei M, Ahmadi SR, Nguyen H, Armaghani DJ (2024) Improved determination of the S-wave velocity of rocks in dry and saturated conditions: application of machine-learning algorithms. Transp Geotech 49:101371
46. Zhou J, Wang Z, Li C, Wei W, Wang S, Armaghani DJ, Peng K (2023) Hybridized random forest with population-based optimization for predicting shear properties of rock fractures. J Comput Sci 72:102097
47. Armaghani DJ, Mohamad ET, Narayanasamy MS, Narita N, Yagiz S (2017) Development of hybrid intelligent models for predicting TBM penetration rate in hard rock condition. Tunn Undergr Space Technol 63:29–43
48. Biau G, Scornet E (2016) A random forest guided tour. Test. https://doi.org/10.1007/s11749-016-0481-7
49. Breiman L (2001) Random forests. Mach Learn 45:5–32

MIX
Papier aus verantwortungsvollen Quellen
Paper from responsible sources
FSC® C105338

If you have any concerns about our products,
you can contact us on
ProductSafety@springernature.com

In case Publisher is established outside the EU,
the EU authorized representative is:
Springer Nature Customer Service Center GmbH
Europaplatz 3, 69115 Heidelberg, Germany

Printed by Libri Plureos GmbH
in Hamburg, Germany